T0291010

PVT Property Correlations

PVT Property Correlations
Selection and Estimation

Ahmed El-Banbi
Professor and Chair of Department, Petroleum Engineering,
AUC, Cairo, Egypt

Ahmed Alzahabi
Assistant Professor, University of Texas of the Permian Basin, Odessa,
TX, United States

Ahmed El-Maraghi
Senior Petroleum Engineering, Qarun Petroleum Company, Cairo, Egypt

Gulf Professional Publishing
An imprint of Elsevier

Gulf Professional Publishing is an imprint of Elsevier
50 Hampshire Street, 5th Floor, Cambridge, MA 02139, United States
The Boulevard, Langford Lane, Kidlington, Oxford, OX5 1GB, United Kingdom

Notices

Knowledge and best practice in this field are constantly changing. As new research and experience broaden
our understanding, changes in research methods, professional practices, or medical treatment may become
necessary.

Practitioners and researchers must always rely on their own experience and knowledge in evaluating and
using any information, methods, compounds, or experiments described herein. In using such information or
methods they should be mindful of their own safety and the safety of others, including parties for whom
they have a professional responsibility.

To the fullest extent of the law, neither the Publisher nor the authors, contributors, or editors, assume any
liability for any injury and/or damage to persons or property as a matter of products liability, negligence or
otherwise, or from any use or operation of any methods, products, instructions, or ideas contained in the
material herein.

British Library Cataloguing-in-Publication Data
A catalogue record for this book is available from the British Library

Library of Congress Cataloging-in-Publication Data
A catalog record for this book is available from the Library of Congress

ISBN: 978-0-12-812572-4

For Information on all Gulf Professional Publishing publications
visit our website at https://www.elsevier.com/books-and-journals

**Working together
to grow libraries in
developing countries**

www.elsevier.com • www.bookaid.org

Publishing Director: Joe Hayton
Senior Acquisitions Editor: Katie Hammon
Senior Editorial Project Manager: Kattie Washington
Production Project Manager: Vijayaraj Purushothaman
Cover Designer: Christian Bilbow

Typeset by MPS Limited, Chennai, India

Contents

For additional information on the topics covered in the book, visit the companion site: https://www.elsevier.com/books-and-journals/book-companion/9780128125724

Biographies

Ahmed El-Banbi

Dr. El-Banbi is currently a Professor of Petroleum Engineering and chair of the department at the American University in Cairo (AUC). He has 25 years of diversified international experience in reservoir and petroleum engineering. He has worked as an engineer, trainer, and a technology developer. Ahmed spent 12 years with Schlumberger, where he held a variety of technical and managerial positions in five countries. He has considerable experience in managing multi-disciplinary teams and performing integrated reservoir studies. In addition to academic research and teaching experience, he also had short assignments with a major oil company and a consulting company. He has authored and co-authored more than eighty technical papers, two book chapters, and holds one US patent. He has been on numerous SPE committees, program chair for the SPE North Africa Technical Conference and Exhibition, and technical reviewer for the SPE Reservoir Engineering and Evaluation Journal and other journals. Ahmed holds BS and MS degrees from Cairo University, and MS and PhD degrees from Texas A&M University; all in petroleum engineering.

Ahmed Alzahabi

Dr. Alzahabi is currently an Assistant Professor at The University of Texas of the Permian Basin. He has a PhD and an MS, both in petroleum engineering; PhD from Texas Tech University and an MS from Cairo University. He previously served as a researcher at the Energy Industry Partnerships, working in the field of energy to solve complex problems for the industry. He is experienced in introducing new technologies in well placement and fracturing in conventional and unconventional oil and gas reservoirs. His research involves developing techniques for Permian Wolfcamp exploitation. He has participated in six US patent applications, edited and reviewed for multiple journals, and is active in SPWLA, SPE, NAGPS, SEG, and AAPG. He has contributed a book chapter and is writing a book on Fracturing Horizontal wells.

Ahmed El-Maraghi

Mr. El-Maraghi is a senior petroleum engineer with Qarun Petroleum Company. He has ten years of experience in reservoir and production engineering. He has extensive experience in well test analysis and has performed several reservoir studies. Ahmed also worked as a trainer and software developer. He is an avid user and developer of artificial intelligence tools in petroleum engineering. Ahmed holds a BS from Suez Canal University, an MS from Cairo University and is currently a PhD candidate in Cairo University, researching neural networks applications in log interpretation. Ahmed has authored and co-authored six papers.

Foreword

As a practicing petroleum engineer and educator, hardly a day goes by without my need for hydrocarbon PVT properties. My initial exposure to hydrocarbon properties used rigorous approaches described by chemical engineers. Subsequently Dr. Marshall Standing, who developed some of the first correlations available for practical applications, taught me, and my student colleagues at Stanford University, how to use correlations to estimate hydrocarbon fluid properties. Before the existence of personal computers, the high-end HP handheld calculator provided programs for engineers to use. The correlations were much more practical to use than the tedious chemical engineering calculations.

By far the best source for hydrocarbon fluid properties is a representative fluid sample from the reservoir. However, when assets change hands, such data may be difficult to find, especially for fields that have been on production for many years. Often reports and journal publications also provide only limited information on hydrocarbon fluid properties. Existing PVT and fluid properties correlations provide engineers with a mechanism to estimate quantities needed for conceptual, screening, and full field simulation models.

PVT Property Correlations, Selection and Estimation provides a thorough collection of existing correlations for hydrocarbon fluid properties and, more importantly, underscores the importance of making accurate fluid property estimations and how to select the model that best represents fluid from a given reservoir. The emphasis on the types of hydrocarbon reservoirs is especially useful for petroleum engineering students. By accompanying the book with programmed correlations, students can focus on broader and essential uses for accurate fluid property data. Further, the scope of the book will serve as a handbook for experienced practitioners.

Dr. Christine Ehlig-Economides
University of Houston

Preface

Estimation of PVT properties is essential in performing almost any engineering calculation in the petroleum industry. We cannot think of any application in this industry that does not require some formation volume factor or viscosity values. PVT properties are derived from laboratory measurements and/or PVT correlations. Errors in PVT data inference can cause wrong estimates of reserves, inaccurate prediction of production rates, suboptimal designs of artificial lift systems, and incorrect parameter estimates from well tests. In absence of PVT laboratory measurements, it is of paramount importance to use appropriate techniques to estimate accurate PVT properties.

The importance of PVT properties is clear from the significant amount of research available on the topic. Numerous PVT correlations have been developed over the years, which shows the importance of the topic. The large number of available correlations may also cause confusion and ambiguity for young engineers. With the development of many correlations and appearance of claims of superiority of one correlation over the others, we felt the need for a book that tackles this very topic. Guidelines are needed to select and properly use the correlations. This book hopefully clarifies the proper ways to use different correlations for different fluid types. The examples in this book were chosen to provide hands-on approaches and correlations applications. In addition, the book discusses techniques for selection of correlations that will better perform in different situations. Many of these PVT correlations are provided through access to a PVT properties calculator.

In today's environment artificial intelligence techniques applications in the petroleum industry are growing to tackle many of our everyday operational activities. A chapter dedicated to discussing artificial neural network applications in deriving PVT properties is therefore worthwhile. With this chapter the book intends to cover a wide spectrum from the early days of simple PVT correlations to the most sophisticated techniques available to date on this topic.

Ahmed El-Banbi

Acknowledgments

A book on PVT cannot be written without extensive referencing to the experts who advanced this topic with great ideas and wealth of information. Among the prominent researchers who contributed immensely to this topic, I must acknowledge Bill McCain, Curtis Whitson, Tarek Ahmed, and Karen Pedersen. Contributions by Muhammad Al-Marhoun, Abhijit Dandekar, Ali Danesh, and many others cannot be overlooked. Earlier contributors such as Donald Katz, Harvey Kenndy, and Marshall Standing shaped the path towards advancing PVT techniques.

Thanks are also due to the professors at both Cairo University and Texas A&M University for numerous discussions and valuable insights over the years. I would also like to acknowledge my graduate students whose enthusiasm kept me eager to continue learning. Immense appreciation goes to the editorial and production staff of Elsevier, especially to Kattie Washington, Katie Hammon, and Vijayaraj Purushothaman.

This book is dedicated to the late Dr. Hamdi El-Banbi whose first technical contribution was in the field of PVT (El-Banbi, Hamdi A. 1963. An Investigation of the Pressure—Volume—Temperature Behavior of Paraffinic Systems in the Presence of Olefins, Aromatics, and Contaminants. PhD dissertation, Texas A&M University, College Station, TX.).

Ahmed El-Banbi

Chapter 1

Introduction

The accuracy of many petroleum engineering calculations (e.g., material balance calculations, reserves estimation, well test analysis, advanced production data analysis, nodal analysis, surface network modeling, surface separation, and numerical reservoir simulations) largely depends on the accuracy of pressure, volume, and temperature (PVT) data. In an ideal situation, PVT data are determined from laboratory experiments performed on representative fluid samples collected from wellhead, surface separators, or wellbore. PVT reports give the results of PVT experiments usually conducted at reservoir temperature. Other routine measurements are usually taken as part of monitoring programs in oil- and gas-field operations. These include frequent measurement of API gravity of stock-tank oil, dead oil viscosity at atmospheric temperature or at different temperatures, specific gravity of separator gas, and composition of separator gas. In some cases, bubble point and dew point pressure can be estimated from production data. In addition, downhole density of reservoir fluids can be estimated from measurements of pressure versus depth (repeat formation tester measurements, RFT).

Common hydrocarbon PVT properties that appear in many calculations in reservoir and production engineering include saturation pressure (bubble point pressure for oils and dew point pressure for gases); solution gas−oil ratio; vaporized oil−gas ratio; formation volume factors for oil and gas; oil and gas density; oil and gas viscosity; single- and two-phase z-factor for gases; and oil and gas isothermal compressibility. The generation of representative values for these properties is the focus of this book.

Other PVT properties that are used in specific applications are reviewed in other texts (McCain, 1990; Whitson and Brule, 2000; Ahmed, 2016; Caroll, 2009; Mullins et al., 2007). These include water PVT properties, surface tension, minimum miscibility pressure, K values, hydrate formation pressure and temperature, asphaltene onset pressure and temperature, and wax appearance temperature.

IMPORTANCE OF ACCURATE PVT PROPERTIES

PVT properties are required for most reservoir, production, and surface-processing calculations. Inaccurate estimation of PVT properties can lead to significant errors in calculation results. Spivey and Pursell (1998) studied the

PVT Property Correlations. DOI: https://doi.org/10.1016/B978-0-12-812572-4.00001-1

effects of errors in PVT data on well test analysis interpretation results. They showed that errors in fluid compressibility affect the interpreted distance to boundaries. Distance to boundaries affects calculation of reserves from well tests (as in the case of reservoir limit tests). Errors in fluid viscosity affect estimates of permeability from well test analysis. Errors in fluid compressibility and formation volume factor have minor effects on estimated skin factor. In dual porosity reservoirs, errors in formation volume factor affect the interporosity flow coefficient (which determines how much fluid is transferred from the matrix to the fracture system with pressure drop between the two systems). Formation volume factor also affects fracture conductivity estimates for hydraulically fractured wells.

Ambastha and van Kruysdijk (1993) performed an error analysis study to quantify the effect of errors in the material balance equation for volumetric gas reservoirs. They concluded that errors in gas PVT data (z-factor and two-phase z-factor) in addition to errors in reservoir pressure can produce significant errors in the calculated original gas-in-place (OGIP) volumes. They used Monte Carlo simulation techniques to generate many cases for investigation of the upper and lower bounds for OGIP estimation errors expected from the errors in input data. The reported errors in OGIP reached 80% or more. The authors also investigated the effect of depletion level on the severity of errors. From the reported data, it was concluded that at any level of depletion, errors in gas PVT properties can lead to significant errors in OGIP estimation.

Baker et al. (2003) studied the effect of PVT data errors on material balance equation results for oil reservoirs. They studied the use of PVT correlations to derive PVT data for material balance analysis. They concluded that deriving PVT data from correlations without tuning the correlation to match the solution gas—oil ratio above the bubble point pressure could lead to significant errors in results of material balance analyses. They also studied the effect of introduction of systematic and random errors into the PVT data for material balance analysis. They concluded that the impact of PVT errors on material balance results could be significant in two cases: (1) if the decrease in reservoir pressure over the production history of the reservoir is small or (2) if the oil is highly volatile.

Hutchinson (1951) reported that many parameters including oil and water compressibility and formation volume factors affect the analysis results in material balance and pressure build-up test analysis calculations. The effect of isothermal oil compressibility in undersaturated oil reservoirs is significant; hence, the importance of using accurate estimates.

Trengove et al. (1991) reported the effects of changing PVT data on simulation results in the case of a gas condensate reservoir. They used an equation of state (EOS) program to match the laboratory-measured PVT data and to produce modified black-oil (MBO) PVT functions for further use in reservoir simulation. They observed that variation in PVT data resulted in variation in MBO properties, leading to errors in recovery predictions from the simulator.

It is generally observed that estimation of correct PVT properties is more challenging for more volatile fluids. Reserves and production calculations are the most sensitive to errors in PVT data for more volatile fluids.

SATURATED AND UNDERSATURATED RESERVOIRS

Undersaturated reservoirs or fluids are reservoirs whose initial pressure is above the saturation pressure of the fluid. The saturation pressure is called bubble point pressure for oils, and dew point pressure for gases. The majority of oil and gas reservoirs are discovered in the undersaturated state. The reservoir or the fluid it contains is said to be "saturated" if the initial reservoir pressure is the same as the saturation pressure.

When a reservoir is discovered as a saturated reservoir, two phases coexist: hydrocarbon liquid or oil phase, and hydrocarbon vapor or gas phase. The two phases are separated by a gas—oil contact (GOC). If the main fluid in the reservoir is oil, the reservoir is called "gas cap oil reservoir." If the dominant fluid is gas condensate, the oil zone is usually termed "oil rim."

Fig. 1.1 shows phase diagrams for a saturated oil reservoir (at reservoir temperature of 200°F). The saturation pressure (bubble point pressure) for the oil is initially 1930 psig. At this pressure, a phase envelope for the oil shows characteristics of black oil (to be discussed in Chapter 2). The gas

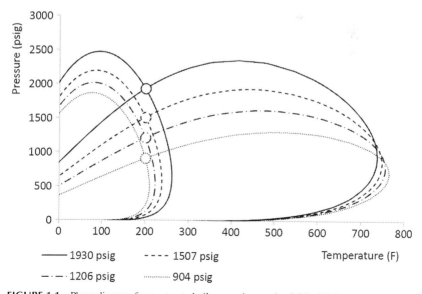

FIGURE 1.1   Phase diagram for a saturated oil reservoir near the GOC. *GOC*, gas—oil contact.

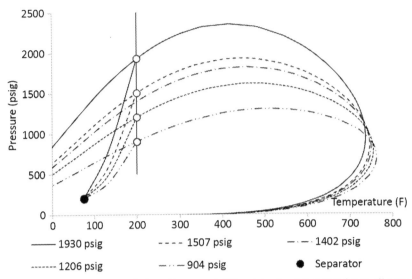

FIGURE 1.2 Development of phase diagrams at different levels of depletion for an oil reservoir (oil near the GOC). *GOC*, gas—oil contact.

phase (gas cap) exists at the same reservoir temperature and pressure. The gas phase diagram is also shown in the graph. At reservoir temperature of 200°F, the gas phase diagram exhibits the behavior of dry or wet gas. With depletion of reservoir fluids, the reservoir pressure declines, and both oil and gas compositions change. The phase diagrams for oil and gas also change. For example, in Fig. 1.1, oil and gas phase diagrams are shown for three more reservoir pressure steps at 1507, 1206, and 904 psig. The different phase diagrams show that both oil and gas phase behavior will differ with depletion of the reservoir. The figure assumes that the fluids are located near the GOC, and that the reservoir is given enough time at every depletion step to reach equilibrium.

Fig. 1.2 shows the depletion path in the reservoir at reservoir temperature (200°F). Oil will move from reservoir conditions for every reservoir pressure level to surface separator conditions. The separator conditions are usually well within the two-phase region. If the reservoir depletes slowly, equilibrium oil and gas are created near the GOC according to the phase diagrams in the figure. In many situations, however, the oil and gas movement in the reservoir can hinder enough contact between gas and oil and can prevent equilibrium. Oil produced at every reservoir pressure level will lose both pressure and temperature on its way to surface through the wellbore, choke, and surface pipes until it reaches the separator. The free (or equilibrium) gas shown in Fig. 1.3 will also move to surface through production wells. If this gas is rich with liquids (as in the case of gas condensates associated with

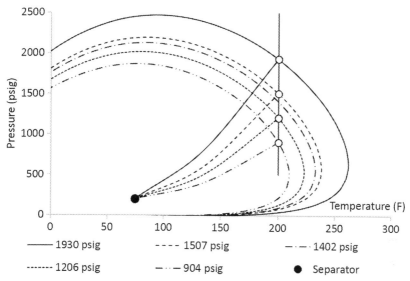

FIGURE 1.3 Phase diagrams of equilibrium gas at different levels of depletion for an oil reservoir (gas near the GOC). *GOC*, gas–oil contact.

volatile oils), it may drop some of the liquids it contains when it reaches the surface. The concept of equilibrium is used in tracking the produced fluids to surface. At every new pressure and temperature on the way of oil and gas to surface, new equilibrium between the oil and gas is assumed for every increment of pipe.

COMMON MODELS FOR PVT MODELING

Three common PVT models are used to handle the changes in PVT properties with the changes in pressure, temperature, and composition of the fluid. These models are usually termed (1) black-oil, (2) MBO, and (3) compositional models.

The choice of model to use in a particular situation depends on fluid type and the process to be modeled in the reservoir and/or the production system (El-Banbi et al., 2000; Fevang et al., 2000). The models differ in data requirements, complexity, and accuracy. The following sections discuss the differences between the three common PVT models.

Black-Oil Models

Black-oil models (not to be confused with the black-oil fluid type, to be discussed in Chapter 2) are fluid models that assume the composition of

	Hydrocarbon vapor phase (gas phase)	Hydrocarbon liquid phase (oil phase)	Water phase
Surface gas	X	X	
Stock-tank oil (surface oil)		X	
Surface water			X

FIGURE 1.4 Relation of produced fluids on surface and reservoir phases for black-oil models.

produced oil (stock-tank oil) and produced gas on surface to remain constant. Despite the fact that, for most fluids, composition of produced fluids can change when the reservoir pressure declines below the saturation pressure (bubble point pressure for oils and dew point pressure for gases), the black-oil model can adequately model many fluid types under reservoir processes such as depletion and secondary recovery (e.g., water and casing-head gas injection).

Fig. 1.4 is a schematic representation of the assumptions of the black-oil model. Most production, reservoir, and surface facilities calculations deal with surface fluids measured in the field (e.g., stock-tank oil, surface gas, and surface water). The traditional black-oil model assumes that surface gas comes from either the reservoir gas phase or reservoir oil phase, or both. Gas exists in the reservoir gas phase as free gas, while it exists in the reservoir oil phase as gas in solution. The black-oil model also assumes that the oil produced on surface (stock-tank oil) comes from the reservoir oil phase; the water produced on surface comes from the reservoir water phase.

For most calculations, the oil PVT properties required to describe phase behavior and PVT properties changes for the black-oil model include solution gas—oil ratio, oil formation volume factor, and oil viscosity. The gas PVT properties include gas z-factor or gas formation volume factor and gas viscosity. Partitioning of the produced gas between reservoir gas phase and reservoir oil phase is accomplished through the solution gas—oil ratio function (R_s). All properties are defined in the respective chapters. The techniques to obtain these properties are also reviewed in different chapters.

Variations of the traditional black-oil model can be considered in certain cases to account for gas solubility in water and water vaporization in gas. These variations of the traditional black-oil model usually have limited applications.

The black-oil model is adequate in modeling phase behavior and PVT properties for low gas—oil ratio oil, black-oil, dry gas, and wet gas fluids. The black-oil model is sufficient for most reservoir and production engineering applications (e.g., material balance, reserves estimation, conventional and advanced decline curve analysis, pressure transient analysis, reservoir

	Hydrocarbon vapor phase (gas phase)	Hydrocarbon liquid phase (oil phase)	Water phase
Surface gas	X	X	
Stock-tank oil (surface oil)	X	X	
Surface water			X

FIGURE 1.5   Relation of fluids produced on surface and reservoir phases for MBO models. *MBO*, modified black-oil.

simulation, nodal analysis, and artificial lift design) and for most common reservoir-depletion processes (e.g., primary and secondary recovery).

Modified Black-Oil (MBO) Models

The MBO model (also called extended black oil) is similar to black-oil models in most respects; it differs, however, in the way it handles oil produced on surface (stock-tank oil) (Spivak and Dixon, 1973). The oil produced on surface is assumed to come from both the reservoir gas and oil phases. The MBO model is therefore capable of handling the oil produced from the gas phase. This capability is important for modeling both volatile oil and gas condensate fluids, as the oil produced from the gas phase can be significant. Fig. 1.5 shows the relation of fluids produced on surface (surface gas and stock-tank oil) and the phases existing in the reservoir (gas and oil phases).

Four PVT functions are required for the MBO model in addition to oil and gas viscosity. Viscosities are defined and obtained in much the same way as in black-oil models. The four functions are (1) oil formation volume factor, (2) solution gas−oil ratio, (3) gas formation volume factor, and (4) vaporized oil−gas ratio. The definitions of these functions are given in the following equations:

$$B_o = \frac{\text{Reservoir oil volume at } T \text{ and } P}{\text{Stock} - \text{tankoil from reservoir oil}} \qquad (1.1)$$

$$R_s = \frac{\text{Surface gas from reservoir oil}}{\text{Stock} - \text{tankoil from reservoir oil}} \qquad (1.2)$$

$$B_g = \frac{\text{Reservoir gas volume at } T \text{ and } P}{\text{Surface gas volume from reservoir gas}} \qquad (1.3)$$

$$R_v = \frac{\text{Stock} - \text{tankoil volume from reservoir gas}}{\text{Surface gas volume from reservoir gas}} \qquad (1.4)$$

The first three functions have the same definitions for both black-oil and MBO models. The partitioning of how much surface gas comes from the reservoir oil phase and how much comes from the reservoir free gas phase is accomplished through R_s function as in black-oil models. Tracking the amount of surface oil (stock-tank oil) that comes from the reservoir gas phase is accomplished through the R_v function.

For gas condensate systems, R_v is the same as surface condensate gas ratio since the condensate dropout in the reservoir is usually assumed to be immobile and will not be produced. For volatile oil systems, R_v is the reciprocal of R_s above the bubble point. Below the bubble point pressure, the reservoir oil produces surface oil and surface gas. B_o tracks the relation of the volume of reservoir oil and the surface oil from the reservoir oil. R_s tracks the gas that comes out of the reservoir oil. In addition, the free gas that is formed in the reservoir below bubble point pressure produces both condensate (oil) and gas on surface, since it is gas condensate fluid. B_g tracks the relation between the gas volume in the reservoir and on surface. R_v tracks the liquid volume on surface to the surface gas produced (both these quantities are generated from the reservoir gas).

Like black-oil models, MBO models assume that surface oil and gas maintain the same composition at any pressure. For most reservoir and production engineering calculations, MBO models are adequate for handling volatile oil and gas condensate calculations under depletion and secondary recovery (El-Banbi et al., 2000; Fevang et al., 2000). Processes that result in significant compositional changes are not suitable to be modeled with the MBO approach.

Compositional Models

Compositional PVT models track the surface production of hydrocarbon components. The components could be hydrocarbon components, impurities (i.e., methane, ethane, hydrogen sulfide, carbon dioxide, etc.), or pseudo (lumped or grouped) components. The main assumption in compositional models is that the components (or pseudocomponents) can come from both the reservoir gas and oil phases. In most compositional models, surface-produced water still comes from the reservoir water phase. The partitioning of each hydrocarbon component between reservoir gas and oil phases is handled with the use of equilibrium constants (K values). Equilibrium constants for any component are defined as

$$K_i = \frac{Y_i}{X_i} \tag{1.5}$$

Equilibrium constants are functions of pressure, temperature, and composition. Correlations exist to calculate K values. However, the most accurate source of K values is tuned EOS models. In general, the application of compositional PVT models requires availability of laboratory-measured PVT experiments and construction of EOS models. Fig. 1.6 schematically depicts

	Hydrocarbon vapor phase (gas phase)	Hydrocarbon liquid phase (oil phase)	Water phase
Component 1	X	X	
Component 2	X	X	
Component 3	X	X	
......	X	X	
......	X	X	
Surface water			X

FIGURE 1.6   Relation of fluid components and reservoir phases for compositional models.

the relations between the surface components, and their existence in the reservoir gas and oil phases. In this model, all hydrocarbon components (or pseudocomponents) produced on surface are assumed to have existed in either the reservoir gas or oil phases, or both.

The preparation of PVT data for compositional models requires the use of EOS programs and is beyond the scope of this book. Compositional PVT models are needed to handle fluids that change composition significantly with different recovery processes. Applications of compositional models include (1) tracking the hydrocarbon components in the production stream (rather than merely volumes of gas and oil produced), (2) gas cycling in rich gas condensate reservoirs, (3) miscible and near miscible processes, (4) cases with severe compositional gradient (e.g., near-critical fluids with large compositional variation versus depth), and (5) separator conditions changes during reservoir depletion. Some commercial reservoir simulators allow changing black-oil PVT data (oil formation volume factor and solution gas-oil ratio) when separator conditions change during the simulation run. The extension of these corrections to MBO simulation were presented by Ibrahim et al. (2011).

NOMENCLATURE

B_g gas formation volume factor, rcf/scf
B_o oil formation volume factor, rbbl/STB
CGR   condensate gas ratio
CVD   constant volume depletion
EOS   equation of state
GOC   gas−oil contact
K_i equilibrium constant for component (i)

MBO   modified black-oil
OGIP   original gas in place
PVT   pressure, volume, temperature
RFT   repeat formation tester
r_s vaporized oil−gas ratio, STB/MMscf
R_s solution gas−oil ratio, scf/STB
R_{si} initial solution gas−oil ratio, scf/STB
R_v vaporized oil−gas ratio, STB/MMscf
X_i mole fraction of component (i) in hydrocarbon liquid phase
Y_i mole fraction of component (i) in hydrocarbon vapor phase

REFERENCES

Ahmed, T., 2016. Equations of State and PVT Analysis, 2nd ed. Gulf Professional Publishing, ISBN: 9780128017524.

Ambastha, A.K., van Kruysdijk, C.P.J.W., 1993. Effects of input data errors on material balance analysis for volumetric, gas and gas-condensate reservoirs. In: Paper Presented at the CIM 1993 Annual Technical Conference, Calgary, May 9−12.

Baker, R.O., Regier, C, Sinclair, R., 2003. PVT error analysis for material balance calculations. In: Paper Presented at the Canadian International Petroleum Conference, Calgary, Alberta, June 10−12.

Caroll, J., 2009. Natural Gas Hydrates—A Guide for Engineers, 2nd ed. Gulf Professional Publishing, Burlington, MA, USA.

El-Banbi, A.H., Forrest, J.K., Fan, L., McCain, W.D. Jr., 2000. Producing rich-gas-condensate reservoirs—case history and comparison between compositional and modified black-oil approaches. In: Paper SPE 58988 Presented at the SPE Fourth International Petroleum Conference and Exhibition, Villahermosa, Mexico. February 1−3.

Fevang, O., Singh, K., Whitson, C.H., 2000. Guidelines for choosing compositional and black-oil models for volatile oil and gas-condensate reservoirs. In: Paper SPE 63087 Presented at the 2000 SPE Annual Technical Conference and Exhibition, Dallas, TX. October 1−4.

Hutchinson, C.A. Jr., 1951. Effect of data errors on typical reservoir engineering calculations. In: Paper Presented at the Oklahoma City Meeting of the Petroleum Branch, AIME, October 3−5.

Ibrahim, M., El-Banbi, A.H., El-Tayeb, S., Sayyouh, H., 2011. Changing separator conditions during black-oil and modified black-oil simulation runs. In: Paper SPE 142462 Presented at the SPE Middle East Oil and Gas Show and Conference, Manama, Bahrain, March 6−9.

McCain, W.D. Jr., 1990. The Properties of Petroleum Fluids, 2nd ed. PennWell Books, Tulsa, Oklahoma, USA.

Mullins, O.C., Sheu, E.Y., Hammami, A., Marshall, A.G. (Eds.), 2007. Asphaltenes, Heavy Oils, and Petroleomics. Springer, ISBN 13: 978-0-387-31734-2. Available from: https://doi.org/10.1007/0-387-68903-6.

Spivak, A., Dixon, T.N., 1973. Simulation of gas condensate reservoirs. In: Paper SPE 4271 Presented at the 3rd Numerical Simulation of Reservoir Performance Symposium, Houston, January 10−12.

Spivey, J.P., Pursell, D.A., 1998. Errors in input data and the effect on well-test interpretation results. In: SPE 39773 Presented at the SPE Permian Basin Oil & Gas Recovery Conference, Midland, Texas, March 23−26, 1998.

Trengove, R.D., Hann, J.H., Skates, J.R., 1991. The impact of PVT data quality on hydrocarbon recovery predictions. In: Paper SPE 22988 Presented at the SPE Asia-Pacific Conference, Perth, Western Australia, November 4–7.

Whitson, C.H., Brule, M.R., 2000. Phase Behavior. SPE Monograph, Vol. 20. SPE Richardson, TX, USA.

PROBLEMS

1.1 Define saturated and undersaturated reservoirs. Explain the differences between saturated and undersaturated reservoirs.

1.2 Explain the main differences between oil and gas reservoir fluids.

1.3 Draw and label completely phase diagrams ($P-T$ diagrams) for undersaturated oil, undersaturated gas condensate, and saturated oil reservoirs.

1.4 What are the main assumptions of black-oil, modified black-oil (MBO), and compositional PVT models?

1.5 What are the main differences between black-oil and MBO models?

1.6 Give example applications for compositional PVT models.

1.7 Define vaporized oil–gas ratio and solution gas–oil ratio.

Chapter 2

Reservoir-Fluid Classification

Petroleum reservoir fluids are hydrocarbons that can exist as gases (hydrocarbon vapor phase) or oils (hydrocarbon liquid phase). If the reservoir temperature is higher than the critical temperature of the fluid, the reservoir fluid is gas. Otherwise, the fluid is oil. Although every fluid has a unique set of characteristics, it is useful to distinguish fluids in terms of typical behavior groups.

Engineers are interested in classifying reservoir-fluid types for practical reasons: reservoir-fluid classification can affect decisions as to fluid sampling, production, and laboratory PVT experiments. Categorizing fluids into distinct groups according to similarity in behavior facilitates the use of PVT models to describe phase behavior and predict fluid properties. More importantly, fluid types affect the choice of model to apply for reservoir behavior prediction (e.g., material balance or reservoir simulation models). Several qualitative (but relatively few quantitative) classifications of reservoir fluids have been suggested.

REVIEW OF RESERVOIR-FLUID CLASSIFICATIONS

Moses (1986) classified oil fluids into ordinary oils (also called black oils) and near-critical or volatile-oils. Black oils are characterized by gas—oil ratio (GOR) up to 2000 scf/STB, stock-tank oil gravities up to 45° API, and oil-formation volume factor of less than 2.0 rbbl/STB; volatile oils usually have GOR of 2000—3000 scf/STB, stock-tank oil gravity is usually 40° API or higher, and formation volume factor is above 2.0 rbbl/STB. C7 + for volatile oils is between 12.5% and 20%. Noting the lack of sharp division between black oils and volatile oils, Moses highlighted the difficulty of classifying the oil as black oil or volatile oil outside of a laboratory setting. Retrograde condensate systems are characterized by GOR of 3000 to 150,000 scf/STB and stock-tank oil gravities between 40 and 60° API. The color of stock-tank fluid cannot be used to distinguish fluid type.

McCain (1994b) classified reservoir fluids into five categories, relying primarily on the criteria of Moses (1986) to distinguish black oils, volatile oils, and gas condensates, and using C7 + mole % and initial producing GOR for wet gases and dry gases. Other criteria (e.g., stock-tank oil API gravity and color of stock-tank fluid) are not decisive in classifying the fluid type. Wet gases are those which produce GOR higher than 15,000 scf/STB

PVT Property Correlations. DOI: https://doi.org/10.1016/B978-0-12-812572-4.00002-3

[condensate gas ratio (CGR) of less than 67 STB/MMscf] and C7 + of less than 4%. Dry gases are those with initial producing GOR of more than 100,000 scf/STB (CGR of less than 10 STB/MMscf) and C7 + mole % of less than 0.7%. The distinction between gas condensate behavior and wet gas behavior rests on the shape of two-phase z-factor. The behavior of the five reservoir fluids is examined in a series of papers (McCain, 1993a,b, 1994a; McCain and Bridges, 1994; McCain and Piper, 1994).

Whitson and Brule (2000) suggest a classification of five fluid types: (1) dry gases, (2) wet gases, (3) gas condensates, (4) volatile oils, and (5) black oils. This study used C7 + mole % to distinguish between fluid types according to the following criteria: gas condensates have C7 + mole % less than 12.5%, volatile oils from 12.5% to 17.5%, and black oils greater than 17.5%. GOR for gas condensates is from 3000 to 150,000 scf/STB (which is equivalent to oil-gas ratios from about 350 to 5 STB/MMscf). The API of stock-tank oil for gas condensate fluids typically ranges between 40 and 60° API. It is noted in the study that although the color of stock-tank oil lightens from volatile oils to gas condensates, color is not a reliable criterion to distinguish reservoir-fluid types.

Pederson et al. (2015) also classify reservoir fluids into five categories: (1) natural gas mixtures, (2) gas condensate mixtures, (3) near-critical mixtures or volatile oils, (4) black oils, and (5) heavy oils. Near-critical mixtures are defined as fluids with reservoir temperature close to the critical temperature of the fluid. They may be volatile oils or gas condensates, depending on the proximity of reservoir temperature to critical temperature. For near-critical fluids, the composition and properties of the gas and liquid inside the two-phase region are similar. The change in composition and properties is rapid with the decline in pressure below saturation pressure.

Ahmed (2016) discusses additional categories in the classification of petroleum reservoir fluids. This classification depends on composition of fluid, location of reservoir temperature relative to critical fluid temperature, initial reservoir pressure and temperature, and pressure and temperature of surface production. Four types of oils can be recognized: (1) ordinary black oil, (2) low-shrinkage crude oil, (3) high-shrinkage (volatile) crude oil, and (4) near-critical crude oil. Ahmed identifies initial GOR as the most important distinguishing criterion of reservoir-fluid type. Color of stock-tank oil is not a defining criterion. Gas reservoirs are divided into four categories as follows: (1) retrograde gas condensates, (2) near-critical gas condensate reservoirs, (3) wet gas reservoirs, and (4) dry gas reservoirs. Ahmed (2016) provided GOR limits for each category.

The criteria of McCain (1994b), offering practical and quantified guidelines to define the fluid type early in the life of the reservoir, seem to be the most widely applied for selection of reservoir-fluid type. Some modifications of these criteria are suggested here for practical purposes.

PRACTICAL CLASSIFICATION OF GASES

Gases exit in reservoirs whose temperatures are higher than the critical fluid temperature. They exhibit dew point pressure at reservoir temperature. Following McCain (1994b), gases can be divided into dry gases, wet gases, and gas condensates. This classification of natural gases is based on practical engineering considerations. The classification should provide guidelines to engineers as to how to derive PVT properties and what equations and models to use in engineering calculations. Fig. 2.1 represents a typical phase diagram for a natural gas. In this phase diagram, the critical point is represented by the point around 75°F and 3200 psia. At this point, both liquid and vapor properties are similar. Inside the phase envelope, two phases (vapor and liquid) coexist. At temperatures above the critical temperature, and outside the two-phase region, the reservoir fluid is single-phase gas. At temperatures below the critical temperature and outside the two-phase region, the reservoir fluid is single-phase oil. During depletion (loss of pressure) of reservoir gas, it may cross the two-phase region and liquid condensate will start to form. The point at which liquid starts to form is termed *dew point pressure*. Inside the two-phase region, the dashed lines are known as "iso-volume lines" or "percent liquid lines." These lines determine how much liquid and vapor coexist at equilibrium at a certain pressure and temperature inside the two-phase region.

Another interesting point on the phase diagram is the cricondentherm. The cricondentherm is the highest temperature seen in the two-phase region

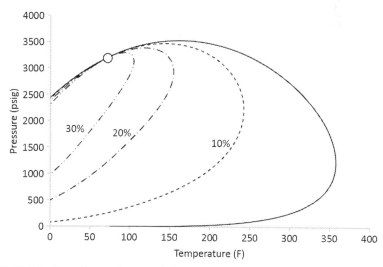

FIGURE 2.1   Phase diagram for reservoir fluid (gas).

(in the example figure, it is approximately 355°F). Three conditions may exist:

1. If the reservoir temperature is above the cricondentherm and the surface conditions are outside the two-phase region, the fluid is determined to be "dry gas."
2. If the reservoir temperature is above the cricondentherm and the surface conditions are inside the two-phase region, the fluid is considered "wet gas."
3. If the reservoir temperature is between the cricondentherm and the critical temperature, the fluid is "gas condensate" or "retrograde gas condensate."

The vast majority of gases will drop some condensate in the reservoir when the reservoir pressure declines below the gas saturation pressure. Fig. 2.2 shows maximum liquid saturation that drops in the reservoir versus C7 + mole % for a large number of gas samples. The presence of larger amounts of C7 + signals a higher degree of gas richness (capability of the gas to drop liquid condensate in the reservoir and on surface). The data is provided by the constant volume depletion experiment that represents the behavior of gas condensate fluid at reservoir temperature. The plot clearly shows that liquid saturation (the amount of condensate dropping out) in the reservoir increases with increasing C7 + mole %. Liquid saturation can be insignificant for C7 + less than 0.7% (McCain and Piper, 1994). At these levels of low liquid saturation, the gas is considered to behave as though it were dry gas. At C7 + mole % above 2%, the liquid dropout in the reservoir

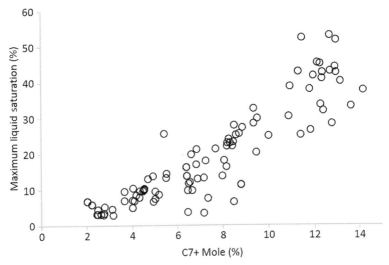

FIGURE 2.2 Maximum liquid saturation versus C7 + mole % for gas samples.

can become significant. At this level, liquid saturation can range from 2% to 7%. At this level, the fluid should be considered gas condensate. Treatment of it, as dry or wet gas, may produce considerable error. The Wet Gases (Chapter 4) and Gas Condensates (Chapter 5) discuss the economic aspects of producing such fluids. The wet gas model can be assumed for fluids with C7 + between 0.7% and 2%.

Fig. 2.3 shows the reservoir pressure at which the maximum liquid dropout occurs. This pressure increases with the increase in C7 + mole %. For the range selected for wet gases (C7 + of 0.7%−2%), the pressure at which maximum liquid dropout occurs is usually low (less than 1500 psia). This level of pressure will usually be reached in the reservoir at a later stage of depletion. At higher pressures, liquid dropout in the reservoir is minimal, thus consistent with the theoretical assumption of wet gas (no liquid condensation in the reservoir).

Fig. 2.4 shows the surface GOR versus C7 + mole %. Producing GOR shows a clear decreasing trend with increasing C7 + mole %. The dry gas concept can be used for C7 + less than 0.7%. This range of C7 + mole % is equivalent to GOR higher than 100,000 scf/STB (or CGR of less than 10 STB/MMscf). The wet gas model can be assumed for C7 + range of 0.7%− 2%. Fig. 2.4 shows that this range is equivalent to GOR range of 100,000 to 33,000 scf/STB (CGR of 10−30 STB/MMscf). For higher levels of C7 + , gas condensate must be assumed to avoid excessive errors in phase behavior prediction. In practice, gas richness of 30 STB/MMscf and above requires that appropriate models be used to forecast the reservoir behavior. At this

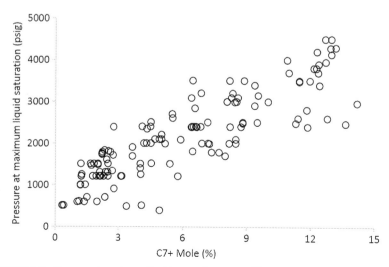

FIGURE 2.3   Pressure at maximum liquid saturation versus C7 + mole %.

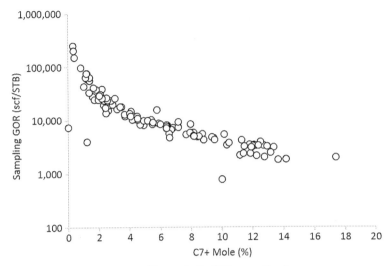

FIGURE 2.4 Surface GOR versus C7 + mole %. *GOR*, gas−oil ratio.

level, the condensate production is significant for the economic value of any field development project.

PRACTICAL CLASSIFICATION OF OILS

When the reservoir temperature is less than the critical fluid temperature, the reservoir fluid is oil. Oil can be classified into three different types: (1) volatile oil, (2) black oil, and (3) low-GOR oil. This classification is based on differences in behavior of the three oils. The handling of each type of oil will differ as to how the PVT properties are obtained and which equations and models are used for forecasting reservoir behavior.

Fig. 2.5 shows a typical phase diagram for a black-oil reservoir. In black oils, the reservoir temperature is far less than the critical fluid temperature. The reservoir temperature for volatile oils is close to the critical temperature of the fluid.

Fig. 2.6 shows initial producing GOR for many oil samples plotted against C7 + mole %. Each point on the graph represents a different fluid. In the PVT laboratory, all these fluids were determined to have bubblepoint pressure at reservoir temperature. The figure shows a clear trend of decreasing initial GOR with increasing C7 + mole %. A similar trend (but not as sharp) is also observed when oil bubblepoint pressure is plotted against the C7 + mole %, as shown in Fig. 2.7.

From these two figures, low-GOR oils can be chosen for oils with C7 + mole % higher than 70%. In this range of C7 +, the oil contains little gas (mostly less than 100 scf/STB with few samples showing slightly higher

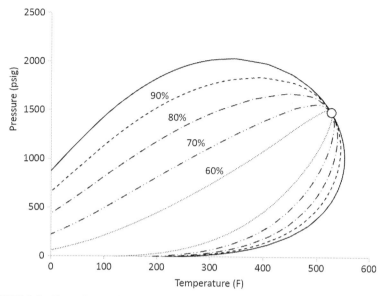

FIGURE 2.5   Phase diagram for an oil fluid.

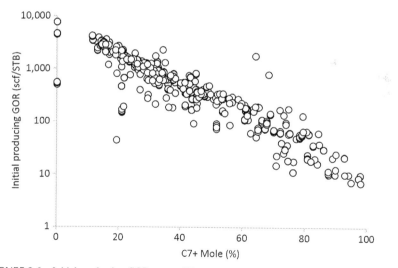

FIGURE 2.6   Initial production GOR versus C7 + mole % for oil fluids. *GOR*, gas–oil ratio.

values of producing GOR). For producing GOR less than 200 scf/STB, gas is usually flared for economic reasons in many oil field operations. Investigating the bubblepoint pressure plot shows that at these high values of C7 + mole %, the bubblepoint is usually low. The bubblepoint is sometimes

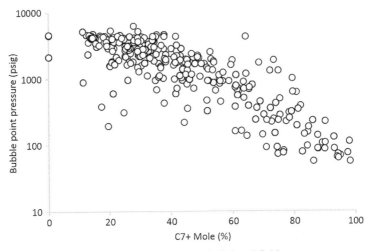

FIGURE 2.7   Bubblepoint pressure versus C7 + mole % for oil fluids.

too low to reach during reservoir depletion. For these low-GOR oils, it is usually sufficient to estimate the PVT properties above the bubblepoint.

The distinction between black oils and volatile oils is not very sharp. The main physical difference between black oils and volatile oils lies in the ability of the associated gas to drop liquids (condensate) on surface. In black oils, the assumption is that its associated gas is dry gas, while the gas associated with volatile oil is gas condensate. This observation makes a difference as to which equations are used to forecast the behavior of black oil and volatile oil fluids (McCain and Piper, 1994). For material balance calculations, the conventional black-oil material balance (Craft and Hawkins 1959) is sufficient for black oils. The material balance for volatile oils requires that oil vaporized in the gas phase be considered (Walsh 1995). Handling volatile oils requires the use of modified black-oil models or compositional models. The use of these models usually requires equation of state modeling. Therefore, preparation of PVT properties for volatile oils is generally more difficult than it is for black oils. The data in the two plots (Figs. 2.6 and 2.7) reveals that an arbitrary distinction (for engineering purposes) can be used as 19% for C7 + mole % and 1750 scf/STB for initial GOR.

Figs. 2.8 and 2.9 show typical phase diagrams for saturated black- and volatile-oil fluids, respectively. For oil reservoirs at their saturation pressure (bubblepoint pressure), both oil and gas coexist in the reservoir. Phase envelopes for gas and oil will intersect at the saturation pressure point, as shown in the two figures. Examination of the phase diagrams for the black oil and its associated gas shows that the gas-phase diagram can be close to dry or wet gas (reservoir temperature is close to the cricondentherm). The volatile

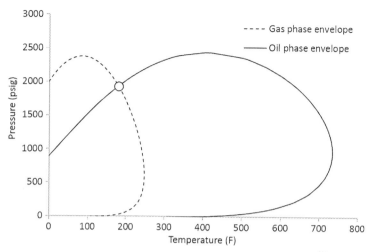

FIGURE 2.8   Saturated black oil and its associated equilibrium gas-phase diagrams.

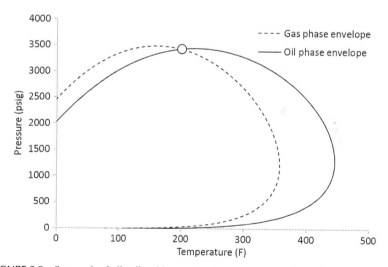

FIGURE 2.9   Saturated volatile oil and its associated equilibrium gas-phase diagrams.

oil and its associated equilibrium gas-phase envelopes show that the reservoir temperature is well far from the cricondentherm (indicating gas condensate behavior).

A similar graph for gas condensate reservoir fluid, at its saturation state, is shown in Fig. 2.10. If a gas condensate reservoir is discovered at its saturation pressure, it is likely that an oil leg exists. The oil associated with the gas condensate fluid is volatile oil.

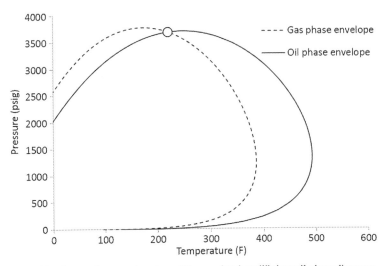

FIGURE 2.10 Saturated gas condensate and its associated equilibrium oil-phase diagrams.

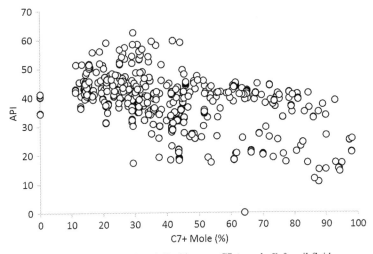

FIGURE 2.11 API gravity of stock-tank liquid versus C7 + mole % for oil fluids.

Some classifications use API gravity of stock-tank liquid as one of the criteria to distinguish reservoir-fluid types. Fig. 2.11 represents a plot of API gravity of stock-tank oil versus C7 + mole %. Although the plot shows a general trend of decreasing API gravity with increasing C7 + mole %, the large variation in API gravity value hinders its use as a distinctive criterion. For low-GOR oils, the API gravity can range from 10 to 40° API. For black

TABLE 2.1 Summary of Field and Laboratory-Measured Criteria for Fluid Classification

Criteria	Oil			Gas		
	Low-GOR Oil	Black Oil	Volatile Oil	Gas Condensate	Wet Gas	Dry Gas
Field evidence (initial producing GOR, scf/STB)	<200	<1750	Up to 3400	Down to 3200	>33,000	>100,000
Field evidence (initial CGR, STB/MMscf)	>5000	>570	>295	30–310	10–30	<10
Composition evidence (C7 + mole %)	>70	>19	12.0–19.0	2–12.75	0.7–2	<0.7

GOR, gas–oil ratio; *CGR*, condensate gas ratio.

oils, the range is wider. API gravity can range from slightly less than 20 to more than 60° API. For volatile-oil fluids, API gravity can range from 35 to 60° API.

SUMMARY OF RESERVOIR-FLUID CLASSIFICATION CRITERIA

The most important criteria in classifying reservoir fluids are initial producing GOR and C7 + mole %. The initial producing GOR is readily available from production data records in the field. C7 + mole % requires the measurement of reservoir-fluid composition. The compositional measurement is usually found in PVT laboratory reports. Other criteria (API gravity, color of stock-tank liquid, and oil-formation volume factor at bubblepoint pressure) can be used only as confirmation, as reported previously. Table 2.1 summarizes the essential criteria for reservoir-fluid classification.

For the vast majority of the fluids, both the GOR and the C7 + criteria are honored. However, Table 2.1 shows some overlap between gas condensate and volatile-oil fluids. Some fluids can be identified as gas condensates in the PVT laboratory even though their C7 + mole % is close to 12.75%, while other fluids may be identified in the laboratory to be volatile oils with C7 + around 12%. On rare occasions, due to the extremely volatile nature of the fluid, the liquid volumes may not be visually confirmed to approach an all-gas or all-liquid system. Presence of high concentrations of H_2S in the fluid sample (above 15%) usually complicates the fluid identification problem.

NOMENCLATURE

CGR   condensate gas ratio
CVD   constant volume depletion
EOS   equation of state
FVF   formation volume factor
GOR   gas−oil ratio

REFERENCES

Ahmed, T., 2016. Equations of State and PVT Analysis, second ed. Gulf Professional Publishing, ISBN: 9780128017524.

Craft, B.C., Hawkins, M.F., 1959. Applied Petroleum Reservoir Engineering. Prentice-Hall, Englewood Cliffs, NJ.

McCain Jr., W.D., 1993a. Chemical composition determines behavior of reservoir fluids. Pet. Eng. Intl. 18−25 (Oct.).

McCain Jr., W.D., 1993b. Black oils and volatile oils − what's the difference? Pet. Eng. Intl. 24−27 (Nov.).

McCain Jr., W.D., 1994a. Revised gas−oil ratio criteria key indicators of reservoir fluid type. Pet. Eng. Int. 57−60 (April).

McCain Jr., W.D., 1994b. Heavy components control reservoir fluid behavior. *JPT* (Sept.), 746–750. Technology Today Series, Trans., AIME, 297.

McCain Jr., W.D., Bridges, B., 1994. Volatile oils and retrograde gases—what's the difference? Pet. Eng. Int. 35–36 (Jan.).

McCain Jr., W.D., Piper, L.D., 1994. Reservoir gases exhibit subtle differences. Pet. Eng. Int. 45–46 (Mar.).

Moses, P.L., 1986. Engineering applications of phase behavior of crude oil and condensate systems. JPT 715 (July).

Pederson, K.S., Christensen, P.L., Shaikh, J.A., 2015. Phase Behavior of Petroleum Reservoir Fluids, 2nd ed. Taylor & Francis Group, FL, USA.

Walsh, M.P., 1995. A generalized approach to reservoir material balance calculations. JCPT, Vol. 34, No. 1, 55–63, (Jan.).

Whitson, C.H., Brule, M.R., 2000. Phase Behavior. SPE Monograph, Vol. 20. SPE Richardson, TX, USA.

PROBLEMS

2.1 The following data is available for a newly discovered reservoir. Initial reservoir pressure is 3750 psia, and reservoir temperature is 210°F. Primary separator pressure and temperature are 590 psia and 65°F. Locate the initial reservoir conditions on the phase envelope and connect the possible path of the fluid from reservoir conditions to separator conditions. Based on the shape of the phase envelope, identify the reservoir fluid type.

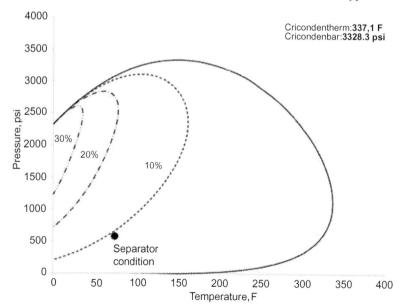

2.2 The following compositional information are available for two reservoir fluids. Identify the fluid type for the two samples. Which fluid is expected to have higher GOR on surface?

Component	Mole %	
	Sample 1	Sample 2
N_2	0.6	0.67
CO_2	5.6	5.13
C1	68.21	75.28
C2	7.1	8.19
C3	3.89	3.99
iC4	0.8	0.7
nC4	1.7	1.5
iC5	0.7	0.5
nC5	0.8	0.6
C6	1.51	0.72
C7 +	9.2	2.66
C7 + molecular weight	242	140
C7 + specific gravity	0.872	0.7832
Temperature, °F	283	293

2.3 The following reservoir parameters and laboratory-measured PVT properties are available for a reservoir fluid produced by a petroleum company. Identify the fluid type. Explain all the criteria used in the fluid identification and discuss whether each criterion agrees or disagrees with your fluid classification.

Reservoir	Parameters
T, F	239
API	41.2
Gas gravity	0.874
Oil FVF	4.08
Rs	7803

Pressure (psig)	Oil Density (g/cm³)	Gas Gravity	Z-Factor	Oil FVF (bbl/STB)	Gas FVF	Rs (ft³/scf)	Oil Viscosity (cp)	Gas Viscosity (cp)	Compressibility (1/psi x 10⁻⁶)
7000	0.65			3.55		7803	0.13		
6500	0.64			3.62		7803	0.12		40.9998
6000	0.62			3.72		7803	0.12		49.9998
5501	0.6			3.83		7803	0.11		58.9818
5001	0.58			3.96		7803	0.11		67.9818
4604	0.56			4.48		7803	0.11		
4001	0.57	0.874	0.809	4.07	0.0039	6862	0.11	0.032	
3500	0.58	0.8846	0.768	3.74	0.0043	6081	0.12	0.029	
3001	0.59	0.899	0.736	3.39	0.0048	5302	0.13	0.026	
2500	0.61	0.912	0.717	3.05	0.0056	4520	0.14	0.023	
2001	0.62	0.9268	0.732	2.72	0.0072	3741	0.15	0.02	
1501	0.64	0.9415	0.7813	2.37	0.0102	2961	0.17	0.016	
1001	0.66	0.9662	0.844	2.03	0.01641	2180	0.21	0.014	
501	0.69	1.002	0.918	1.69	0.0352	1400	0.42	0.013	
0	0.74	1.0888	0.997	1.1	1.3404	0	1.036	0.012	

2.4 Identify the fluid type from the simple information given below for each one of the following five fluids:

- 43,478 scf/STB primary separator gas−oil ratio
- 20% C7 + mole %
- 1.2% C7 + mole %
- Fluid with field stock-tank oil gravity of 53° API and initial producing gas−oil ratio of 5440 scf/STB
- Fluid with API gravity of 17 and oil-formation volume factor at the bubblepoint pressure of 1.01 bbl/STB

2.5 Identify the fluid type for the following fluid samples based on the given compositional data and initial producing gas-oil ratio.

	Sample 1	Sample 2	Sample 3	Sample 4	Sample 5
CO_2	5.6	5.13	7.65	0.92	0.01
N_2	0.6	0.67	0.27	0.22	0.35
C1	68.21	75.28	84.11	58.77	34.61
C2	7.1	8.19	5.79	7.57	4.12
C3	3.89	3.99	1.19	4.09	1
iC4	0.8	0.7	0.18	0.91	0.75
nC4	1.7	1.5	0.26	2.09	0.49
iC5	0.7	0.5	0.09	0.77	0.43
nC5	0.8	0.6	0.07	1.15	0.21
C6	1.51	0.72	0.07	1.76	1.62
C7 +	9.2	2.66	0.32	21.75	56.3
C7 + MW	242	140	157.7	227	276
GOR, scf/STB	5000	3300	120,000	1489	300

2.6 Consider an existing producing gas reservoir whose initial reservoir pressure is 5000 psi. Initial producing gas−oil ratio is 99,000 scf/STB. Identify the fluid type. Also, assess if there could be any expected liquid dropout in the reservoir or on the surface.

2.7 Well X (shown below) is a discovery well in a field in north Africa. It was discovered in 2012 and has production data given in the following two figures. The produced stock-tank liquid was initially yellowish with gravity of 55° API.

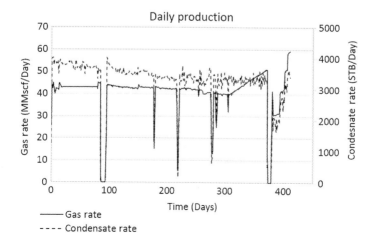

Daily production

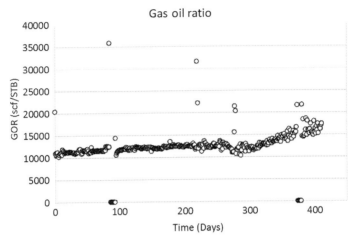

a. Classify this reservoir fluid.
b. What is the initial CGR?
c. During the production history of this field, the stock-tank liquid gravity increased to 62° API and the producing gas—oil ratio increased according to the GOR plot. Does this information confirm your classification? Why or why not?
d. Laboratory analysis of a sample from this reservoir gave the following composition. Does this information confirm your classification? Why or why not?

Component	Composition (Mole Fraction)
CO_2	0.0118
N_2	0.0267
H_2S	0.03500
C1	0.5827
C2	0.0952
C3	0.0624
C4'S	0.0312
C5'S	0.0297
C6	0.0158
C7+	0.1095
	1.000

Properties of heptanes plus	
Specific gravity	0.834
Molecular weight	216 lb/lb mole

e. During the depletion of the reservoir, do you think the saturation pressure was reached? If yes, when was it reached? Is it bubblepoint or dew point pressure?

Chapter 3

Dry Gases

Dry gases represent gases that when produced drop no condensate at the surface. Fig. 3.1 represents a typical phase diagram for what can be considered dry gas. Dry gases drop condensate neither at the surface nor in the reservoir, and so the reservoir gas composition is the same as the surface gas composition. The reservoir temperature is higher than the cricondentherm (maximum temperature of the two-phase region of the phase diagram). This behavior is theoretically the behavior of pure methane. In practice, the majority of gas reservoirs contain small amounts of higher hydrocarbons in addition to methane. Such natural gas reservoirs produce a small, inappreciable amount of condensate when the gas is brought to surface. As long as this amount of condensate is very small, the behavior of these reservoirs can usually be approximated with the dry gas model.

According to McCain and Piper (1994), the dry gas concept pertains to gases that contain C7 + mole percent less than 0.8%, or the gases with producing condensate gas ratio less than 10 STB/MMscf. This ratio is equivalent to initial gas-oil ratio (GOR) of higher than 100,000 scf/STB. In practice, a large portion of the gas produced worldwide fits these criteria and can be considered dry gas for engineering purposes.

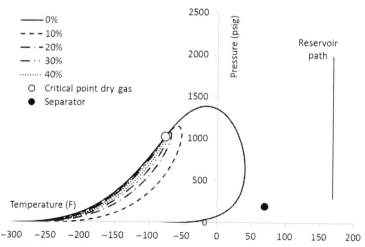

FIGURE 3.1   Theoretical dry gas phase diagram.

PVT Property Correlations. DOI: https://doi.org/10.1016/B978-0-12-812572-4.00003-5

29

PVT PROPERTIES FOR DRY GASES

In fact, the most widely available types of hydrocarbon gases can be considered dry gases. Although some of these gases can produce some condensate, the dry gas concept can be used to handle them. The engineering of dry gas reservoirs requires multiple calculations, including material balance, reservoir and well production forecasts, flow in pipes calculations, and surface facilities. All such calculations require estimation of gas PVT properties at different pressures and temperatures. The PVT properties necessary for engineering dry gases include z-factor, gas formation volume factor, gas density, gas viscosity, and isothermal gas compressibility.

Although all these gas PVT properties can be obtained in the laboratory by performing experiments on a representative sample, the number of full PVT reports for dry gases is not significant. It is usually sufficient to measure the composition of the dry gas. The composition can also be measured with simple equipment at the well site. Because dry gases are not expected to drop condensate at reservoir or surface, it is usually accurate to take the sample from the wellhead or at any point in the surface production system (e.g., separator).

Specific Gravity of Gas

Gas-specific gravity is defined as the ratio of the gas density to the density of dry air. Both densities are measured at the same temperature and pressure. This relation is given by the following equation:

$$\gamma_g = \frac{\rho_g}{\rho_a} \tag{3.1}$$

At standard conditions, the densities of gas and air can be represented by the ideal gas law, where the ideal gas law is given by the following equation:

$$p\, V = n\, R\, T \tag{3.2}$$

Both temperature and pressure must be at absolute conditions (i.e., temperature in °R and pressure in psia are used instead of °F and psig). The number of moles (n) can be represented by the mass of the gas divided by the molecular weight as in the following equation:

$$n = \frac{m}{M} \tag{3.3}$$

Recall that density is given by mass over volume, then

$$p\, V = \frac{m}{M} R\, T \tag{3.4}$$

and

$$\rho = \frac{m}{V} \tag{3.5}$$

Then,

$$p = \rho \frac{R\,T}{M} \tag{3.6}$$

Solving for density,

$$\rho = \frac{p\,M}{R\,T} \tag{3.7}$$

Substituting in the specific gravity definition, we have

$$\gamma_g = \frac{\rho_g}{\rho_a} = \frac{pM_g/R\,T}{pM_a/R\,T} = \frac{M_g}{M_a} = \frac{M_g}{29} \tag{3.8}$$

The molecular weight of the gas (also called apparent molecular weight) is calculated from the following equation:

$$M_g = \sum_{i=1}^{nc} y_i \times M_i \tag{3.9}$$

where y_i is the mole fraction of the components forming the gas mixture, and nc is the number of components. The mole fractions of the gas are measured in the compositional measurement procedure.

Gas Deviation Factor

Also called gas compressibility factor, z-factor, or supercompressibility, gas deviation factor is defined as the ratio between the actual (or real) volume of the gas and the ideal volume of the gas. It is given by the following equation:

$$z = \frac{V_{real}}{V_{ideal}} \tag{3.10}$$

Gas deviation factor is a measure of how the gas behavior deviates from the ideal behavior (given by the ideal gas law). It is determined in the laboratory by comparing the volume occupied by the gas under pressure and temperature with the volume of the gas calculated from the ideal gas law under the same pressure and temperature.

Gas deviation factor has been correlated with parameters that are readily available so it can be computed in absence of laboratory data. It approaches unity at pressures and temperatures close to standard pressure and temperature. The gas deviation factor is a property essential to determine for any

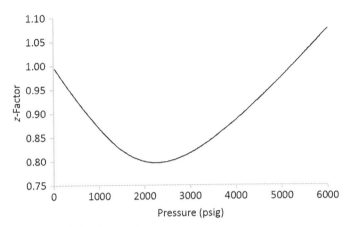

FIGURE 3.2 Gas deviation factor (z-factor).

further calculations involving dry gases. Fig. 3.2 shows a typical shape of z-factor for a dry gas at isothermal conditions.

Gas Formation Volume Factor

Formation volume factor of gas is defined as the volume of gas at reservoir pressure and temperature required to produce one standard volume of gas at the surface. Gas formation volume factor has units of volume/volume and can be represented in a variety of units (e.g., rcf/scf, rbbl/scf, and rbbl/Mscf). Gas formation volume factor is given by the following equation under the assumption that the z-factor is equal to 1 at standard conditions.

$$B_g = \frac{V_{res}}{V_{sc}} = \frac{znRT/p}{nRT_{sc}/p_{sc}} = \frac{zTp_{sc}}{T_{sc}p} \tag{3.11}$$

It will be noted that gas formation volume factor is fully defined from the real gas law. Therefore, no correlations are available to compute gas formation volume factor. Engineers must compute the z-factor (from correlations if not available from laboratory measurements) and then use it in the calculation of gas formation volume factor. Fig. 3.3 shows a typical gas formation volume factor plot for a dry gas.

Gas Density

Gas density is defined as the mass of the gas occupying a certain volume at specified pressure and temperature. The density is usually represented in units of lbm/ft^3. Another common density representation is the "gas gradient" that is expressed in units of psi/ft. Gas density is a function of the pressure and temperature conditions for the gas. Due to its high compressibility,

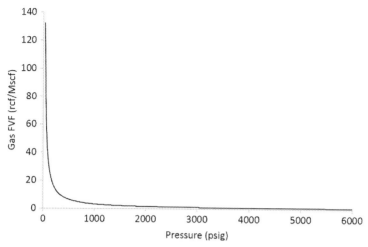

FIGURE 3.3   Gas formation volume factor.

gas can change its volume significantly with change in pressure. Therefore, density changes (at low pressure) can be significant. At high pressures, the gas molecules are packed together and may approach the behavior of liquid (small variation of density with pressure). The following two equations are used to calculate gas density in the two commonly used units.

$$\rho = \frac{p\,M}{z\,R\,T}\,\text{lbm}/\text{ft}^3 \qquad (3.12)$$

$$\rho = \frac{1}{144}\frac{pM}{zRT}\,\text{psi}/\text{ft} \qquad (3.13)$$

Gas density is not calculated from correlations. However, correlations can be used to calculate the z-factor, which is used in the above equations to calculate the density of the gas at a specified pressure and temperature. Fig. 3.4 shows a typical density plot for a dry gas.

Problem 1—Calculation of Gas Density and Specific Gravity

For the gas composition given in Table 3.1, calculate the gas density in lbm/ft^3 and gas gradient in psi/ft at 2000 psia and 60°F. Assume z-factor to be 0.97 at the specified pressure and temperature. Also calculate the gas-specific gravity at the same pressure and temperature.

Solution of Problem 1

We first calculate the apparent molecular weight of the gas (Table 3.2).

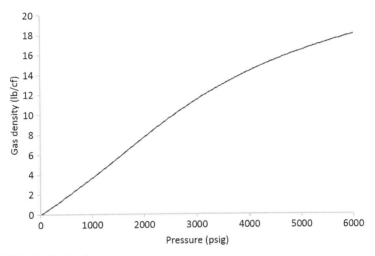

FIGURE 3.4 Gas density.

TABLE 3.1 Problem 1 Gas Composition

Component	Mole %
N_2	2.92
CO_2	1.07
C1	87.63
C2	4.77
C3	1.85
i-C4	0.49
n-C4	0.58
i-C5	0.20
n-C5	0.20
C6	0.29
Total	100

The gas density is given by

$$\rho = \frac{p\,M}{z\,R\,T} = \frac{(2000\text{psi})(18.76\text{lbm}/\text{lbm mol})}{(0.97)(10.732\text{psift}^3/\text{lbm molR})(520\text{R})}$$
$$= 6.93\text{lbm}/\text{ft}^3$$

TABLE 3.2 Problem 1 Calculations

Component	Mole %	Mole Fraction	Molecular Weight	$Y_i \times M_i$
N_2	2.92	0.0292	28.013	0.818
CO_2	1.07	0.0107	44.010	0.471
H_2S	0.00	0.0000	34.080	0.000
C1	87.63	0.8763	16.043	14.058
C2	4.77	0.0477	30.070	1.434
C3	1.85	0.0185	44.097	0.816
i-C4	0.49	0.0049	58.123	0.285
n-C4	0.58	0.0058	58.123	0.337
i-C5	0.20	0.0020	72.150	0.144
n-C5	0.20	0.0020	72.150	0.144
C6	0.29	0.0029	86.177	0.250
Total	100	1		18.76

$$\rho = \frac{1}{144}\frac{pM}{zRT} = \frac{6.93}{144} = 0.048 \text{psi/ft}$$

Apparent molecular weight of the gas:

$$M_g = \sum_{i=1}^{nc} y_i \times M_i = 18.76$$

Specific gravity of the gas:

$$\gamma_g = \frac{M_g}{29} = \frac{18.76}{29} = 0.6469$$

Gas Viscosity

Viscosity is defined as the resistance to flow exerted by a fluid. The viscosity is usually measured in centipoise (cp = g/100 s cm). The centipoise is a unit for measuring dynamic viscosity. Dynamic viscosity is of particular interest to reservoir and production engineers as it is a reflection of resistance of the fluid to flow. Kinematic viscosity, or static viscosity, is calculated by dividing the dynamic viscosity by density. Kinematic viscosity is measured in centistokes [centistokes = cp/(g/cm^3)].

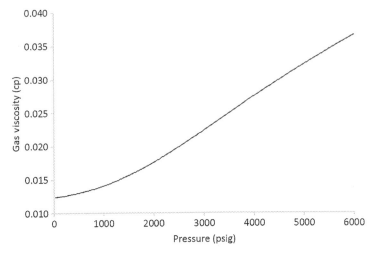

FIGURE 3.5 Gas viscosity.

Fig. 3.5 is a typical viscosity versus pressure plot for a dry gas at constant temperature. As can be noticed from the figure, the viscosity of gas increases with increasing pressure.

Values of gas viscosity are low. They increase with increasing density of gas. Viscosity of gas is difficult to measure in the laboratory, and so engineers typically rely on correlations to estimate it (Comings et al., 2007).

Gas Compressibility

Gas compressibility, or isothermal gas compressibility, is also called the coefficient of isothermal compressibility of gas. It is defined as the relative change in the volume of the gas with respect to the change in pressure at constant temperature. It is expressed in units of reciprocal pressure (usually psi^{-1}). Gas compressibility is defined by the following equation:

$$c_g = -\frac{1}{V}\left(\frac{\partial V}{\partial p}\right)_T \tag{3.14}$$

The compressibility of an ideal gas is given by the following equation:

$$c_g = \frac{1}{p} \tag{3.15}$$

For real gas, compressibility is given by the following equation:

$$c_g = \frac{1}{p} - \frac{1}{z}\left(\frac{\partial z}{\partial p}\right)_T \tag{3.16}$$

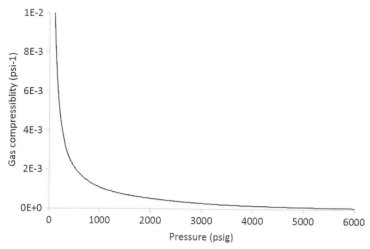

FIGURE 3.6 Gas compressibility.

At low pressures, when gas behaves as if it were an ideal gas, the z-factor derivative with respect to pressure is equal to zero, and the gas compressibility is given by the same form as the ideal gas compressibility. The ideal gas compressibility simple equation can be also used as an approximation for the gas compressibility.

Gas compressibility is measured during the laboratory procedure known as constant composition expansion (CCE). It can also be estimated from correlations or from the definition equation. If the definition equation is used to compute the gas compressibility, a correlation for z-factor must be used to compute the derivative of z-factor with respect to pressure at the given temperature.

Fig. 3.6 shows the typical behavior of gas compressibility. At low pressures, gas compressibility is very high, which means that the gas can expand significantly to occupy a large volume at low pressure. This property is responsible for the high recovery factor from volumetric dry gas reservoirs. At high pressures, gas compressibility decreases and eventually approaches the levels of liquid compressibility.

HANDLING PVT PROPERTIES FOR DRY GASES

PVT properties for gases can be determined in the PVT laboratory by the reservoir fluid study known as the *gas-condensate study*. The gas-condensate study consists of three main experiments (composition, CCE, constant volume depletion). In addition, gas viscosity is often estimated from correlations. A small number of gas-condensate PVT reports contain separator test(s). When a representative fluid sample or PVT laboratory report is not

TABLE 3.3 Gas Formation Volume Factor

Gas Formation Volume Factor	General Equation

TABLE 3.4 Gas Compressibility

Gas Compressibility	General Equation	McCain (1990)

TABLE 3.5 Gas Density

Gas Density	General Equation

TABLE 3.6 Gas Viscosity

Gas Viscosity	Al-Nasser and Al-Marhoun (2012)	Bicker and Katz (1943)	Carr et al. (1954)
De Ghetto et al. (1995)	Dean and Stiel (1965)	Dempecy (1965)	ElSharkawy (2006)
Lee et al. (1966)	Londono et al. (2002)	Londono et al. (2005)	Lucas (1981)
McCain (1990)	Standing (1947)	Sutton (2007)	

available, correlations are used to calculate the PVT properties for gases. A review of laboratory experiments for gases is given in other texts (Ahmed, 2007; Whitson and Brule, 2000).

The inputs to dry gas correlations are rather straightforward. One needs to know either the specific gravity of the gas or the gas composition to be able to calculate all PVT properties for dry gas at any pressure and temperature.

Several correlations were developed over the years to estimate dry gas PVT properties. The calculation of the gas deviation factor is therefore one of the first steps in preparing PVT properties of dry gases. Tables 3.3–3.9 summarize the majority of the known correlations for gas PVT properties.

Estimation of Critical Properties

Critical pressure and temperature are defined as the pressure and temperature at the critical point. For a gas mixture, the critical point is defined as the

TABLE 3.7 z-Factor

	Brill and Beggs (1978)	Burnett (1979)	Carlile and Gillett (1971)
Dindoruk and Christman	Dranchuk and Abou-Kassem (1975)	Dranchuk et al. (1974)	Elechi et al. (2015)
Gopal (1977)	Gray and Sims (1959)	Hall and Yarborough (1973)	Hankinson et al. (1969)
Leung (1983)	Mahmoud (2013)	Nishiumi and Saito (1975)	Obuba et al. (2013)
Papay (1985)	Papp (1979)	Said and El-Banbi (2005)	Sarem (1961)
Standing (1977)	Standing and Katz (1942)		

TABLE 3.8 Critical Pressure and Temperature Using Specific Gravity

T_c and P_c Using Specific Gravity	Elsharkawy et al. (2000)	Londono et al. (2005)	Piper et al. (1993)
Said and El-Banbi (2005)	Standing (1977)	Sutton (1985)	Sutton (2005)
Sutton (2007)	Whitson and Brule (2000)	Wichert and Aziz (1972)	

TABLE 3.9 Critical Pressure and Temperature Using Composition

T_c and P_c Using Composition	Corredor et al. (1992)	Elsharkawy et al. (2000)	Kay (1936)
Piper et al. (1993)	Stewart et al. (1959)	Sutton (1985)	

point (pressure and temperature) at which all properties of the liquid and the gas become identical. It is also the point at which the bubble point line joins the dew point line. For a pure substance, the critical pressure is defined as the pressure above which liquid and gas cannot coexist at any temperature. The critical temperature for a pure substance is the temperature above which the gas cannot become liquid, regardless of the applied pressure.

Critical pressures and temperatures for pure hydrocarbon components are known and can be obtained from many sources. Table 3.10 lists the critical

TABLE 3.10 Critical Properties of Pure Components

ID	Compound	Formula	Mo.Wt. (lb mol)	P_c (psia)	T_c (F)	V_c (ft^3/lbm)	Density (60°F, 14.696 psia)	
							Specific Gravity	lbm/ft^3
1	Methane	CH_4	16.04	667.8	−116.7	0.0988	0.3000	18.701
2	Ethane	C_2H_6	30.07	707.8	90.1	0.0788	0.3562	22.214
3	Propane	C_3H_8	44.10	616.3	206.0	0.0737	0.5070	31.619
4	n-Butane	C_4H_{10}	58.12	550.7	305.6	0.0703	0.5629	35.104
5	Isobutene	C_4H_{10}	58.12	529.1	275.0	0.0724	0.5840	33.430
6	n-Bentane	C_5H_{12}	72.15	488.6	385.6	0.0674	0.6247	38.960
7	Isopentane	C_5H_{12}	72.15	490.4	369.0	0.0679	0.6311	39.360
8	Neopentane	C_5H_{12}	72.15	464.0	321.1	0.0673	0.5967	37.211
9	n-Hexane	C_6H_{14}	86.18	436.9	453.6	0.0689	0.6638	41.400
10	2-Methylpentane	C_6H_{14}	86.18	436.6	435.7	0.0682	0.6579	41.028
11	3-Methylpentane	C_6H_{14}	86.18	453.1	448.2	0.0682	0.6690	41.723
12	Neohexane	C_6H_{14}	86.18	446.9	420.0	0.0668	0.6539	40.778
13	2,3-Dimethylbutane	C_6H_{14}	86.18	453.5	440.0	0.0665	0.6663	41.524
14	n-Heptane	C_7H_{16}	100.21	396.8	512.7	0.0690	0.6882	42.920
15	2-Methylhexane	C_7H_{16}	100.21	396.5	494.9	0.0673	0.6831	42.602
16	3-Methylhexane	C_7H_{16}	100.21	408.1	503.7	0.0646	0.6917	43.136
17	3-Ethylpentane	C_7H_{16}	100.21	419.3	513.4	0.0665	0.7028	43.828
18	2,2-Dimethylpentane	C_7H_{16}	100.21	402.2	477.1	0.0665	0.6783	42.302

19	2,4-Dimethylpentane	C_7H_{16}	100.21	397.0	475.8	0.0668	0.6773	42.242
20	3,3-Dimethylpentane	C_7H_{16}	100.21	427.1	505.7	0.0662	0.6977	43.514
21	Triptane	C_7H_{16}	100.21	428.4	496.3	0.0636	0.6946	43.317
22	n-Octane	C_8H_{18}	114.23	360.6	564.1	0.0690	0.7070	44.090
23	Diisobutyl	C_8H_{18}	114.23	360.6	530.3	0.0676	0.6979	43.527
24	Isooctane	C_8H_{18}	114.23	372.5	519.3	0.0657	0.6962	43.421
25	n-Nonane	C_9H_{20}	128.26	331.8	610.5	0.0684	0.7219	45.020
26	n-Decane	$C_{10}H_{22}$	142.29	304.4	651.6	0.0679	0.7342	45.790
27	Carbon monoxide	CO	28.01	507.5	-220.4	0.0532	0.7894	49.231
28	Carbon dioxide	CO_2	44.01	1071.0	87.9	0.0342	0.8180	51.016
29	Hydrogen sulfide	H_2S	34.08	1306.0	212.6	0.0460	0.8014	49.983
30	Sulfur dioxide	SO_2	64.06	1145.0	315.8	0.0306	1.3974	87.148
31	Ammonia	NH_3	17.03	1636.0	270.4	0.0681	0.6183	38.562
32	Air	N_2O_2	28.96	546.9	-221.4	0.0517	0.8748	54.555
33	Hydrogen	H_2	2.02	188.1	-399.9	0.5164	0.0711	4.432
34	Oxygen	O_2	32.00	736.9	-181.2	0.0367	1.1421	71.230
35	Nitrogen	N_2	28.01	493.0	-232.7	0.0516	0.8094	50.479
36	Chlorine	CL_2	70.91	1118.4	291.0	0.0280	1.4244	88.831
37	Water	H_2O	18.02	3207.9	705.5	0.0509	1.0000	62.366

properties of the more common hydrocarbon and nonhydrocarbon components normally found in reservoir fluids.

Critical pressures and temperatures are not usually measured for hydrocarbon mixtures or for the plus fraction in a composition experiment and must therefore be estimated from correlations. Critical pressure and temperature correlations require that either the specific gravity of the gas or the full composition be known. If the critical properties are estimated for a mixture (rather than a pure component), they are termed pseudocritical pressure and pseudocritical temperature.

If the gas-specific gravity is known or estimated, several correlations can be used for computing pseudocritical pressure and temperature. Examples of these correlations include Standing (1977) and Sutton (2007). If impurities exist (nonhydrocarbon components such as CO_2, H_2S, and N_2), a correlation such as Wichert and Aziz (1972) will be used to correct the pseudocritical properties for the presence of nonhydrocarbons. Note that the Wichert and Aziz correlation requires the composition of the nonhydrocarbons. Other correlations that correct directly for the presence of nonhydrocarbons also exist (e.g., Piper et al., 1993; Elsharkawy et al., 2000).

If the full composition of the gas is known, other sets of correlations can be used to estimate the pseudocritical pressure and temperature for the gas mixture. The most straightforward are Kay's rules (Kay, 1936). These mixing rules are calculated as follows:

$$p_{pc} = \sum_{i=1}^{nc} y_i p_{ci} \qquad (3.17)$$

$$T_{pc} = \sum_{i=1}^{nc} y_i T_{ci} \qquad (3.18)$$

Calculation of pseudocritical properties from composition requires that critical pressure and temperature be known for the plus fraction. Numerous correlations exist for calculating the pseudocritical properties for the plus fraction such as C7 + (Riazi, 2005). These correlations depend on different inputs. The common inputs are molecular weight and/or specific gravity of the C7 + (both are usually measured in the PVT laboratory for the plus fraction) and normal boiling temperature. In general, for dry gases, the C7 + mole fraction is a small value. Therefore, the weight of C7 + critical properties is also small in the summations given by Eqs. (3.17) and (3.18). Any C7 + critical properties correlations can therefore be used. Wichert and Aziz (1972) correlation can then be used to correct for nonhydrocarbon impurities.

Other correlations that use composition and correct for nonhydrocarbon impurities also exist (e.g., Piper et al., 1993; Elsharkawy et al., 2000). These correlations and others are given in Appendix B. Their ranges of applicability are given in Appendix D.

Estimation of Reduced Pressure and Temperature

Some PVT properties are correlated with reduced pressure and temperature. The most critical is the z-factor. When defined for a gas mixture, reduced pressure and temperature are usually referred to as *pseudoreduced pressure* and *pseudoreduced temperature*. They are defined by the following equations:

$$p_{pr} = \frac{p}{p_{pc}} \tag{3.19}$$

$$T_{pr} = \frac{T}{T_{pc}} \tag{3.20}$$

Estimation of Gas Deviation Factor

Gas deviation factor (z-factor) is used in the calculation of almost all PVT properties of gas. There are more than 20 known (and different) correlations for calculating z-factor. All of them compute z-factor from known pseudoreduced pressure and pseudoreduced temperature.

Estimation of Other PVT Properties of Gas

Several gas properties can be directly calculated if the z-factor is known at specific pressure and temperature. These properties include gas formation volume factor, density, and compressibility of the gas. Viscosity of the gas, however, is computed from correlations. Several correlations for gas viscosity exist.

Sequence of Calculations

The main input that allows engineers to calculate all dry gas PVT properties is either the gas-specific gravity or the gas composition. In either case, impurities (nonhydrocarbon) mole fractions must be known for greater accuracy. The first step is to calculate the pseudocritical pressure and temperature using one of the correlations. The second step is to calculate the pseudoreduced pressure and temperature at the specified system pressure and temperature. Then z-factor is calculated from any of the available correlations. After that, other PVT properties are calculated using the z-factor values.

Fig. 3.7 is a flow diagram that describes the sequence of calculations for computing dry gas PVT properties.

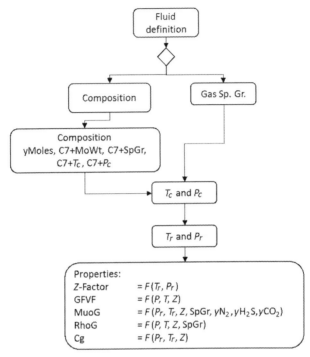

FIGURE 3.7 Sequence of calculations for computing dry gas PVT properties.

APPLICATIONS OF TECHNIQUES TO COMPUTE PVT PROPERTIES FOR DRY GASES

Many gas reservoirs worldwide can be treated as dry gas for engineering purposes. These reservoirs do not drop appreciable amounts of condensate when the gas is produced at surface. Engineering these reservoirs requires the calculation of their PVT properties. The following solved examples illustrate use of the correlations to arrive at estimates of gas PVT properties.

Problem 2—Estimation of Critical Properties

For the gas composition given in Table 3.11, calculate critical pressure and temperature from both gas-specific gravity and gas composition by the following methods:

- Sutton (2007) using the gas-specific gravity.
- Kay's rules (Kay, 1936) and Elsharkawy (2000) using the gas composition.

TABLE 3.11 Problem 2 Gas Composition

	y_i (mol%)
Hydrogen sulfide	0.50%
Carbon dioxide	1.25%
Nitrogen	0.00%
Methane	75.75%
Ethane	8.25%
Propane	6.35%
i-Butane	2.15%
n-Butane	2.25%
i-Pentane	1.15%
n-Pentane	1.10%
Hexanes	0.50%
Heptanes plus	0.75%
Sum	100%

C7 + molecular weight and specific gravity are estimated in the laboratory to be 143 lb/lb mol and 0.752, respectively. Use Riazi and Daubert (1987) to calculate critical properties for C7 + .

Solution of Problem 2

From the composition, the gas-specific gravity would be calculated as in Table 3.12.

Using Gas-Specific Gravity to Calculate the Gas Mixture Critical Properties

Application of Sutton (2007)

$$P_{PcHC} = 744 - 125.4\gamma_{gHC} + 5.9\gamma_{gHC}^2 \tag{3.21}$$

and

$$T_{PcHC} = 164.3 + 357.7\gamma_{gHC} - 67.7\gamma_{gHC}^2 \tag{3.22}$$

where

$$\gamma_{gHC} = \frac{\gamma_g\left(y_{H_2S}M_{H_2S} + C_{CO_2}M_{CO_2} + y_{N_2}M_{N_2}\right)/M_{air}}{y_{HC}} \tag{3.23}$$

TABLE 3.12 Gas-Specific Gravity

	Y_i (mol%)	M_w (Mo.Wt.)	$y_i \times M_w$
Hydrogen sulfide	0.50%	34.08	0.17
Carbon dioxide	1.25%	44.01	0.55
Nitrogen	0.00%	28.01	0.00
Methane	75.75%	16.04	12.15
Ethane	8.25%	30.07	2.48
Propane	6.35%	44.10	2.80
i-Butane	2.15%	58.12	1.25
n-Butane	2.25%	58.12	1.31
i-Pentane	1.15%	72.15	0.83
n-Pentane	1.10%	72.15	0.79
Hexanes	0.50%	86.18	0.43
Heptanes plus	0.75%	143.00	1.07
Sum	1		23.84

Gas Sp. Gr. = 23.84/29.1 = 0.82.

and

$$y_{HC} = 1 - y_{N_2} - y_{CO_2} - y_{H_2S} \qquad (3.24)$$

Then:

$$y_{HC} = 1 - 0.0 - 0.0125 - 0.005 = 0.9825$$

$$\gamma_{gHC} = \frac{0.82(0.17 + 0.55 + 0.0)/29.1}{0.9825} = 0.021$$

$$P_{PcHC} = 744 - 125.4 \times 0.021 + 5.9 \times 0.021^2 = 741.4$$

$$T_{PcHC} = 164.3 + 357.7 \times 0.021 - 67.7 \times 0.021^2 = 171.7F(631.4R)$$

Using Gas Composition to Calculate the Gas Mixture Critical Properties

First, we need to calculate the critical properties for C7 +. We use Riazi and Daubert (1987).

$$T_c = 544.4 \times Mo.Wt.^{0.2998} \times \gamma^{1.0555} \times e^{-1.3478 \times 10^{-4} \times Mo.Wt. - 0.61641 \times \gamma} \qquad (3.25)$$

$$P_c = 4.5203 \times 10^4 \times \text{Mo.Wt.}^{-0.8063} \times \gamma^{1.6015} \times e^{-1.8078 \times 10^3 \times \text{Mo.Wt.} - 0.3084 \times \gamma} \tag{3.26}$$

Then:

$$T_c = 1099.18 \ R \ (639.5 \ F) \ \text{and} \ P_c = 319.5 \ \text{psia}$$

Kay's Rules (Kay 1936) With Wichert and Aziz Correction

$$T_{pc} = T_{pc}^* - \epsilon \tag{3.27}$$

$$P_{pc} = \frac{P_{pc}^*(T_{pc}^* - \epsilon)}{T_{pc}^* + y_{H_2S}(1 - y_{H_2S})\epsilon} \tag{3.28}$$

$$\epsilon = 120\left[\left(y_{CO2} + y_{H_2S}\right)^{0.9} - \left(y_{CO2} + y_{H_2S}\right)^{1.6} + 15\left(y_{CO2}^{0.5} - y_{H_2S}^4\right)\right] \tag{3.29}$$

$$P_{pc}^* = \sum_{i=1}^{N} y_i P_{ci} \tag{3.30}$$

$$T_{pc}^* = \sum_{i=1}^{N} y_i T_{ci} \tag{3.31}$$

Then

Table 3.13 summarizes the T_c^* and P_c^* calculation using Kay's rules (1936):

Then $T_c^* = -37.3$ F (427.7 R) and $P_c^* = 652.2$ psia

$$\epsilon = 120\left[(0.0125 + 0.005)^{0.9} - (0.0125 + 0.005)^{1.6} + 15\left(0.0125_{CO2}^{0.5} - 0.005_{H_2S}^4\right)\right]$$
$$= 4.82$$

Then $T_c = 423$ R and $P_c = 646$ psia

Elsharkawy (2000)

$$T_{pc} = \frac{K^2}{J} \tag{3.32}$$

and

$$P_{pc} = \frac{T_{pc}}{J} \tag{3.33}$$

TABLE 3.13 Critical Properties Calculation by Kay's Rules

	y_i (mol%)	T_c	P_c	$y_i \times T_c$	$y_i \times P_c$
Hydrogen sulfide	0.50%	212.40	1291.34	1.06	6.46
Carbon dioxide	1.25%	87.69	1058.26	1.10	13.23
Nitrogen	0.00%	−233.10	477.33	0.00	0.00
Methane	75.75%	−116.52	658.38	−88.26	498.72
Ethane	8.25%	89.80	693.65	7.41	57.23
Propane	6.35%	206.01	602.68	13.08	38.27
i-Butane	2.15%	274.69	514.36	5.91	11.06
n-Butane	2.25%	305.29	535.96	6.87	12.06
i-Pentane	1.15%	369.81	468.36	4.25	5.39
n-Pentane	1.10%	385.59	474.83	4.24	5.22
Hexanes	0.50%	454.10	425.01	2.27	2.13
Heptanes plus	0.75%	639.50	319.50	4.80	2.40
Sum	100%			−37.3	652.2

where

$$J = \left[\sum y_i \left(T_c/P_c \right) \right]_{C_1-C_6}$$
$$+ \left\{ a_0 + \left[a_1 \left(yT_c/P_c \right) \right]_{C_{7+}} + \left[a_2 \left(yT_c/P_c \right) \right]_{N_2} \right.$$
$$\left. + \left[a_3 \left(yT_c/P_c \right) \right]_{CO_2} + \left[a_4 \left(yT_c/P_c \right) \right]_{H_2S} \right\}$$
(3.34)

$$K = \left[\sum y_i \left(T_c/P_c^{0.5} \right) \right]_{C_1-C_6}$$
$$+ \left\{ b_0 + \left[b_1 \left(yT_c/P_c^{0.5} \right) \right]_{C_{7+}} + \left[b_2 \left(yT_c/P_c^{0.5} \right) \right]_{N_2} \right.$$
$$\left. + \left[b_3 \left(yT_c/P_c^{0.5} \right) \right]_{CO_2} + \left[b_4 \left(yT_c/P_c^{0.5} \right) \right]_{H_2S} \right\}$$
(3.35)

Where the equations parameters are given in Table 3.14.
 By using Elsharkawy (2000)

$J = -0.0791$
$K = -2.0102$

Then

$T_c = -51.07$
$P_c = 645.53$

TABLE 3.14 Elsharkawy (2000) Constants

i	a	b
0	−0.040279933	−0.776423332
1	0.881709332	1.030721752
2	0.800591625	0.734009058
3	1.037850321	0.909963446
4	1.059063178	0.888959152

TABLE 3.15 Results Comparison

	Sutton	Elsharkawy	Kay's Rules
T_c (R)	631.4	408.6	422.8
P_c (psia)	741.4	645.5	644.8

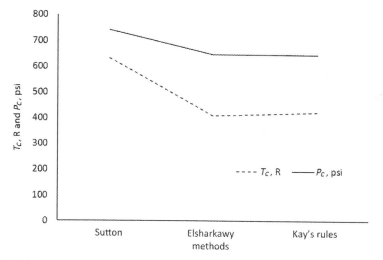

FIGURE 3.8 Comparison plot for the critical properties using the three methods.

Table 3.15 and Fig. 3.8 summarize the results for the critical properties using the three methods:

Problem 3—Estimation of Gas Deviation Factor

For the gas in Problem 2, estimate the z-factor at 230°F and for a pressure range from 100 to 6000 psia using Dranchuk and Abou-Kassem (1975) and

TABLE 3.16 Results Comparison

Pressure (psia)	z-Factor	
	Dranchuk and Abou-Kassem	Hall and Yarborough
100	0.986	0.986
521	0.930	0.929
943	0.879	0.877
1364	0.836	0.834
1786	0.805	0.804
2207	0.788	0.790
2629	0.784	0.792
3050	0.794	0.807
3471	0.817	0.831
3893	0.849	0.863
4314	0.889	0.899
4736	0.933	0.939
5157	0.977	0.980
5589	1.020	1.022
6000	1.062	1.065

Hall and Yarborough (1973). Compare the two calculated z-factors. Use the pseudocritical properties estimated by Kay's rules.

Solution of Problem 3

Table 3.16 shows the calculation results.
   Fig. 3.9 shows the comparison plot.

Problem 4—Estimation of Other Gas PVT Properties

For the above gas (Problem 3) and using the estimated z-factor with the two correlations, compute the gas formation volume factor, gas density, and gas compressibility. Compare the two z-factor calculations to show the effect of z-factor on different PVT properties.
   Also estimate gas viscosity using Lee et al. (1966).

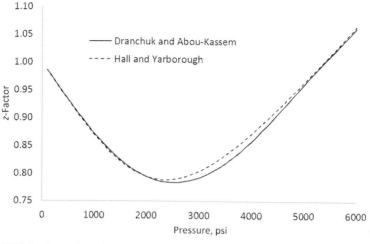

FIGURE 3.9   Comparison plot for the z-Factor calculation.

Solution of Problem 4

Tables 3.17 and 3.18 present the calculation results using the two different z-factors.

Figs. 3.10−3.17 show the comparison plots for gas formation volume factor, gas density, gas compressibility, and gas viscosity. In each plot, the two z-factors were used to calculate each gas property. The difference between the two calculations is shown in the figure following each property comparison plot.

An interesting observation from these plots is that small differences in z-factor yield even smaller differences in other properties such as gas formation volume factor, gas density, gas compressibility, and gas viscosity.

ENGINEERING DRY GAS RESERVOIRS

PVT properties of the gas must be estimated so reserves estimation, production forecast, and gas flow in pipes can be calculated. In addition, heat content is usually necessary for pricing of the gas. The following section details the heat content calculations for any gas.

Heat Content

Heat content of the gas is defined as the amount of heat produced when gas is burned completely to carbon dioxide and water. It is usually measured in BTU/scf. The net heating value (also known as the lower calorific value) of a gas is defined as the amount of heat released by combusting a specified quantity and returning the temperature of the combustion products to 150°C,

TABLE 3.17 DAK Results

Dranchuk and Abou-Kassem

Pressure (psia)	z-Factor	B_g (cf/SCF)	Gas Gradient (psi/ft)	c_g (psi^{-1})	Gas Viscosity (cp)
100	0.99	0.1925	0.002	1.01E − 02	0.041
521	0.93	0.0348	0.013	2.05E − 03	0.042
943	0.88	0.0182	0.024	1.15E − 03	0.043
1364	0.84	0.0120	0.036	7.66E − 04	0.044
1786	0.81	0.0088	0.050	5.23E − 04	0.046
2207	0.79	0.0070	0.063	3.57E − 04	0.048
2629	0.78	0.0058	0.075	2.48E − 04	0.050
3050	0.79	0.0051	0.086	1.81E − 04	0.052
3471	0.82	0.0046	0.095	1.41E − 04	0.054
3893	0.85	0.0043	0.102	1.17E − 04	0.056
4314	0.89	0.0040	0.108	1.01E − 04	0.058
4736	0.93	0.0038	0.113	8.89E − 05	0.059
5157	0.98	0.0037	0.118	7.98E − 05	0.061
5579	1.02	0.0036	0.122	7.24E − 05	0.062
6000	1.06	0.0035	0.126	6.63E − 05	0.064

which assumes that the latent heat of vaporization of water in the reaction products is not recovered. The gross heating value (also known as the higher calorific value, gross energy, or total heating value) is defined as the amount of heat released by a specified quantity once it is combusted and the products have returned to a temperature of 25°C, which takes into account the latent heat of vaporization of water in the combustion products.

For the conditions of the gas before combustion, the heating value of the gas is referred to as either wet or dry. Wet heating value means that the gas is saturated with water vapor before combustion. Dry heating value means that the gas contains no water vapor.

The heating value of an ideal gas at standard conditions is calculated from the following equation:

$$L_{c,ideal} = \sum_{i=1}^{nc} y_i L_{ci} \tag{3.36}$$

TABLE 3.18 Hall and Yarborough Results

Pressure (psia)	z-Factor	B_g (cf/SCF)	Gas Gradient (psi/ft)	c_g (psi^{-1})	Gas Viscosity (cp)
100	0.99	0.1925	0.002	1.01E − 02	0.041
521	0.93	0.0348	0.013	2.05E − 03	0.042
943	0.88	0.0181	0.024	1.15E − 03	0.043
1364	0.83	0.0119	0.037	7.65E − 04	0.044
1786	0.80	0.0088	0.050	5.22E − 04	0.046
2207	0.79	0.0070	0.062	3.58E − 04	0.047
2629	0.79	0.0059	0.074	2.51E − 04	0.050
3050	0.81	0.0052	0.084	1.81E − 04	0.052
3471	0.83	0.0047	0.093	1.45E − 04	0.054
3893	0.86	0.0043	0.101	1.19E − 04	0.055
4314	0.90	0.0041	0.107	1.02E − 04	0.057
4736	0.94	0.0039	0.113	8.94E − 05	0.059
5157	0.98	0.0037	0.118	8.00E − 05	0.061
5579	1.02	0.0036	0.122	7.25E − 05	0.062
6000	1.06	0.0035	0.126	6.64E − 05	0.064

The following equation is used to convert the calculated heating value from ideal gas to real gas at standard conditions.

$$L_c = \frac{L_{c,\text{ideal}}}{z} \tag{3.37}$$

The compressibility factor at standard conditions is calculated using the individual hydrocarbon components' z-factors at standard conditions and using the following equation:

$$z = 1 - \left(\sum_{i=1}^{nc} y_i \sqrt{1 - z_j} \right)^2 \tag{3.38}$$

The correction for heat content from ideal gas behavior to real gas behavior is minor and does not make significant difference in the overall heat content for any natural gas. Table 3.19 lists the net and gross heating values for the hydrocarbon components of natural gases at standard conditions.

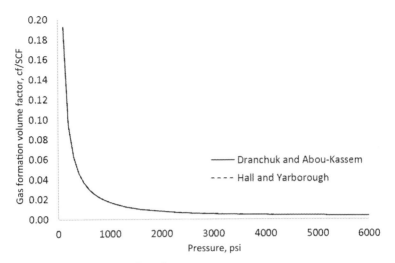

FIGURE 3.10 Gas formation volume factor.

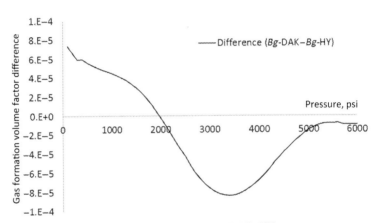

FIGURE 3.11 Gas formation volume factor difference (DAK−HY).

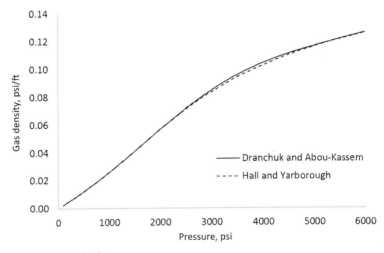

FIGURE 3.12 Gas density.

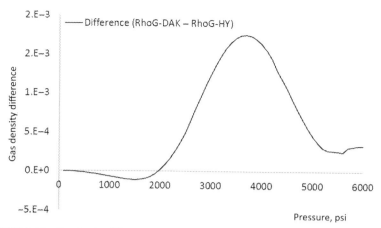

FIGURE 3.13 Gas density difference (DAK−HY).

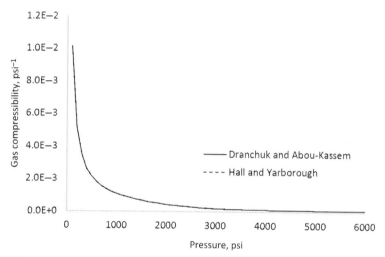

FIGURE 3.14 Gas compressibility.

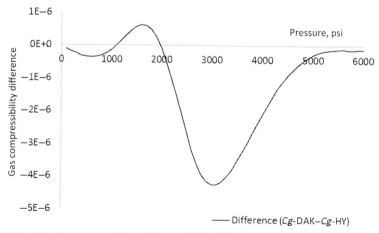

FIGURE 3.15 Gas compressibility difference (DAK−HY).

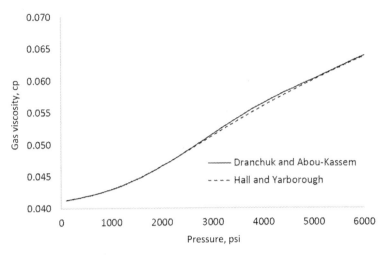

FIGURE 3.16 Gas viscosity.

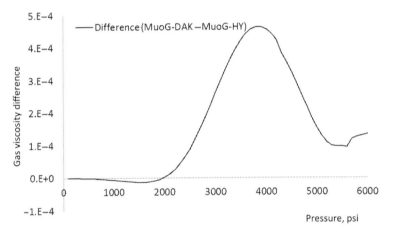

FIGURE 3.17 Gas viscosity difference (DAK−HY).

Problem 5—Estimation of Heating Value of Natural Gas

Calculate the net heat content for the gas given in Problem 1.

Solution of Problem 5

Following the equations for calculating the heating value of the gas from its composition, we obtain the following results:

$$L_{c,\text{ideal}} = 1080.7 \, \text{BTU/scf}$$

$$z = 0.9973$$

$$L_c = 1084 \, \text{BUT/scf}$$

The details of the calculations are given in Table 3.20.

TABLE 3.19 Net and Gross Heating Values for the Hydrocarbon Components

Gas	Gross Heating Value (BTU/scf)	Net Heating Value (BTU/scf)	z-Factor at the Standard Conditions
Methane	1012	911	0.998
Ethane	1783	1631	0.9919
Propane	2557	2353	0.9825
Isobutene	3354	3094	0.9711
n-Butane	3369	3101	0.9667
Isopentane	4001	3698	0.948
n-Pentane	4009	3709	0.942
Neopentane	3987	3685	–
n-Hexane	4755.9	4403.8	0.91
2-Methylpentane	4747.3	4395.2	–
3-Methylpentane	4750.3	4398.2	–
Neohexane	4736.2	4384	–
2,3-Dimethylbutane	4745	4392.9	–
n-Heptane	5502.5	5100	–
2-Methylhexane	5494.6	5092.2	–
3-Methylhexane	5498.6	5096	–
3-Ethylpentane	5500.7	5098.3	–
2,2-Dimethylpentane	5481.9	5079.6	–
2,4-Dimethylpentane	5486.7	5084.2	–
3,3-Dimethylpentane	5488.8	5086.4	–
Triptane	5483.5	5081.2	–
n-Octane	6248.9	5796.1	–
Diisobutyl	6233.5	5780.5	–
Isooctane	6231.7	5778.8	–
n-Nonane	6996.5	6493.2	–
n-Decane	7742.9	7189.6	–

(Continued)

TABLE 3.19 (Continued)

Gas	Gross Heating Value (BTU/scf)	Net Heating Value (BTU/scf)	z-Factor at the Standard Conditions
Cyclopentane	3763.7	3512.1	–
Methylcyclopentane	4501.2	4199.4	–
Cyclohexane	4481.7	4179.7	–
Methylcyclohexane	5215.9	4863.6	–
Ethene (ethylene)	1599.8	1499.1	–
Propene (propylene)	2332.7	2181.8	–
1-Butene (butylene)	3079.9	2878.7	–
cis-2-Butene	3072.2	2871	–
trans-2-Butene	3068	2866.8	–
Isobutene	3061.1	2859.9	–
1-Pentene	3826.5	3575	–
1,2-Butadiene	2939.9	2789	–
1,3-Butadiene	2879.9	2729	–
Isoprene	3612.1	3410.8	–
Acetylene	1475.5	1423.2	–
Carbon monoxide	320.5	320.5	0.9997
Hydrogen sulfide	637.1	586.8	0.9943

TABLE 3.20 Heating Value Calculation for Problem 5

Component	Mole %	L_{cj}, Heating Value (BTU/scf)	$y_j L_{cj}$	z_j	$\frac{y_j}{(1-z_j)^{0.5}}$
N_2	2.92	0	0.0000	0.9943	0.002205
CO_2	1.07	0	0.0000	0.9997	0.000185
C1	87.63	1010	885.0630	0.9980	0.093189
C2	4.77	1769.6	84.4099	0.9919	0.004293
C3	1.85	2516.1	46.5479	0.9825	0.002447

(Continued)

TABLE 3.20 (Continued)

Component	Mole %	L_{cj}, Heating Value (BTU/scf)	$y_j L_{cj}$	z_j	$\frac{y_j}{(1 - z_j)^{0.5}}$
i-C4	0.49	3251.9	15.9343	0.9711	0.000833
n-C5	0.58	3262.3	18.9213	0.9667	0.001058
i-C5	0.20	4000.9	8.0018	0.9480	0.000456
n-C5	0.20	4008.9	8.0178	0.9420	0.000482
C6	0.29	4755.9	13.7921	0.9100	0.000870
C7+	0.00	5502.5	0.0000	0.8520	0.000000
			1080.6881	BTU/ scf	0.052019
		$z = 0.9972941$			
		$L_c = 1084$ BTU/scf			

NOMENCLATURE

CCE	constant composition expansion
CVD	constant volume depletion
B_g	gas formation volume factor (rcf/scf)
m	mass of gas (lbm)
$L_{c,\text{ideal}}$	heating value of ideal gas (BTU/scf)
L_c	heating value of real gas (BTU/scf)
L_{ci}	heating value of component i (BTU/scf)
M_a	molecular weight of air (lbm/lbm mol)
M_g	molecular weight of gas (lbm/lbm mol)
n	number of moles of gas
nc	number of components of the gas mixture
p	pressure (psia)
P_c	critical pressure (psia)
P_{pc}	pseudocritical pressure (psia)
P_r	reduced pressure (dimensionless)
P_{pr}	pseudoreduced pressure (dimensionless)
R	universal gas constant (psiaft^3/lbm mol °R)
T	temperature (°R)
T_c	critical temperature (°R)
T_{pc}	pseudocritical temperature (°R)
T_r	reduced temperature (dimensionless)
T_{pr}	pseudoreduced temperature, dimensionless
V	volume of gas (ft^3)
V_{res}	volume of gas at reservoir conditions (ft^3)

V_{sc} volume of gas at standard conditions (ft^3)
y_i mole fraction of the components in the gas (fraction)
z gas deviation factor (dimensionless)
γ_g specific gravity of gas (air = 1.0)
ρ_a density of air (lbm/ft^3)
ρ_g Density of gas (lbm/ft^3)

REFERENCES

Ahmed, T., 2007. Equations of State and PVT Analysis Applications for Improved Reservoir Modeling. Gulf Publishing Company, Houston, TX.

Al-Nasser, K.S., Al-Marhoun, M.A., 2012. Development of New Gas Viscosity Correlations. Society of Petroleum Engineers, Richardson, TX, USA. Available from: https://doi.org/10.2118/153239-MS.

Brill, J.P., Beggs, H.D., 1974. Two-Phase Flow in Pipes. University of Tulsa, INTERCOMP Course, The Hague, The Netherlands.

Bicker, L.A., Katz, D.L., 1943. Viscosity of natural gases. Trans. AMIE 155, 754.

Burnett, R.R., 1979. Calculator gives compressibility factors. Oil Gas J. 70−74.

Carlile, R.E., Gillett, B.E., 1971. Digital solutions of an integral. Oil Gas J. 68−72.

Carr, N.L., Kobayashi, R., Burrows, D.B., 1954. Viscosity of hydrocarbon gases under pressure. Trans. AIME 201, 264−272.

Comings, E.W., Mayland, B.J., Egly, R.S., 2007. The Viscosity of Gases at High Pressure. University of Illinois, Chicago, USA, Bulletin Series No. 354.

Corredor, J.H., Piper, L.D., McCain Jr., W.D., 1992. Compressibility Factors for Naturally Occurring Petroleum Gases (1992 Version). Society of Petroleum Engineers. Available from: https://doi.org/10.2118/24864-MS.

De Ghetto, G., Paone, F., Villa, M. 1995. Pressure-volume-temperature correlations for heavy and extra heavy oils. In: Paper SPE 30316-MS Presented at the International Heavy Oil Symposium, Calgary, 19−21 June.

Dean, D.E., Stiel, L.I., 1965. The viscosity of nonpolar gas mixtures at moderate and high pressures. AICHE J. 526−532.

Dempsey, J.R., 1965. Computer routine treats gas viscosity as a variable. Oil Gas J. 141.

Dindoruk, B., 2012. Development of a correlation for the estimation of condensate to gas ratio (CGR) and other key gas properties from density data. Society of Petroleum Engineers. Available from: https://doi.org/10.2118/160170-MS.

Dranchuk, P.M., Abou-Kassem, J.H., 1975. Calculation of z-factors for natural gases using equations of state. J. Can. Pet. Tech. 14, 34−36.

Dranchuk, P.M., Purvis, R.A., Robinson, D.B., 1974. Computer calculation of natural gas compressibility factors using the Standing and Katz correlations. Institute of Petroleum Technical Series, pp. 1−13, No. IP74-008.

Elechi, V.U., Ikiensikimama S.S., Azubuike I.I. 2015. A correlation for estimating gas compressibility factor in the Niger delta. In: Festschrift for J. A. Ajienka, University of Port Harcourt, Nigeria, pp. 137−148.

Elsharkawy, A.M., 2006. Efficient methods for calculations of compressibility, density and viscosity of natural gases. JCPT 45 (6), 55−61.

Elsharkawy, A.M., Hashem, Y.S.K.S., Alikhan, A.A., 2000. Compressibility Factor for Gas Condensates. Society of Petroleum Engineers. Available from: https://doi.org/10.2118/59702-MS.

Gopal, V.N., 1977. Gas Z-factor equations developed for computer. Oil Gas J. 58−60.

Gray, E.H., Sims, H.L., 1959. Z-factor determination in a digital computer. Oil Gas J. 80−81.

Hall, K.R., Yarborough, L., 1973. A new equation of state for Z-factor calculations. Oil Gas J. 82−92.

Hankinson, R.W., Thomas, L.K., Phillips, K.A., 1969. Predict natural gas properties. Hydrocar. Process. 106−108.

Kay, W., 1936. Gases and vapors at high temperature and pressure − density of hydrocarbon. Ind. Eng. Chem. 28 (9), 1014−1019.

Lee, A.L., Gonzalez, M.H., Eakin, B.E., 1966. The viscosity of natural gas. Trans. AIME 237, 997−1000.

Leung, L.C., 1983. Numerical evaluation of the effect of simultaneous steam and carbon dioxide injection on the recovery of heavy oil. JPT 35 (9), 1591−1599. SPE-10776-PA.

Londono, F.E., Archer, R.A., Blasingame, T.A. 2002. Simplified correlations for hydrocarbon gas viscosity and gas density validation and correlation behavior using a large scale database. In: Paper SPE 75721 Presented at the SPE GasTechnology Symposium, Calgary, Canada, April 30−May 2. doi:10.2118/75721-MS.

Londono, F.E., Archer, R.A., Blasingame, T.A., 2005. Correlations for hydrocarbon-gas viscosity and gas density-validation and correlation of behavior using a large-scale database. SPEREE 8 (6), 561−572.

Lucas, K., 1981. Chem. Ing. Tech. 53 (12), 959−960 (in German).

Mahmoud, M.A. 2013. Development of a new correlation of gas compressibility factor for high pressure gas reservoirs. In: Paper SPE 164587 Presented at the North Africa Technical Conference and Exhibition held in Cairo, Egypt, April 15−17.

McCain Jr., W.D., 1990. The Properties of Petroleum Fluids, 2nd ed PennWell Books, Tulsa, OK, USA.

McCain Jr., W.D., 1991. Reservoir-fluid property correlations − state of the art. SPERE 266−272.

McCain Jr., W.D., Piper, L.D., 1994. Reservoir gases exhibit subtle differences. Pet. Eng. Int. 45−46.

Nishiumi, H., Saito, S., 1975. An improved generalized BWR equation of state applicable to low reduced temperatures. J. Chem. Eng. Jpn. 5, 356−360.

Obuba, J., Ikiesnkimama, S.S., Ubani, C.E., Ekeke, I.C., 2013. Natural gas compressibility factor correlation evaluation for Niger delta gas fields. IOSR J. Electr. Electron. Eng. 6 (4), 01−10 (IOSR-JEEE). e-ISSN: 2278-1676, p-ISSN: 2320-3331.

Papay, J. 1968. A termelestechnologiai parameterek valtozasa a gazlelep muvelese soran. OGIL Musz, Tud, Kuzl. Budapest, pp. 267−273.

Papp, I. 1979. Uj modszer Ioldgazok elteresi tenvezojenek szamirasara. Koolaj es Foldgaz, November, pp. 345-47.

Piper, L.D., McCain, W.D., Jr., Corredor, J.H. 1993. Compressibility factors for naturally occur-ring petroleum gases (1993 version). In: Paper SPE 26668 Presented at the SPE Annual Technical Conference and Exhibition, Houston, TX, USA, October 3−6. doi:10.2118/26668-MS.

Riazi, M.R. 2005. Characterization and Properties of Petroleum Fractions. ASTM Manual Series: MNL50. 100 Barr Harbor, West Conshohocken, PA, USA.

Riazi, M.R., Daubert, T.E., 1987. Characterizing parameters for petroleum fractions. Ind. Eng. Chem. Res. 26 (24), 755−759.

Said, S., El-Banbi, A.H., 2005. A new and simple z-factor correlation. In: Paper Presented at the 9th International Mining, Petroleum, and Metallurgical Engineering Conference, Cairo University, Cairo, Egypt, 21−24 February.

Sarem, A.M., 1961. Z-factor equation developed for use in digital computers. OGL 118.

Standing, M.B., 1974. Petroleum Engineering Data Book. Norwegian Inst. of Technology, Trondheim, Norway.

Standing, M.B., 1977. Volumetric and Phase Behavior of Oil Field Hydrocarbon Systems. Society of Petroleum Engineers, ISBN-13: 978-0895203007.

Standing, M.B., Katz, D.L., 1942. Density of natural gases. Trans. AIME 146, 159−169.

Stewart, W.F., Burkhardt, S.F., Voo, D. 1959. Prediction of pseudo-critical parameters for mixtures. In: AIChE Meeting, Kansas City, MO, USA.

Sutton, R.R. 1985. Compressibility factors for high-molecular weight reservoir gases. In: Paper SPE14265 Presented at the SPE Annual Technical Meeting and Exhibition, Las Vegas, USA.

Sutton, R.P., 2005. Fundamental PVT calculations for associated and gas/condensate natural gas systems. In: Paper SPE 97099 Presented at the SPE Annual Technical Conference and Exhibition, Dallas, TX, USA. October 9−12.

Sutton, R.P., 2007. Fundamental PVT calculations for associated and gas/condensate natural-gas systems. SPE Res. Eval. J. 10 (3), 270−284.

Whitson, C.H. Brule, M.R. 2000. *Phase Behavior*, vol. 20. SPE Monograph Series, SPE, Richardson, TX, USA.

Wichert, E., Aziz, K., 1972. Calculate Z's for sour gases. Hydrocar. Process 51, 119−122.

FURTHER READING

Beggs, H.D., 1984. Gas Production Operations. OGCI Publications, Tulsa, OK.

PROBLEMS

3.1 For the gas composition given below, calculate the gas density in lbm/ft^3 and psi/ft. at 2000 psi and 60°F. Assume z-factor to be 0.97 at the specified pressure and temperature. Also, calculate the gas-specific gravity at the same pressure and temperature.

Component	Mole %
N_2	0.55
CO_2	8.71
C1	82.73
C2	5.49
C3	1.24
i-C4	0.21
n-C4	0.29
i-C5	0.12
n-C5	0.09
C6	0.11
C7 +	0.46
Total	100

3.2 For the gas composition given below, calculate critical pressure and temperature from both gas-specific gravity and gas composition using the following methods:

- Sutton (2007) from the gas-specific gravity.
- Kay's rules (Kay, 1936) and Elsharkawy (2000) from the gas composition.

Component	Mole %
H_2S	0
CO_2	7.65
N_2	0.27
C1	84.11
C2	5.79
C3	1.19
i-C4	0.18
n-C4	0.26
i-C5	0.09
n-C5	0.07
C6	0.07
C7+	0.32

C7+ molecular weight and specific gravity are estimated in the laboratory to be 157.7 lb/lb mol and 0.693, respectively.

3.3 For the gas in the above problem, estimate the z-factor at 230°F and for a pressure range from 100 to 5000 psia using Dranchuk and Abou-Kassem (1975), Gopal (1977), and Hall and Yarborough (1973) correlations. Compare between the calculated z-factors. Use the pseudocritical properties estimated by Kay's rules and by Piper et al. (1993).

3.4 For the above gas (Problem 3) and using the estimated z-factor with any of the correlations, calculate the gas formation volume factor, gas density, and gas compressibility. Also estimate gas viscosity using Lee et al. (1966) and Londono et al. (2005) correlations. Compare between the two calculated viscosity curves.

Chapter 4

Wet Gases

Wet gases lie between dry gases and gas condensates. This reservoir fluid type covers a large portion of the gases produced from many basins. The designation wet gas refers to the behavior of gases that condense some liquids on surface. Hydrocarbon liquids produced on surface from wet gases are less than those produced from gas condensate reservoirs. The phase behavior of wet gases is described in Fig. 4.1. Theoretically, wet gases do not drop out condensate in the reservoir. This means that the reservoir gas composition does not change at any stage of depletion. Also theoretically, the reservoir temperature is higher than the cricondentherm (i.e., the reservoir temperature line does not cross the two-phase envelope). Almost any natural gas we produce drops some condensate at least on surface. When the amount of the condensate produced on surface (relative to the amount of gas) is low, we can use the wet gas "model" or wet gas "concept" to describe the fluid.

According to McCain, we can use the wet gas concept for gases that contain C7+ from 0.8% to 4%. If reservoir gas composition is not available, the initial gas–oil ratio (GOR) (or condensate gas ratio, CGR) can be used

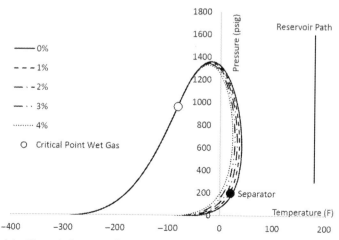

FIGURE 4.1   Theoretical wet gas phase diagram.

PVT Property Correlations. DOI: https://doi.org/10.1016/B978-0-12-812572-4.00004-7
65

to define the wet gas and generate its PVT properties. Initial CGR between 10 and 66 STB/MMscf is associated with wet gases. The ranges suggested by McCain are modified in this book based on analysis of many gas samples and based on considering the economic value of condensate relative to the produced gas. Details on the modified criteria are given in the Reservoir-Fluid Classification (Chapter 2).

PVT PROPERTIES FOR WET GASES

The engineering of wet gases includes reservoir, production, and process calculations. Almost all calculations required to forecast the reservoir behavior or estimate production rates of a well require PVT properties. The required PVT properties for engineering wet gases include the same set of PVT properties required for engineering dry gases (z-factor, gas formation volume factor, gas viscosity, gas density, and gas compressibility) and for estimating dew point pressure and natural gas liquids (NGL) derived from gas plants.

HANDLING PVT PROPERTIES FOR WET GASES

As the wet gases lie between dry gases and gas condensates, the prediction of wet gas properties can be approached in two different ways. Wet gases can be treated as gas condensates or as dry gases. We recommend the latter approach when condensation in the reservoir is not significant. Chapter 2 contains additional explanation as to which gases can be approximated as wet gas.

If the wet gas is treated as if it were a gas condensate, the best approach will be to run complete gas condensate PVT experiments (including CVD experiment) on a representative reservoir fluid sample. The laboratory observations should be used to tune an equation of state (EOS) model. The resultant EOS model can then be utilized to generate a compositional or modified black oil PVT model.

If the wet gas is treated as dry gas, then the composition or specific gravity of the gas will be utilized in deriving all required PVT properties from correlations. While the application of gas composition is straightforward (as with dry gases), the application of specific gravity is not. In dry gases, the composition of surface gas is similar to the composition of reservoir gas, and therefore, the surface gas specific gravity can be used as input to the dry gas PVT correlations. In wet gases, however, surface gas and reservoir gas do not have the same composition. In wet gases, gas drops liquid condensate in surface facilities. Therefore, if the reservoir gas specific gravity is to be included in any PVT property calculations, surface gas specific gravity will have to be adjusted to account for the condensate drop out on surface. Several approaches have been developed to incorporate the surface condensate into the calculation of reservoir gas specific gravity. The following summarizes these different approaches.

Option 1—From Reservoir Gas Composition

If the reservoir gas composition is available, we can use the standard procedure (explained in the Dry Gases Chapter) to calculate the pseudocritical properties and follow the dry gas procedure to calculate z-factor and other PVT properties. This option is essentially the same as calculating PVT properties for dry gases from reservoir gas composition. The main difference will be that both the mole fraction for C7+ and C7+ molecular weight in wet gases are higher than that of dry gases. Practically, since the C7+ mole fraction is usually significant in wet gases, it is necessary to have a reasonably accurate estimate for its molecular weight.

Option 2—From Reservoir Gas Specific Gravity

This calculation option requires that we find out the specific gravity of the reservoir gas (not the surface gas). The specific gravity of reservoir gas is calculated from the composition of reservoir gas. If separator gas and liquid compositions are the only compositions available, then recombination calculations are required to compute the reservoir gas composition. If separator gas, stock-tank gas, and stock-tank liquid compositions are available, recombination calculations are required to first calculate the reservoir gas composition. Once the reservoir gas composition is available, the specific gravity of reservoir gas can be computed. The PVT properties calculation procedure is the same as the calculation procedure for dry gas properties from specific gravity of the gas (explained in the Dry Gases Chapter). The main difference is that we can consider the measured specific gravity on surface for dry gases to represent the reservoir gas specific gravity. However, for wet gases, the surface gas does not generally represent reservoir gas. The reservoir gas specific gravity for wet gases must be calculated from reservoir gas composition.

Option 3—From Knowledge of Surface CGR

When either reservoir gas composition or surface fluids (gas and condensate) compositions are not available, we can use measured surface GOR or CGR to estimate the reservoir gas specific gravity. Eq. 4.1 is used to estimate the specific gravity of reservoir gas. The equation requires the surface GOR (or CGR), specific gravity of separator gas, and API gravity of stock-tank condensate.

$$\gamma_{\mathrm{gw}} = \frac{R\gamma_{g\mathrm{SP}} + 4600\gamma_o}{R + 133,300\left(\gamma_o/M_o\right)} \tag{4.1}$$

While

$$R = \frac{1,000,000}{\text{CGR}} \tag{4.2}$$

Specific gravity of oil is related to API gravity by Eq. 4.3, and the molecular weight of stock-tank oil can be calculated from Eqs. 4.4 or 4.5.

$$\gamma_o = \frac{141.5}{\text{API} + 131.5} \tag{4.3}$$

$$M_o = \frac{5954}{\text{API} - 8.811} \tag{4.4}$$

or

$$M_o = \frac{42.43\gamma_o}{1.008 - \gamma_o} \tag{4.5}$$

Option 4—Two-Stage Separation With Full Definition

Wet gases are usually processed with either two (one separator and stock-tank) or three (two separators and stock-tank) stage separation system. If two-stage separation is utilized, we can use Eq. 4.6 to estimate the reservoir gas specific gravity. The equation requires knowledge of both the separator GOR and the stock-tank GOR (which is usually not available). It requires also that the specific gravity of both the first stage separator gas and the stock-tank gas be known. Practically, the stock-tank gas specific gravity is seldom available. Other quantities in the equation such as specific gravity of stock-tank condensate can be evaluated from the API gravity and molecular weight of stock-tank condensate can be estimated from the correlations above.

$$\gamma_{gw} = \frac{R_1\gamma_1 + 4602\gamma_o + R_3\gamma_3}{R_1 + (133, 310\gamma_o/M_o) + R_3} \tag{4.6}$$

Option 5—Three-Stage Separation With Full Definition

As with the option above, for the three-stage separation system (two separators and stock-tank), the specific gravity of the reservoir gas can be calculated from the following equation. GORs and specific gravity of separators and stock-tank gases must be known.

$$\gamma_{gw} = \frac{R_1\gamma_1 + 4602\gamma_o + R_2\gamma_2 + R_3\gamma_3}{R_1 + (133, 310\gamma_o/M_o) + R_2 + R_3} \tag{4.7}$$

TABLE 4.1 Parameter Values for Two-Stage Separation Equations

A	Value	B	Value
1	1.4599	0	635.53
2	1.3394	1	0.36182
3	7.0943	2	1.0544
4	1.1436	3	5.0831
5	−0.93446	4	1.5812
		5	−0.7913

Option 6—Two-Stage Separation (Practical)

In many wet gas systems, two-stage separation is applied. In general, it is difficult to have stock-tank GOR and stock-tank gas specific gravity available. Gold et al. (1989) developed correlations to calculate the additional gas production from the stock-tank oil and the vapor equivalent of primary separator liquid. The use of these correlations allows engineers to estimate the reservoir gas specific gravity from the knowledge of primary separator GOR and specific gravity of primary separator gas, density of stock-tank liquid, and primary separator conditions. Estimates of these inputs are usually readily available. The correlations are given in Eqs. 4.8 to 4.10. Table 4.1 gives the correlations parameter values.

$$\gamma_{\text{gw}} = \frac{R_1 \gamma_1 + 4602 \gamma_o + G_{pa}}{R_1 + V_{\text{eq}}} \tag{4.8}$$

$$G_{pa} = A_1 (p_{s1} - 14.65)^{A_2} \gamma_1^{A_3} \text{API}^{A_4} T_{s1}^{A_5} \tag{4.9}$$

$$V_{\text{eq}} = B_0 + B_1 (p_{s1})^{B_2} \gamma_1^{B_3} \text{API}^{B_4} T_{s1}^{B_5} \tag{4.10}$$

Option 7—Three-Stage Separation (Practical)

As in the two-stage separation system above, if the three-stage separation system (two separators and stock-tank) is applied, correlations similar to those given above can be used (also developed by Gold et al., 1989). In these correlations, the additional gas production accounts for the gas production from the secondary separator and the stock-tank. The vapor equivalent of primary separator liquid accounts for the stock-tank liquid production and the gas production from the secondary separator and the stock tank. The equation to calculate the reservoir gas specific gravity is similar to that of the previous option; however, the correlations for the additional gas and vapor

TABLE 4.2 Parameter Values for Three-Stage Separation Equations

C	Value	D	Value
1	2.9922	0	535.92
2	0.9705	1	2.6231
3	6.8049	2	0.79318
4	1.0792	3	4.6612
5	−1.1960	4	1.2094
6	0.55367	5	−0.84911
		6	0.26987

equivalent are given with different correlations. The correlations parameters are given in Table 4.2.

$$G_{pa} = C_1(p_{s1} - 14.65)^{C_2} \gamma_1^{C_3} \mathrm{API}^{C_4} T_{s1}^{C_5} T_{s2}^{C_6} \tag{4.11}$$

$$V_{eq} = D_0 + D_1(p_{s1})^{D_2} \gamma_1^{D_3} \mathrm{API}^{D_4} T_{s1}^{D_5} T_{s2}^{D_6} \tag{4.12}$$

Sequence of Calculations

If the full composition of reservoir gas is used (Option 1), then the pseudo-critical temperature and pressure are calculated for the reservoir gas. There are several options for calculating the pseudocritical temperature and pressure from composition. The calculations require knowledge of the molecular weight and specific gravity of C7 +. The pseudoreduced temperature and pressure are then calculated, and the standard procedure to compute z-factor is followed. Gas deviation factor (z-factor) can be computed with several correlations. The rest of the PVT properties (gas formation volume factor, gas density, gas viscosity, and gas compressibility) are computed from z-factor and other readily available parameters. For each PVT property, more than one correlation is usually available for computing the property. The details of all these correlations are presented in the Dry Gases Chapter.

If any of the other options to calculate the reservoir gas specific gravity are used (Options 2−7), the pseudocritical temperature and pressure are calculated from the reservoir gas specific gravity. Several correlations are available to calculate the pseudocritical temperature and pressure. Some correlations correct directly for nonhydrocarbon impurities and others must be corrected afterwards. The details of these correlations are given in the Dry Gases Chapter. Once the pseudocritical temperature and pressure are

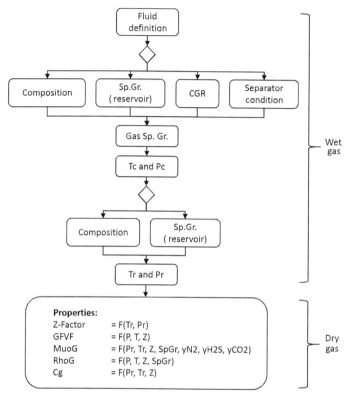

FIGURE 4.2 Flowchart for wet gas pressure, volume, and temperature (PVT) calculations options.

computed, the standard procedure outlined above is followed to compute z-factor and other PVT properties. The flow diagram presented in Fig. 4.2 is a schematic of the calculation procedure of wet gas properties. Notice that the sequence of calculations for PVT properties after calculating pseudoreduced temperature and pressure is the same as in dry gas calculations procedure.

APPLICATIONS OF TECHNIQUES TO COMPUTE PVT PROPERTIES FOR WET GASES

Two wet gas samples were selected to illustrate the computation of specific gravity using more than one technique. The calculated specific gravities of each gas were then used to compute z-factor and gas formation volume factor at initial reservoir pressure. The two wet gases were carefully selected to represent one with low liquid content (lean gas) and another with relatively high liquid content (close to being considered gas condensate fluid). Table 4.3 lists the compositions of the two gases and the properties of C7 + measured

TABLE 4.3 Compositional Information Data for the Two Example Fluids

Component	Mole %	
	Sample 1	Sample 2
N_2	0.49	2.3
CO_2	6.36	7.82
C1	72.79	77.6
C2	9.28	7.41
C3	3.81	2.08
i-C4	0.78	0.44
n-C4	1.16	0.48
i-C5	0.44	0.18
n-C5	0.37	0.17
C6	1	0.16
C7 +	3.53	1.36
C7 + molecular Weight	163.48	144.7
C7 + specific Gravity	0.81	0.78

in the laboratory. This information (except for C7 + properties) is usually readily available in the field. Notice that Sample 1 is considered gas condensate by the criteria presented in the Reservoir-Fluids Classification Chapter.

Problem 1—Calculation of Wet Gas Specific Gravity

For the two gases given in Table 4.3, calculate the gas specific gravity from gas composition, CGR, and separator conditions. Additional information is provided in Table 4.4.

Solution of Problem 1

Investigating the data in Tables 4.3 and 4.4 shows that we can calculate the gas specific gravity from composition, CGR, and knowledge of separator conditions using the practical approach.

TABLE 4.4 Field Measured Parameters for the Two Example Fluids

Parameter	Unit	Sample 1	Sample 2
Initial reservoir pressure	psia	10,000	5498
Reservoir temperature	°F	204	290
CGR	STB/MMscf	69.81	15.7
Stock-tank gravity	API	43.2	45.2
First stage separator gas gravity	(Air = 1)	0.78	0.72
First stage separator GOR	scf/STB	13,079	64,780
First stage separator pressure	psia	575	540
First stage separator temperature	°F	74	107
Second stage separator gas gravity	(Air = 1)	NA	NA
Second stage separator GOR	scf/STB	NA	NA
Second stage separator pressure	psia	250	NA
Second stage separator temperature	°F	74	NA

CGR, condensate gas ratio; GOR, gas–oil ratio.

Calculation of Reservoir Gas Specific Gravity From Composition

The procedure described in the Dry Gases Chapter applies here. The molecular weight of the $C7+$ is used in addition to the mole fractions and molecular weight of lighter hydrocarbons and impurities, and the following results are obtained:

Sample 1: Gas specific gravity $= 27.43/28.96 = 0.95$
Sample 2: Gas specific gravity $= 22.71/28.96 = 0.78$

Calculation of Reservoir Gas Specific Gravity From CGR

Molecular weight of the condensate is estimated from stock-tank liquid gravity using a correlation. Other information is obtained from Table 4.4. Table 4.5 summarizes the results of the calculations.

Calculation of Reservoir Gas Specific Gravity From Separator Data

Full information of first stage separators are available for both gases in Table 4.4. However, the full information on other stages of separation is not available for other separator(s). We cannot therefore use the option that requires knowledge of full information of separator data. However, we can use the practical method to calculate reservoir gas specific gravity from partial separator information. Gold et al. (1989) procedures (Options 6 and 7)

TABLE 4.5 Calculation of Reservoir Gas Specific Gravity from CGR for the Two Example Fluids

Parameter	Unit	Sample 1	Sample 2
CGR	STB/MMscf	69.81	15.7
Separator gas specific gravity	(Air = 1)	0.78	0.72
Stock-tank gravity	API	43.2	45.2
Oil specific gravity	(Water = 1)	0.81	0.8
Molecular weight of stock-tank oil	lbm/lbm mol	173.14	163.62
Reservoir gas specific gravity	(Air = 1)	0.997	0.770

CGR, condensate gas ratio.

TABLE 4.6 Calculation of Reservoir Gas Specific Gravity from Separator Data for the Two Example Fluids

Parameter	Unit	Sample 1	Sample 2
Reservoir gas calculations option		3 stages practical	2 stages practical
V_{eq}	scf/STB	1535.87	1167.06
G_{pa}	scf/STB-gravity	940.31	619.93
Oil specific gravity	(Water = 1)	0.81	0.80
Reservoir gas specific gravity	(Air = 1)	1.02	0.77

are used to calculate the reservoir gas specific gravity for Samples 1 and 2, respectively. Table 4.6 summarizes the important results of these calculations, and Table 4.7 compares the results of the reservoir gas specific gravity calculated by the different approaches.

Problem 2—Calculation of Wet Gas z-Factor and Formation Volume Factor

For the gas specific gravity obtained in Problem 1, calculate the gas z-factor and gas formation volume factor at initial reservoir pressure and reservoir temperature.

TABLE 4.7 Reservoir Gas Specific Gravity Comparison for the Two Example Fluids

Method	Sample 1	Sample 2
Composition data	0.950	0.780
CGR data	0.997	0.770
Separator data	1.020	0.770

CGR, condensate gas ratio.

TABLE 4.8 Results of z-Factor and Gas Formation Volume Factor for the Two Example Fluids

Parameter	Unit	Option	Sample 1	Sample 2
z-Factor	Dimensionless	From gas composition	1.503	1.052
		From CGR	1.510	1.067
		From separator conditions	1.549	1.067
Gas formation volume factor	rcf/scf	From gas composition	0.00282	0.00406
		From CGR	0.00284	0.00412
		From separator conditions	0.00273	0.00412

Solution of Problem 2

Using the three reservoir gas specific gravity values obtained for the two gases, the procedures of the Dry Gases Chapter are followed to compute pseudocritical pressure and temperature with the procedure of Piper et al. (1993). The pseudoreduced pressure and temperature are calculated using the initial reservoir pressure and reservoir temperature. The technique of Dranchuk and Abou-Kassem (1975) for z-factor calculation is then applied. The gas formation volume factor is calculated by the equation from the Dry Gases Chapter. The important results for these calculations are summarized in Table 4.8.

The variation of the two gases z-factors and formation volume factors are not significant between the different approaches. The z-factor variation for

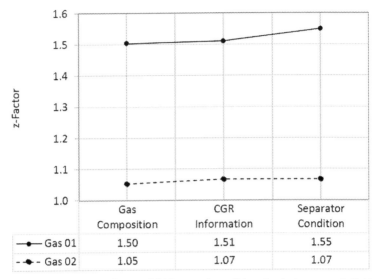

	Gas Composition	CGR Information	Separator Condition
●── Gas 01	1.50	1.51	1.55
- ◆ - Gas 02	1.05	1.07	1.07

FIGURE 4.3 Gas deviation factor at initial reservoir conditions for the two example fluid samples calculated from different options.

Samples 1 and 2 are shown in Fig. 4.3, while the calculated reservoir gas formation volume factors are shown in Fig. 4.4.

ENGINEERING WET GAS RESERVOIRS

Engineering wet gas reservoir fluid requires discussion of the main differences between wet gases on one hand, and dry gases and gas condensates on the other hand. There are differences associated with the importance of estimating dew point pressure, heat content, gallons per mcf (GPM), and calculation of fluids in place and their economic value. We will use the same two wet gases presented earlier to illustrate some of the key calculations.

The above two samples were modeled with three-parameter SRK EOS (Soave, 1972; Peneloux et al., 1982), and their phase envelopes were calculated. Fig. 4.5 shows the computed phase diagram for Sample 1 (can be considered gas condensate), while Fig. 4.6 presents the phase diagram for Sample 2 (which follows the expected theoretical behavior of wet gases). Sample 1 phase diagram shows that the reservoir temperature lies in the two-phase region (indicating theoretical gas condensate behavior). The same plot also shows a low amount of liquid drop out in the reservoir. It can be regarded as a lean gas condensate sample, or simply a wet gas sample (according to McCain's criteria). Sample 2, however, shows the reservoir temperature to be higher than the cricondentherm. This sample follows the theoretical behavior of wet gases as no liquid condensation at reservoir

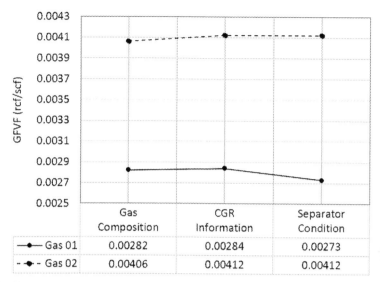

	Gas Composition	CGR Information	Separator Condition
Gas 01	0.00282	0.00284	0.00273
Gas 02	0.00406	0.00412	0.00412

FIGURE 4.4   Gas formation volume factor at initial reservoir conditions for the two example fluid samples calculated from different options.

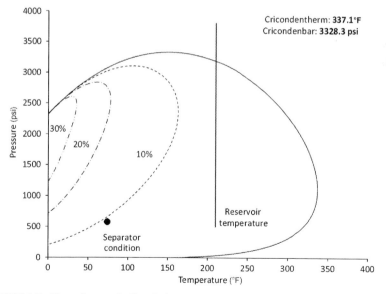

FIGURE 4.5   Phase diagram for Sample 1.

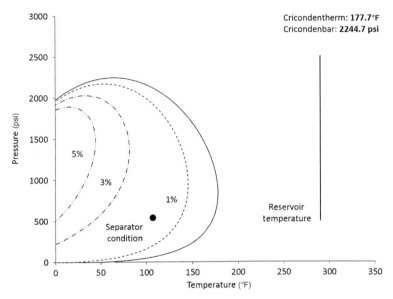

FIGURE 4.6   Phase diagram for Sample 2.

conditions is expected. The separator conditions in both samples are well within the two-phase region, indicating condensation of liquid hydrocarbon at surface.

Dew Point Pressure

In dry gas reservoirs, the assumption is that the gas does not exhibit a dew point at reservoir temperature or at surface. In wet gases, it is assumed that the vaporized liquid in the gas under reservoir conditions will condense on surface. This means that the fluid will exhibit two-phase conditions at the separator or on the way to the separator. The assumption is also that a dew point pressure is not reached at reservoir temperature. Practically, however, and for engineering purposes, we can approximate some gas condensate reservoirs to follow the wet gas model for its simplicity. In this regard, we would like to be able to estimate the dew point pressure and the liquid drop out in the reservoir. Dew point pressure can be estimated at reservoir temperature from the correlations given in the Gas Condensates Chapter, if not available from a PVT report. If the liquid drop out at reservoir temperature is not significant, the wet gas model assumption is usually sufficient.

Heat Content

In general, the heat content for wet gases is higher than the heat content for dry gases due to the presence of greater amounts of higher molecular weight hydrocarbons. The same procedure explained to calculate heat content for dry gases is followed to calculate the heat content for wet gases. Sometimes, gas may be sold before further processing in a gas plant. In this case, it will usually have higher heat content and be sold for a premium. When gas contains significant amounts of liquids, it may sometimes make economic sense to process the gas to extract the NGL. The heat content of the two wet gases presented in this chapter is calculated in the following example.

Problem 3—Calculating Heat Content for Wet Gases

Calculate heat content for the two gases with the composition given above.

Solution of Problem 3

Following the same procedure for calculating the heating value in the Dry Gases Chapter, Tables 4.9 and 4.10 show the detailed calculations for the heat content for the two example gases.

TABLE 4.9 Heating Value Calculations for Sample 1

Component	Mole %	L_{cj}	$y_j \times L_{cj}$	z_j	$y_j \times (1 - z)^{0.5}$
N_2	0.49	0.0	0.000	0.994	0.000370
CO_2	6.36	0.0	0.000	1.000	0.001102
C1	72.79	101.0	73.518	0.998	0.032553
C2	9.28	1769.6	164.219	0.992	0.008352
C3	3.81	2516.1	95.863	0.983	0.005040
i-C4	0.78	3251.9	25.365	0.971	0.001326
n-C4	1.16	3262.3	37.843	0.967	0.002117
i-C5	0.44	4000.9	17.604	0.948	0.001003
n-C5	0.37	4008.9	14.833	0.942	0.000891
C6	1.00	4755.9	47.559	0.910	0.003000
C7 +	3.53	5502.5	194.238	0.852	0.013580
			1332.703		0.069334
	$z = 0.9952$				
	$L_c = 1339$				

TABLE 4.10 Heating Value Calculations for Sample 2

Component	Mole %	L_{cj}	$y_j \times L_{cj}$	z_j	$y_j \times (1-z)^{0.5}$
N_2	2.30	0.0	0.000	0.994	0.001736
CO_2	7.82	0.0	0.000	1.000	0.001354
C1	77.60	101.0	78.376	0.998	0.034704
C2	7.41	1769.6	131.127	0.992	0.006669
C3	2.08	2516.0	52.335	0.983	0.002752
i-C4	0.44	3251.9	14.308	0.971	0.000748
n-C4	0.48	3262.3	15.659	0.967	0.000876
i-C5	0.18	4000.9	7.202	0.948	0.000410
n-C5	0.17	4008.9	6.815	0.942	0.000409
C6	0.16	4755.9	7.609	0.910	0.000480
C7 +	1.36	5502.5	74.834	0.852	0.005232
			1093.65		0.055371
	$z = 0.9969$				
	$L_c = 1097$				

Volumetric Calculations of Fluids in Place for Wet Gas Reservoirs

The main difference between dry gases and wet gases is the production of reasonable (sometimes significant) condensate amounts in the latter. Therefore, the original gas in place (OGIP) and prediction of reservoir performance calculations for wet gas reservoirs differ from those for dry gas reservoirs. In volumetric calculations, it is necessary to consider the volume of liquid (vaporized in the gas phase under reservoir conditions) that will condense at surface. The gas under reservoir conditions is found as gas with vaporized condensate. When it is brought to the surface, both gas and condensate will be produced. The following section details the required calculations to estimate the expected amounts of both gas and condensate that will be produced from the reservoir gas.

The fraction of OGIP that will be produced in the gaseous phase is given by Eq. 4.13.

$$G = f_g G_T \tag{4.13}$$

The amount of condensate present in the reservoir gas (as vaporized liquid) when it is brought to the surface is calculated from Eq. 4.14.

TABLE 4.11 Reservoir and Fluid Parameters for the Two Samples

Parameter	Unit	Reservoir 1	Reservoir 2
Fluid sample		Sample 1	Sample 2
Initial reservoir pressure	psia	10,000	5498
Reservoir temperature	°F	204	290
Total producing GOR	scf/STB	14,325	63,695
Reservoir area	ac	12,000	10,000
Average reservoir net pay thickness	ft	35	40
Average porosity	Fraction	0.18	0.23
Average initial water saturation	Fraction	0.32	0.19

GOR, gas–oil ratio.

$$N = \frac{f_g G_T}{R_t} \tag{4.14}$$

The fraction of the total initial gas in place that will be produced in the gaseous phase at the surface is calculated from Eq. 4.15.

$$f_g = \frac{R_t}{R_t + (132,800 \gamma_o / M_o)} \tag{4.15}$$

Problem 4—Calculating Hydrocarbon in Place for Wet Gases

The two fluid samples above were taken from two different reservoirs with the basic reservoir data given in Table 4.11. Estimate the total initial gas in place, the fraction of the total initial gas in place that will be produced in the gaseous phase, and the initial oil (condensate) in place.

Solution of Problem 4

By use of the following volumetric calculation for total initial gas in place (Eq. 4.16), the total (gas with the vaporized condensate) gas in place can be calculated.

$$G_T = \frac{7758\, A\, h\, \phi\, (1 - S_{wi})}{B_{gi}} \tag{4.16}$$

As an example calculation for Reservoir 1:

$$G_T = \frac{(7758)(12,000)(35)(0.18)(1 - 0.32)}{0.00284} = 1.4 \times 10^{11} \text{ scf} = 0.14 \text{ tcf}$$

TABLE 4.12 Results of Volumetric Calculations for the Two Example Fluids

Parameter	Unit	Reservoir 1	Reservoir 2
Total OGIP from gas composition	scf	1.41427E + 11	1.42396E + 11
Total OGIP from CGR	scf	1.40431E + 11	1.40322E + 11
Total OGIP from separator conditions	scf	1.46089E + 11	1.40322E + 11
Fraction of total OGIP produced as gas (f_g)	Fraction	0.958432621	0.98990883
OOIP from gas composition	STB	9462,335	2213,026
OOIP from CGR	STB	9395,699	2180,797
OOIP from separator conditions	STB	9774,280	2180,797

CGR, condensate gas ratio. OGIP, original gas in place. OOIP, original oil in place.

$$f_g = \frac{R_t}{R_t + (132,800\gamma_o/M_o)} = \frac{14,325}{14,325 + ((132,800)(0.81)/173.14)} = 0.958$$

$$N = \frac{f_g G_T}{R_t} = \frac{(0.958)(1.4 \times 10^{11})}{14,325} = 9.36 \times 10^6 \text{STB}$$

The calculations of total gas in place and oil in place for both reservoirs and with the different assumptions for reservoir gas specific gravities are summarized in Table 4.12.

Economic Considerations for Wet Gas Reservoirs

In dry gas reservoirs, condensate production may not be very important from an economic point of view. However, for many wet gas reservoirs, condensate production constitutes an important revenue stream in the project. Failure to quantify the economic impact of the expected condensate production may lead to overlooking some potentially promising wet gas development projects. The condensate production from some wet gas reservoirs (which are close to being considered gas condensate) may be as high as half the value of the entire project.

For the above two reservoirs, and assuming recovery factor for both reservoir of 85%, gas price of $3/Mscf, and condensate price of $50/STB, the results are given in Table 4.13.

The average OGIP and OOIP are used in the calculations shown above. In the first reservoir, the condensate value is even higher than the value of the gas. The richness of the wet gas plays a significant role in development

TABLE 4.13 Results of Economic Calculations for the Two Example Fluids

Parameter	Unit	Reservoir 1	Reservoir 2
Recovery factor	Fraction	0.85	0.85
Value of produced gas	$	363,754,545	359,583,467
Value of produced condensate	$	405,624,459	93,140,448
Fraction of condensate value to total	%	52.7	20.6

project economics. It is safe to assume that the same recovery factor can be applied for both gas and condensate as theoretically no condensate will drop out in the reservoir. All produced condensate on surface is initially vaporized with the gas.

Plant Products (GPM)

In some wet gas reservoirs, and when production rates are high, it makes economic sense to process the produced gas for additional recovery of liquids and plant products (GPM). The calculations for plant products are covered in the Gas Condensates Chapter.

NOMENCLATURE

API   density of stock-tank liquid (API)
B_{gi} gas formation volume factor at initial pressure (rcf/scf)
CGR   condensate gas ratio (STB/MMscf)
CVD   constant volume depletion
EOS   equation of state
f_g the fraction of the total initial gas in place that will be produced in the gaseous phase at surface, fraction
G_T total original gas in place (scf)
G portion of gas in place that will be produced in the gaseous phase (scf)
GOR   gas−oil ratio (scf/STB)
GPM   gallons per mcf
G_{pa} additional gas production (scf/STB gravity)
h reservoir net pay thickness (ft)
M_o molecular weight of stock-tank liquid (lbm/lbm mol)
N oil in place (STB)
NGL   natural gas liquids
p_{s1} primary separator pressure (psi)
R gas−oil ratio (scf/STB)
R_1 primary separator GOR (scf/STB)
R_2 secondary separator GOR (scf/STB)

R_3 stock-tank GOR (scf/STB)
R_t total producing gas−oil ratio (scf/STB)
S_{wi} initial water saturation (fraction)
T_{s1} primary separator temperature (°F)
T_{s2} secondary separator temperature (°F)
V_{eq} vapor equivalent of primary separator liquid (scf/STB)
$\varnothing$ porosity (fraction)
γ_1 specific gravity of primary separator gas (air = 1.0)
γ_2 specific gravity of secondary separator gas (air = 1.0)
γ_3 specific gravity of stock-tank gas (air = 1.0)
γ_{gw} reservoir gas specific gravity (water = 1.0)
γ_o specific gravity of stock-tank liquid (water = 1.0)

REFERENCES

Dranchuk, P.M., Abou-Kassem, J.H., 1975. Calculation of z-factors for natural gases using equations of state, J. Can. Pet. Tech., 14, pp. 34−36.
Gold, D.K., McCain Jr., W.D., Jennings, J.W., 1989. An improved method for the determination of the reservoir-gas specific gravity for retrograde gases, J. Pet. Tech., 41, pp. 747−752.
Peneloux, A., Rauzy, E., Freze, R., 1982. A consistent correction for Redlich−Kwong−Soave volumes, Fluid Phase Equilib., 8, pp. 7−23.
Piper, L.D., McCain Jr., W.D., Corredor, J.H., 1993. Compressibility factors for naturally occurring petroleum gases. In: Paper SPE 26668 Presented at the 1993 SPE Technical Conference and Exhibition, Houston, October 3−6.
Soave, G., 1972. Equilibrium constants from a modified Redlich−Kwong equation of state, Chem. Eng. Sci., 27, pp. 1197−1203.

FURTHER READING

Beggs, H.D., 1984. Gas Production Operations. OGCI Publications, Tulsa, OK.
Craft, B.C., Hawkins, M.F., Terry, R.E., 1991. Applied Petroleum Reservoir Engineering. Prentice Hall, Englewood Cliffs, NJ, USA.
McCain Jr., W.D., 1990. The Properties of Petroleum Fluids, 2nd ed. PennWell Books, Tulsa, OK, USA.
McCain Jr., W.D., Piper, L.D., 1994. Reservoir gases exhibit subtle differences. Pet. Eng. Int. (March), pp. 45−46.
Moses, P.L., 1986. Engineering applications of phase behavior of crude oil and condensate systems, J. Pet. Tech., 38, p. 715.

PROBLEMS

4.1 Two sets of gas samples were selected for further analysis from two exploratory wells. Identify the fluid type for each sample. Also, calculate the heating value of the two gas samples.

Component	Mole %	
	Sample 1	Sample 2
N_2	0.6	0.67
CO_2	5.6	5.13
C1	68.21	75.28
C2	7.1	8.19
C3	3.89	3.99
i-C4	0.8	0.7
n-C4	1.7	1.5
i-C5	0.7	0.5
n-C5	0.8	0.6
C6	1.51	0.72
C7+	9.2	2.66
C7+ molecular weight	242	140
C7+ specific gravity	0.872	0.7832
Temperature (°F)	283	293

4.2 Calculate wet gas specific gravity, given gas composition, CGR, and separator conditions for the following three samples. Additional information is given in the following table:

	Sample 1	Sample 2	Sample 3
Initial reservoir pressure (psia)	7500	8902	5915
Reservoir temperature (°F)	210	236.5	300
CGR (STB/MMscf)	39	48.9	68.5
Stock-tank gravity (API)	51.1	48.98	44
First stage separator gas gravity	0.631	0.614	0.811
First stage separator GOR (scf/STB)	25,133	20,439	14,603
First stage separator pressure (psia)	498	335	617.7
First stage separator temperature (°F)	99	104	176

GOR, gas−oil ratio; *CGR*, condensate gas ratio.

4.3 Draw and label completely a wet gas phase diagram. Explain the differences between a typical wet gas and a gas condensate fluid.

4.4 The two fluid samples in Problem 1 were taken from two different reservoirs with the basic reservoir data given in the following table. Estimate the total initial gas in place, the fraction of the total initial gas in place that will be produced in the gaseous phase, and the initial oil (condensate) in place.

	Sample 1	Sample 2
Initial reservoir pressure (psia)	4614	3365
Reservoir temperature (°F)	283	293
Total producing GOR (scf/STB)	4885	10,333
Reservoir area (ac)	13,000	10,000
Average reservoir net thickness	40	50
Average porosity	0.22	0.25
Average initial water saturation	0.29	0.21

GOR, gas−oil ratio.

Chapter 5

Gas Condensates

Gas condensate fluids (also called retrograde gas condensates) are usually found in deeper reservoirs. From the phase behavior point of view, gas condensates lie between wet gases and volatile oils reservoir fluids. They represent reservoirs that predominantly contain gas, but produce significant amounts of liquids when the gas reaches the surface. The liquids are originally vaporized in the gas phase, and condense on surface due to the inability of the gas to retain those liquids when it reaches a surface pressure and temperature below reservoir pressure and temperature. Fig. 5.1 represents a typical gas condensate phase diagram. Both reservoir temperature and separator conditions lie within the two-phase envelope.

Gas condensate reservoirs are found either as under-saturated (initial reservoir pressure is higher than dew point pressure) or saturated (initial reservoir pressure is equal to dew point pressure). The reservoir temperature is higher than the critical temperature. Above the dew point pressure, the gas condensate fluid has constant composition, and the composition of both the produced gas and condensate is constant. When production occurs below the dew point pressure, gas and condensate change composition. Whether the initial pressure lies above or on the dew point curve, the reservoir pressure decreases (at the reservoir temperature) with reservoir depletion. This decrease causes some condensate to fall (drop out) from the gas phase in the reservoir. With further reduction in reservoir pressure, the condensate (liquid) saturation in the reservoir increases until it reaches a pressure point (varying according to the gas condensate fluid) after which liquid saturation decreases. This phenomenon is termed "retrograde condensation." At these low pressures, the reservoir gas (vapor) composition becomes leaner and can accept revaporized condensate. In general, the liquid that drops out in the reservoir does not move and can be recovered only by means of expensive gas cycling operations. Fig. 5.2 shows the surface (producing) condensate−gas ratio (CGR) with reservoir pressure for a typical gas condensate fluid.

According to McCain (1994), reservoirs that initially produce with CGR between 66 and 312 STB/MMscf are considered gas condensate reservoirs (equivalent to GOR between 3200 and 15,000 scf/STB). Another criterion

PVT Property Correlations. DOI: https://doi.org/10.1016/B978-0-12-812572-4.00005-9

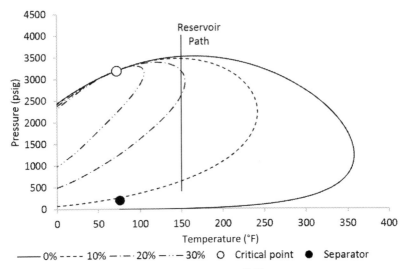

FIGURE 5.1   Typical phase diagram for gas condensate fluid.

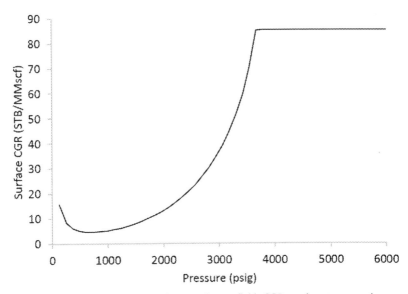

FIGURE 5.2   Surface CGR for typical gas condensate fluid. *CGR*, condensate−gas ratio.

that can be used to define gas condensate reservoirs is the C7 + mole percent. McCain reports values between 4% and 12.7% for gas condensates. In
Chapter 2, Reservoir-Fluid Classification, these criteria are revisited and
slightly modified.

PVT PROPERTIES FOR GAS CONDENSATES

Gas condensates carry significant economic value over dry gas reservoirs. They also pose challenges in performance management. The two main challenges in producing gas condensate reservoirs are (1) significant loss of economic value due to condensation of liquids in the reservoir and (2) loss of well productivity due to the formation of high liquid saturation around the wellbore (condensate banking).

In general, prediction of reservoir behavior can be challenging in gas condensate reservoirs. Understanding phase changes requires the use of appropriate PVT properties that are capable of capturing the variations that occur in solubility and revaporization in addition to volume changes of both oil and gas (El-Banbi et al., 2000a; Fevang et al., 2000). The phase behavior changes can also lead to production problems such as condensate banking (El-Banbi et al., 2000b).

Several PVT properties are required for engineering gas condensate reservoirs. These PVT properties include the same set of PVT properties required for engineering dry gases (specific gravity of gas, z-factor, gas formation volume factor, gas viscosity, gas density, and gas compressibility), in addition to dew point pressure, two-phase z-factor, CGR, and natural gas liquids (NGL).

PVT properties for gas condensate fluids are required in almost all reservoir, production, and surface facilities calculations. The economic value of gas condensate reservoirs depends largely on the richness of the gas (how much condensate drops out from the gas when it reaches the surface) and the management of the reservoir. In some cases, additional value can be obtained when the gas stream is further processed in a gas plant to produce additional liquids. The following sections summarize the PVT properties that are needed in calculations related to management of gas condensate reservoirs.

Specific Gravity of Gas Condensates

At dew point pressure and above, gas composition is constant. The specific gravity of the gas above the dew point pressure is defined and obtained in the same way as for wet gas fluids. Chapter 4, Wet Gases, contains the details on the options available to calculate the specific gravity of the reservoir gas.

Dew Point Pressure

Dew point pressure is defined as the pressure at which the first droplet of liquid appears at a specific temperature. The phase envelope for a typical gas condensate shows that the dew point pressure is a function of the temperature. Dew point pressure usually refers to dew point pressure at reservoir temperature, unless another temperature is specified.

Knowledge of dew point pressure (at reservoir temperature) is necessary in production of gas condensate reservoirs. The composition of the produced well stream changes when reservoir pressure falls below dew point pressure. In general, the production stream becomes leaner with lower amounts of heavier hydrocarbon components.

Several correlations are available to calculate dew point pressure from basic information. These correlations can be grouped in three categories: (1) correlations that depend on full composition information; (2) correlations that depend on gas specific gravity and field information; and (3) correlations that depend on downhole composition information. The next section summarizes the use of these distinct correlation types.

Gas Deviation Factor (z-Factor)

Single-phase gas deviation factor is measured in the laboratory for gas condensate fluids (in the constant composition expansion procedure). Single-phase gas deviation factor for gas condensates has the same definition as for dry gases. For pressures above the dew point pressure in gas condensates, z-factor is adequate for material balance (MB) calculations. The z-factor for gas condensates can also be obtained in the same way as in dry gases if gas composition is used. If gas specific gravity is used, the condensate drop out on surface must be taken into consideration. The same calculations presented in Chapter 4, Wet Gases, are followed to calculate the reservoir gas specific gravity. Then, the standard procedure is followed to find the pseudo-critical properties and the pseudo-reduced properties. The single-phase z-factor is then calculated from any of the available z-factor correlations.

Other PVT Properties for Gas Condensates

As reviewed in Chapter 3, Dry Gases, the PVT properties dependent on z-factor (e.g., gas formation volume factor, gas density, gas viscosity, and gas compressibility) have the same definitions for gas condensates as dry gases. These properties can be calculated from single-phase z-factor according to the methods explained in Chapter 3, Dry Gases.

Two-Phase z-Factor

The two-phase z-factor is a quantity that is back-calculated from the constant volume depletion (CVD) experiment procedure performed on gas condensate fluids. The following equation is used to calculate two-phase z-factor in the laboratory (Whitson and Brule, 2000).

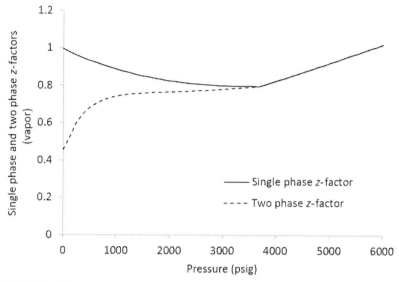

FIGURE 5.3 Single-phase and two-phase z-factors.

$$z_{2\emptyset} = \frac{p}{(p_d/z_d)\left[1 - (n_p/n)\right]} \tag{5.1}$$

Fig. 5.3 shows the difference between single-phase and two-phase z-factors for a moderately rich gas condensate sample. The two-phase z-factor is concave downward for gas condensates. The importance of the two-phase z-factor arises from the fact that it should be used in material balance (MB) calculations for gas condensate fluids when pressure data extend to pressures above and below the dew point pressure (Arukhe and Mason, 2012).

Rayes et al. (1992) developed a correlation to compute two-phase z-factor from pseudo-reduced pressure and temperature. Their correlation is recommended in absence of laboratory data of two-phase z-factor. The correlation form is given in Appendix B, Gas Correlations Formulae.

Condensate–Gas Ratio and Condensate Yield

CGR is the ratio of liquid volume produced from the gas stream in gas condensate reservoirs to the amount of gas produced. The volumes of both liquid and gas are usually expressed under standard conditions. In rich gas condensate reservoirs, the produced gas is usually further processed in a gas plant to produce NGL. However, the NGL produced in gas plants are not usually added to the CGR.

Condensate yield, however, is defined as the volume of condensate produced in the stock-tank divided by the volume of well stream production (Tarek, 2016). Again, both volumes are measured at standard conditions.

The well stream volume is the volume of fluid produced from the well before undergoing separation in the first stage separator.

When reservoir pressure exceeds the dew point pressure, both CGR and condensate yield are theoretically constant. With reservoir depletion, the pressure declines below the dew point pressure, and producing CGR and condensate yield decline as well. Few correlations are available to estimate the changes in producing CGR as function of reservoir pressure. These correlations usually depend on initial CGR, specific gravity of first stage separator gas, API gravity of the stock-tank condensate, and reservoir pressure. One correlation (England, 2002) estimates the condensate yield at initial conditions (or at dew point pressure and above) from simple data. The correlation was developed for particular regions. Other correlations presented in Appendix B, Gas Correlations Formulae are used to calculate the CGR decline with decreasing pressures below the dew point pressure.

Although not in agreement with the correct definition, in practice, CGR is sometimes reported as the volume of condensate collected in the stock-tank to the volume of gas produced from the first stage separator. We notice, however, that the majority of the gas is produced in the first stage separator. This is the amount of gas that is usually measured accurately (because it is often sold or used in the field for gas lift or power generation). Chapter 4, Wet Gases, contains methods to calculate the additional gas produced from second stage separator and stock-tank by use of correlations, when not available from field measurements.

Natural Gas Liquids

NGL are also called plant products. In operation of gas condensate reservoirs, field separation can recover high value condensates in the stock-tank. Additional liquids can also be recovered if the separator gas is processed through a gas plant. NGL refers to the liquids that are extracted from natural gas after further processing of the gas in a gas plant. It is a process of fractionation that results in production of simple hydrocarbons from the hydrocarbon mixture known as natural gas. NGL refers to ethane, propane, normal butane, iso-butane, and pentane and heavier hydrocarbons. These quantities are measured in "Gallons Per Thousand" or GPM. They can be also measured in barrels per million cubic feet of gas (bbl/MMscf).

To calculate how much GPM can be obtained from a specific gas, the composition of the gas must be known. The GPM of every component is calculated from any of the following equations:

$$\text{GPM}_i = 19.65 \frac{y_i M_i}{\rho_{oi}} \text{ gallons/Mscf} \qquad (5.2)$$

TABLE 5.1 Recovery Efficiency of NGL Components from Different Plants

NGL Components	Absorption or Lean Oil Plants	Refrigeration Plants	Cryogenic Plants
Ethane	0.15–0.30	0.80–0.85	0.85–0.90
Propane	0.65–0.75	1	1
Butanes	0.99	1	1
Pentanes and C5 +	0.99	1	1

$$GPM_i = 0.3151 \frac{y_i M_i}{\gamma_{oi}} \text{ gallons/Mscf} \qquad (5.3)$$

The GPM can be converted to bbl/MMscf according to the following equation:

$$GPM \text{ in } \frac{bbl}{MMscf} = GPM \text{ in } \frac{gal}{Mscf} \times 1000 \frac{Mscf}{MMscf} \times \frac{1 \text{ bbl}}{42 \text{ gal}}$$

$$= 23.81 \, GPM \text{ in bbl/MMscf} \qquad (5.4)$$

Complete recovery of these liquids is not possible. Recovery efficiency depends on the technology used in the gas plant. Table 5.1 summarizes the NGL component recovery efficiency of different types of plants.

The lean oil plants are the simplest type of plants. Kerosene type oil is circulated through the plant to absorb light hydrocarbon components from the gas. The light components are separated from the rich oil and the lean oil is recycled. Refrigeration plants use propane to chill the gas and remove more liquids. Cryogenic plants use turbo expander technology to cool the gas to subzero temperatures to remove liquids.

Modified Black Oil Approach for Gas Condensate

Modified black oil (MBO) approach can be adequately used in many applications to model gas condensate behavior (El-Banbi et al., 2000a; Fevang et al., 2000). If MBO approach is used to model the gas condensate fluid behavior, it is preferable to derive the MBO PVT properties from an equation of state (EOS). Several techniques are available to derive the MBO properties from an EOS (Fattah et al., 2006). Four PVT properties are needed as a function of pressure in the MBO approach. These are (1) oil formation volume factor; (2) gas formation volume factor; (3) solution gas–oil ratio; and (4) vaporized oil–gas ratio.

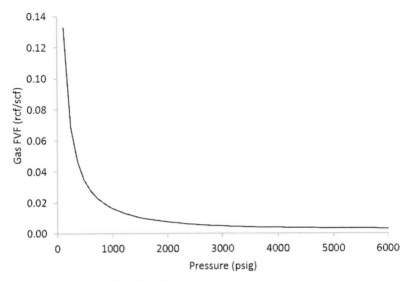

FIGURE 5.4 Gas formation volume factor.

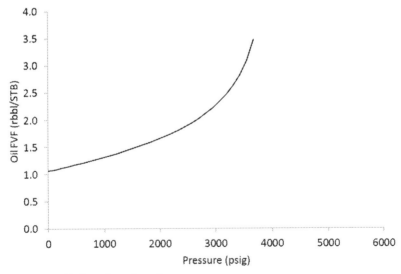

FIGURE 5.5 Oil formation volume factor.

Figs. 5.4–5.7 show the four MBO PVT functions versus pressure for a typical gas condensate fluid. These PVT functions were calculated by use of Whitson and Torp (1983) procedure applied on CVD data from an EOS model.

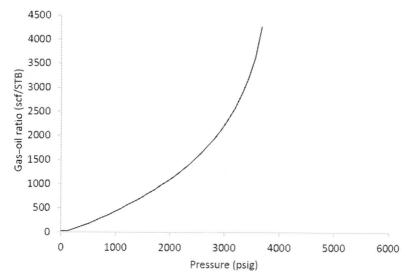

FIGURE 5.6 Solution gas−oil ratio.

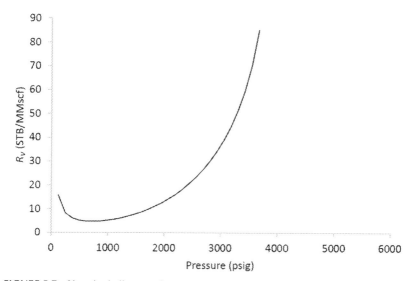

FIGURE 5.7 Vaporized oil−gas ratio.

Two correlations exist for computing the four MBO PVT parameters from readily available field data (El-Banbi et al., 2006; Nassar et al., 2013). The correlation forms are given in Appendix A (for oil) and Appendix B (for gases). The first set of correlations (El-Banbi et al., 2006) depend on knowledge of average specific gravity of gas. The other required parameters include reservoir temperature, initial CGR, stock-tank oil gravity, and separator conditions. In the case of two or three-stage separation, the average gas specific gravity must be calculated as a volume-weighted average of the gases released from the different separators and stock-tank. The second set of correlations (Nassar et al., 2013) requires the use of reservoir temperature, primary separator gas gravity, stock-tank API gravity of oil, and separator conditions. Since specific gravity of secondary separator gas and stock-tank gas are not usually available, we find Nassar et al. (2013) correlations to be easier to implement in practice.

In general, the MBO correlations can yield significant error (in the order of 20% absolute average percent error or slightly more when compared with laboratory observed values) in the case of highly rich gas condensate fluids and near critical volatile oils. These errors may be due to the volatility of the fluids. It is always advisable to use these correlations only if an EOS model is not available for the fluid.

HANDLING PVT PROPERTIES FOR GAS CONDENSATE

PVT properties for gas condensate fluids can be determined from PVT laboratory experiments performed on a representative fluid sample. Other texts (e.g., Ahmed, 2016; Whitson and Brule, 2000) explain gas condensate laboratory experiments in full detail. The other available option to calculate the PVT properties of gas condensate fluids is to use correlations.

Gas condensate specific gravity can be obtained by use of the techniques presented in Chapter 4, Wet Gases. The critical properties (critical pressure and temperature) for the gas condensate can be computed with the techniques presented in Chapter 3, Dry Gases. Critical properties correlations either use the gas specific gravity or the gas composition. Some of the important critical properties correlations are given in Appendix B, Gas Correlations Formulae. In general, for gas condensates, the use of more recent critical properties correlations is recommended over the older ones. The older correlations were developed from fluid databases that were dominated by dry gas samples. Appendix D, Gas Correlations Range of Applicability, contains the ranges of applicability of the major correlations. One can consult the Appendix for the best correlation to apply according to the validity ranges of input data.

The PVT properties such as single-phase z-factor, gas formation volume factor, gas density, gas compressibility, and gas viscosity can be computed with the techniques presented in Chapter 3, Dry Gases. Many of these correlations are given in Appendix B, Gas Correlations Formulae.

The PVT properties that are specific to gas condensates (e.g., dew point pressure, two-phase z-factor, and CGR) are calculated using the techniques presented in this chapter. Tables 5.2 and 5.3 list the available dew point pressure correlations from specific gravity and from gas composition, respectively. Tables 5.4 and 5.5 summarize the two-phase z-factor and CGR correlations. Table 5.6 shows the gas condensate specific gravity

TABLE 5.2 Dew Point Pressure Using Gas Specific Gravity

Dew Point Pressure Using Specific Gravity	Al-Dhamen and Al-Marhoun (2011)	Humoud and Al-Marhoun (2001)
Marruffo et al. (2001)	Nassar et al. (2013)	Ovalle et al. (2007)

TABLE 5.3 Dew Point Pressure Using Composition

Dew Point Pressure Using Composition	Ahmadi and ElSharkawy (2017)	Alzahabi et al. (2017)	Elsharkawy (2001)
Elsharkawy (2011)	England (2002)	Godwin (2012)	Kamaria et al. (2016)
Nemeth and Kennedy (1967)	Olds et al. (1945)	Organick and Golding (1952)	Ovalle et al. (2007)
Shokir (2008)			

TABLE 5.4 Two-Phase z-Factor

Two-Phase z-Factor	Jones et al. (1989)	Rayes et al. (1992)

TABLE 5.5 Condensate−Gas Ratio Correlations

CGR Correlations	Dindoruk (2012)	Ovalle et al. (2007)

TABLE 5.6 Gas Condensate Specific Gravity

Gas Condensate Specific Gravity	Gold et al. (1989)	Ovalle et al. (2007)

TABLE 5.7 MBO Correlations

Oil Formation Volume Factor	Gas Formation Volume Factor	Solution Gas–Oil Ratio	Vaporized Oil–Gas Ratio
B_o	B_g	R_s	R_v
El-Banbi et al. (2006)	El-Banbi et al. (2006)	El-Banbi et al. (2006)	El-Banbi et al. (2006)
Nassar et al. (2013)	Nassar et al. (2013)	Nassar et al. (2013)	Nassar et al. (2013)

correlations. The most important correlations for these properties are also given in Appendix B, Gas Correlations Formulae, with applicability ranges given in Appendix D, Gas Correlations Range of Applicability.

If the MBO approach is to be used to describe the gas condensate phase behavior (and EOS model is not available), the MBO correlations can be applied. These correlations are used to calculate the PVT functions required for the MBO modeling approach (oil and gas formation volume factors, solution gas–oil ratio, and vaporized oil–gas ratio). Table 5.7 lists the MBO PVT properties correlations, and the forms of the correlations are given in Appendix B, Gas Correlations Formulae.

Sequence of Calculations

If the gas condensate PVT properties are needed, the flow chart of Chapter 4, Wet Gases, is followed to compute the specific gravity of the gas (with the desired option), followed by critical properties, z-factor, then other PVT properties. If the MBO approach is to be used, the flow chart of Fig. 5.8 is applied. In this sequence of calculations; pressure, temperature, initial vaporized oil–gas ratio, and separator conditions need to be specified. The dew point can be either specified by the user if known, or estimated from correlations. The inputs will be used to calculate the four functions of the MBO model.

ENGINEERING GAS CONDENSATE RESERVOIRS

To extract the maximum value from gas condensate reservoirs, engineering calculations should determine GPM, heat content of the gas, and dew point pressure. Above the dew point pressure, gas condensate can be treated as wet gas, and the methods employed in engineering wet gas reservoirs are adequate for gas condensates. The behavior of gas condensate reservoirs below dew point pressure change significantly from above the dew point.

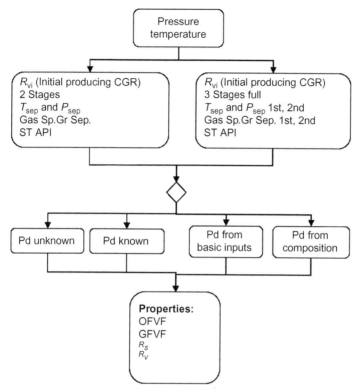

FIGURE 5.8 Flow chart for MBO PVT properties' calculations for gas condensates. *MBO*, modified black oil.

Accurate description of the behavior of gas condensate reservoirs below dew point pressure is important, as the economic value of the gas decreases with reservoir depletion. This decrease is due to the fact that gas becomes leaner with less producing CGR. The following sections discuss the properties required for engineering gas condensate reservoir fluids.

Dew Point Pressure

Dew point pressure can be estimated at any temperature from the dew point correlations of Appendix B, Gas Correlations Formulae. Several correlations are available and these require either the surface gas specific gravity or gas composition. All dew point correlations use surface information except Alzahabi et al. (2017) correlation, which uses the compositional information from downhole fluid analyzer tools to calculate dew point pressure.

Problem 1—Calculation of Dew Point Pressure for a Gas Condensate

Calculate dew point pressure for the gas with the composition given in Table 5.8. Use dew point correlations from Appendix B, Gas Correlations Formulae (e.g., Elsharkawy, 2001; Al-Dhamen and Al-Marhoun, 2011) to show the variation of the calculated dew point. Notice that Elsharkawy (2001) requires the gas composition as input, while Al-Dhamen and Al-Marhoun (2011) require gas specific gravity and basic field information. Compare your calculated dew point pressure with the laboratory-measured value of 3245 psia. The reservoir temperature is 323°F, separator gas specific gravity is 0.786, the separator-gas-to-separator-liquid ratio is 1910 scf/STB at 60°F, separator liquid to stock-tank−liquid ratio is 1.255 rbbl/STB and the condensate specific gravity is 0.7823. Separator conditions are given as follows:

1st stage separator pressure = 800 psi, 1st stage separator temperature = 110°F

2nd stage separator pressure = 60 psi, 2nd stage separator temperature = 75°F

Solution of Problem 1

Calculating the dew point using the basic field data by Al-Dhamen and Al-Marhoun (2011):

$$\ln(P_d) = a_1 + a_2\ln\left(\frac{\text{GOR} \times \gamma_g}{\gamma_{7+}}\right) + a_3\ln(T_R) + a_4\gamma_g + \frac{a_5}{\gamma_{\text{cond.}}} + a_6 e^{a_7 + a_8\ln(\text{GOR})}$$

(5.5)

where the correlation parameters are given in Table 5.9.

TABLE 5.8 Gas Composition (in mole %) for Problem 1

N_2	CO_2	C1	C2	C3	i-C4	n-C4
3.16%	6.42%	50.26%	9.68%	6.20%	2.19%	3.73%

i-C5	n-C5	C6	C7+	C7 + Mw	C7 + SpGr.
1.88%	1.85%	3.04%	11.59%	146	0.782

TABLE 5.9 Dew Point Correlation Constants (Al-Dhamen and Al-Marhoun, 2011)

a1	a2	a3	a4	a5	a6	a7	a8
18.60	− 0.15	− 0.17	0.07	− 5.90	− 0.06	8.50	− 0.75

Then

$$\ln(P_d) = 18.6012$$

$$- 0.1520 \ln\left(\frac{3486 \times 0.786}{0.782}\right) - 0.1674 \ln(323) + 0.0685 \times 0.786$$

$$+ \frac{-0.0559}{0.7823} - 0.0559 \times e^{8.496 + -0.7466 \ln(3486)}$$

$$\ln(P_d) = 8.287613588$$

$$P_d = 3974.3 \text{ psia}$$

Calculating the dew point using gas composition by Elsharkawy (2001):

$$P_d = A_0 + A_1 Tf + A_2 x H_2 S + A_3 x CO_2$$
$$+ A_4 x N_2 + A_5 x C_1 + A_6 x C_2 + A_7 x C_3 + A_8 x C_4 + A_9 x C_5 + A_{10} x C_6$$
$$+ A_{11} x C_7 + A_{12} MW_{C7+} + A_{13} \gamma_{C7+} + A_{14}(x C_{7+} MW_{c7+})$$
$$+ A_{15}\left\{\frac{MW_{c7+}}{\gamma_{C7+}}\right\} + A_{16}\left\{\frac{x C_{7+} MW_{c7+}}{\gamma_{C7+}}\right\} + A_{17}\left\{\frac{x C_{7+}}{x C_1 + x C_2}\right\}$$
$$+ A_{18}\left\{\frac{x C_{7+}}{x C_2 + x C_3 + x C_4 + C_5 + C_6}\right\}$$

$$(5.6)$$

where the correlation parameters are given in Table 5.10.

Table 5.11 summarizes the results of calculating each term of Elsharkawy's (2001) equation, while the dew point pressure is calculated to be:

$$P_d = 3473 \text{ psia}$$

TABLE 5.10 Dew Point Correlation Constants (Elsharkawy, 2001)

A0	A1	A2	A3	A4	A5
4268.85	0.09	−7157.87	−4540.58	−4663.55	−1357.56
A6	A7	A8	A9	A10	A11
−7776.10	−9967.99	−4257.10	−1417.10	691.53	40,660.36
A12	A13	A14	A15	A16	A17
205.26	−7260.32	−352.41	−114.52	8.13	94.92
A18					
238.25					

TABLE 5.11 Dew Point Correlation Terms (Problem 1 Solution)

Term1	Term2	Term3	Term4	Term5	Term6
4268.85	30.38	0.00	− 291.51	− 147.37	− 682.31
Term7	**Term8**	**Term9**	**Term10**	**Term11**	**Term12**
− 752.73	− 618.02	− 252.02	− 52.86	21.02	4712.54
Term13	**Term14**	**Term15**	**Term16**	**Term17**	**Term18**
29,967.96	− 5677.57	− 5963.32	− 21,380.79	175.99	18.35
Term19					
96.65					

TABLE 5.12 Results Comparison

Method	Pressure, psia
Pd from laboratory data	3245.0
Pd from field data (Al-Dhamen and Al-Marhoun, 2011)	3974.3
Pd from composition (Elsharkawy, 2001)	3473.3

Table 5.12 summarizes the dew point calculation results.

Heat Content

The heat content calculations for any gas are the same. Therefore, the techniques presented in Chapter 3, Dry Gases, are used to calculate the heat content of gas condensates. However, it should be noted that in gas condensate operations, the gas that is sold is usually the separator gas (and not the well stream gas). Therefore, the well stream composition or reservoir gas composition is not typically used in heat content calculations. Although the produced gas composition changes when the reservoir pressure declines below the dew point pressure, the heat content of surface separator gas does not change significantly. Fig. 5.9 is a plot of net heat content of a typical first stage gas produced from a gas condensate reservoir as a function of reservoir pressure.

Gas in Place Calculations for Gas Condensate Reservoirs

The volumetric calculations for gas in place in gas condensate reservoirs follow exactly the same procedure for wet gases (explained in detail in

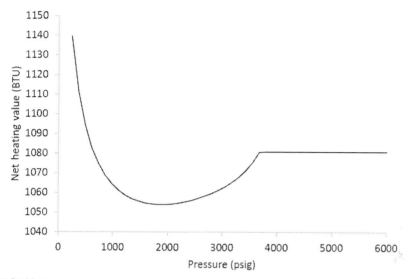

FIGURE 5.9   Net heating value for produced separator gas from a gas condensate reservoir.

Chapter 4, Wet Gases). However, the estimation of original gas in place (OGIP) in gas condensates from MB calculations is different from calculations for dry and wet gases. In dry gases, the MB uses single-phase z-factor in the p/z versus cumulative gas production (Gp) plot. In gas condensates, two modifications must be considered: (1) Two-phase z-factor is used instead of single-phase z-factor and (2) surface condensate production should be added to the cumulative gas production (because the condensate produced on surface is associated with the gas phase in the reservoir). The following equations are used to convert the surface condensate and water production to their equivalent gas (Craft et al., 1991).

$$GE = 133,000 \frac{\gamma_o}{M_o} \tag{5.7}$$

Molecular weight can be estimated from the following correlation, if not measured in the laboratory.

$$M_o = \frac{5954}{\text{API} - 8.811} \tag{5.8}$$

$$M_o = \frac{42.43\gamma_o}{1.008 - \gamma_o} \tag{5.9}$$

$$GE_w = 7390 \text{ scf/surface bbl} \tag{5.10}$$

In a three-stage separation system; and if gas produced from primary separator, secondary separator, and the stock-tank is measured, the following equation is used to estimate total reservoir gas production

$$G_p = G_{p(\text{surf})} + \text{GE} \times N_p = G_{p,\text{ps}} + G_{p,\text{ss}} + G_{p,\text{st}} + \text{GE} \times N_p \qquad (5.11)$$

For a two-stage separation system (primary separator and stock-tank), the following equation applies:

$$G_p = G_{p(\text{surf})} + \text{GE} \times N_p = G_{p,\text{ps}} + G_{p,\text{st}} + \text{GE} \times N_p \qquad (5.12)$$

In practice, the gas production from secondary separator and stock-tank is not usually available. In this case, the following equation is used to estimate the total reservoir gas production.

$$G_p = G_{p,\text{ps}} + V_{\text{eq}} \times N_p \qquad (5.13)$$

where V_{eq} is the vapor equivalent of one barrel of primary separator liquid and is estimated from the correlations given by Gold et al. (1989).

Problem 2—Calculation of Original Gas in Place for Gas Condensate Reservoir Using Material Balance Equations

The gas condensate fluid of Problem 1 is contained in a volumetric reservoir and is produced under depletion. Use the available CVD data to compare the calculated OGIP with the following assumptions: (1) dry gas assumption, and (2) gas condensate assumption. Compare the p/z versus cumulative gas produced in the two cases.

Table 5.13 shows the CVD experiment data where the stock-tank condensate has gravity of 58.6 API (specific gravity of 0.77).

Solution of Problem 2

Calculating the equivalent molecular weight: $M_o = 5954/(\text{API} - 8.811) = 5954/(58.6 - 8.811) = 119.6$ lbm/lbm mol

Calculating the condensate equivalent: $\text{GE} = 133\gamma_o/M_o = 133 \times (0.77/119.6) = 0.856381$ scf/STB

We use the gas equivalent (GE) to calculate the GE to the surface condensate and add it to the total dry gas (i.e., primary, secondary and stock-tank gas) in every pressure step. Table 5.14 summarizes the results of the calculated wet gas values.

Fig. 5.10 shows the comparison between p/z versus cumulative gas produced when single-phase z-factor and two-phase z-factor are used. Extrapolation of the pressure over two-phase z-factor line reaches the correct OGIP for the CVD experiment (1000 Mscf), while this is not the case if single phase z-factor is used.

Plant Products (Gallons Per Thousand)

As explained above, the NGL can be of significant value in gas condensate reservoirs. The economics of gas condensate reservoirs depend on the cost of

TABLE 5.13 CVD Experiment Data for Problem 2

P, psig	Initial	3245	2600	1900	1200	700	700
C7+ Mo. wt, lbm/lbm.mol		145	124	117	114	112	160
C7+ Sp. Gr.		0.782	0.763	0.756	0.753	0.751	0.795
z-Factor		0.823	0.839	0.863	0.893	0.923	
Two-Phase z-Factor		0.823	0.785	0.75	0.701	0.628	
Stock-tank oil, STB/MMscf	230.1	0.0	19.6	38.3	56.1		
1st Separator gas, Mscf	636.5	0.0	119.4	278.4	445.1		
2nd Separator gas, Mscf	123.9	0.0	13.6	27.9	42.8		
Stock-tank gas, Mscf	41.1	0.0	5.4	11.4	17.8		

TABLE 5.14 Calculation Results for Problem 2

Pressure, psig	3245	2600	1900	1200	700
p/z	3943	3099	2202	1344	758
$p/z2P$	3943	3312	2533	1712	1115
GPd, dry gas, Mscf	0	138	318	506	
GPc, condensate equivalent, Mscf	0	17	33	48	
GPw, wet gas, Mscf	0	155	350	554	

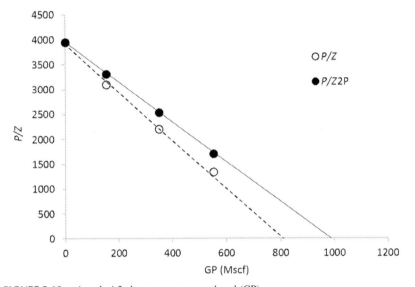

FIGURE 5.10 p/z and $p/z2$ phase versus gas produced (GP).

transportation and processing of gas to extract NGL. The following example details the calculations of GPM for a typical gas.

Problem 3—Calculation of Gallons Per Thousand for Gas Condensate

The separator gas from a gas condensate field with the composition given in Table 5.15 is processed through a lean oil plant. Calculate the NGL recovery for each component. Molecular weight of C7+ is 120 lbm/lbm mole and specific gravity of C7+ is 0.72. Assuming 25% recovery for ethane, 70%

TABLE 5.15 Composition of Separator Gas of Problem 1

N₂	CO₂	H₂S	C1	C2	C3
4.90%	63.60%	0.00%	728.50%	92.80%	38.10%
i-C4	n-C4	i-C5	n-C5	C6	C7+
0.78%	1.06%	0.44%	0.37%	0.99%	3.57%

TABLE 5.16 Calculation of Maximum Available Liquids from Separator Gas of Problem 1

Component	Mole Fraction, Y_i	Molecular Weight, M_i	Liquid-Specific Gravity	Liquid Content, GPM
N₂	0.0049	28.01		
CO₂	0.0636	44.01		
H₂S	0.0000	34.08		
C1	0.7285	16.04		
C2	0.0928	30.07	0.3562	2.4685
C3	0.0381	44.10	0.5070	1.0442
i-C4	0.0078	58.12	0.5629	0.2538
n-C4	0.0106	58.12	0.5840	0.3324
i-C5	0.0044	72.15	0.6247	0.1601
n-C5	0.0037	72.15	0.6311	0.1333
C6	0.0099	86.18	0.6638	0.4050
C7+	0.0357	120.00	0.7200	1.8748
				6.6721

recovery for propane, and 100% recovery for heavier components, how much liquid recovery is expected in bbl/MMscf?

Solution of Problem 3

The maximum available liquid products are calculated at the last column of Table 5.16. The maximum available liquids are 6.6721 GPM or 158.9 bbl/MMscf.

TABLE 5.17 Calculations of Liquids Recovery from Separator Gas of Problem 1

	Efficiency (%)	Liquid Content Recovery	
		GPM	STB/MMscf
Ethane	25	0.617	14.7
Propane	70	0.731	17.4
Butanes	100	0.586	14.0
C5 +	100	2.573	61.3

Using the assumed efficiencies for ethane, propane, butanes, and pentanes plus recovery, the liquid recovery is calculated and Table 5.17 summarizes the results.

Estimation of Modified Black Oil PVT Properties for Gas Condensates

As explained above, MBO approach can be adequate for modeling phase changes in many gas condensate fluids. The following example gives the details of estimating MBO properties from correlations for a typical gas condensate fluid.

Problem 4—Calculation of Modified Black Oil PVT Properties for Gas Condensate

The field information is available for the gas condensate of Problem 1, where the initial producing oil−gas ratio is 230 STB/MMscf. First and second separator gas specific gravities are 0.786 and 0.788, respectively. Use Nassar et al. (2013) gas condensate correlations to calculate dew point pressure, oil and gas formation volume factors, solution gas−oil ratio, and vaporized oil−gas ratio as functions of pressure. Plot the calculated MBO PVT properties.

Solution of Problem 4

Using the gas condensate saturation pressure (dew point) correlation,

$$P_{sat} = \left[A_0 \times R_{vi}^{A1}\right] \times \left[(X+Y)^{A2}\right] \times \left[STO^{A3}\right] \times \left[Tr^{A4}\right]$$

$$X = \frac{SG1}{P_{sep1}}$$

$$Y = \frac{SG2}{P_{sep2}}$$

(5.14)

A0 = 762, A1 = − 0.06, A2 = − 0.017, A3 = 0.96, and A4 = 0.38

Then dew point pressure is calculated to be 6475 psia

$$R_{vi} = \left[A0 \times e^{(A1 \times (X+Y))}\right] + \left[(A2 \times STO^2) + (A3 \times STO) + A4\right] + \left[A5 \times Tr\right]$$

(5.15)

$$R_v = \left[(A0 \times P^2) + (A1 \times P) + A2\right] + \left[e^{(A3 \times X + A4 \times Y)}\right] + \left[e^{(A5 \times V)}\right] \times R_{vi}$$

(5.16)

$$X = SG1 \times P_{sep1}$$
$$Y = SG2 \times P_{sep2}$$

$$V = \frac{STO}{Tr}$$

where

A0 = − 2.0E − 09, A1 = 2.3E − 04 A2 = 2.9E − 01, A3 = − 1.4E − 03,
A4 = 2.4E − 03, and A5 = − 2.3E + 02

$$B_g = \left[A0 \times P^{A1}\right] \times \left[e^{A2 \times X + A3 \times Y}\right] \times \left[e^{A4 \times V}\right]$$

(5.17)

$$X = SG1 \times Psep1$$
$$Y = SG2 \times Pse2$$
$$V = STO \times Tr$$

where

A0 = 404, A1 = − 0.77, A2 = 3.3E − 05,
A3 = − 7.5E − 05, and A4 = 0.002

The results are summarized in Table 5.18. The calculated dew point is highlighted in bold.

Figs. 5.11−5.14 show the calculation results against pressure.

TABLE 5.18 MBO PVT Properties Calculation Results for Problem 4

Pressure, psia	R_s, scf/STB	R_v, STB/Mscf	B_o, rbbl/STB	B_g, rbbl/Mscf
475	409	0.024	1.13	5.81
775	816	0.028	1.22	3.98
1075	1219	0.033	1.32	3.09
1375	1618	0.037	1.40	2.55
1675	2013	0.041	1.49	2.19
1975	2405	0.045	1.57	1.93
2275	2793	0.049	1.65	1.73
2575	3176	0.053	1.73	1.57
2875	3556	0.057	1.80	1.44
3175	3932	0.061	1.87	1.34
3475	4304	0.065	1.94	1.25
3775	4672	0.069	2.00	1.17
4075	5037	0.073	2.06	1.10
4375	5397	0.077	2.12	1.04
4675	5754	0.081	2.17	0.99
5275	6455	0.089	2.27	0.90
5575	6800	0.093	2.32	0.87
5875	7141	0.097	2.36	0.83
6175	7478	0.101	2.40	0.80
6475	**7812**	**0.104**	**2.43**	**0.77**
6975	7812	0.104	2.37	0.73
7475	7812	0.104	2.30	0.69
7975	7812	0.104	2.24	0.66
8475	7812	0.104	2.17	0.63
8975	7812	0.104	2.10	0.60
9475	7812	0.104	2.04	0.57
9975	7812	0.104	1.97	0.55
10,000	7812	0.104	1.97	0.55

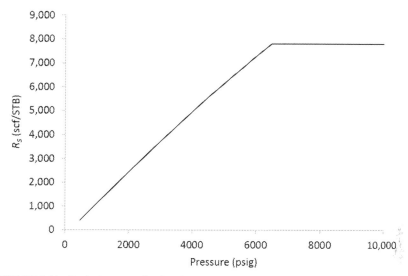

FIGURE 5.11   Producing gas−oil ratio using the MBO approach. *MBO*, modified black oil.

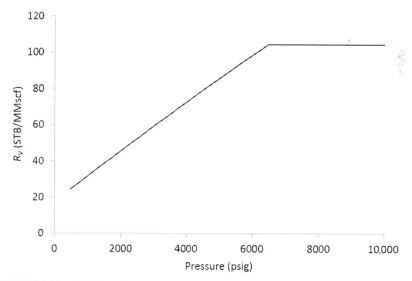

FIGURE 5.12   Producing oil−gas ratio using the MBO approach. *MBO*, modified black oil.

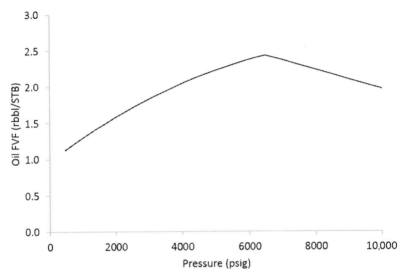

FIGURE 5.13 Oil formation volume factor using the MBO approach. *MBO*, modified black oil.

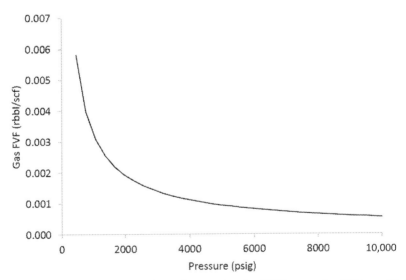

FIGURE 5.14 Gas formation volume factor using the MBO approach. *MBO*, modified black oil.

NOMENCLATURE

API	density of stock-tank liquid, API
B_g	gas formation volume factor, rcf/scf
B_o	oil formation volume factor, rbbl/STB
CGR	condensate−gas ratio
CVD	constant volume depletion
EOS	equation of state
GE	gas equivalent to one stock-tank barrel of condensate, scf/STB
GE_w	gas equivalent to one barrel of surface water, scf/bbl
GPM	gallons per thousand
G_p	total reservoir gas production, scf
$G_{p,ps}$	gas production from primary separator, scf
$G_{p,ss}$	gas production from secondary separator, scf
$G_{p,st}$	gas production from stock-tank, scf
MBO	modified black oil
M_i	molecular weight of component "i", lbm/lbm mol
M_o	molecular weight of stock-tank liquid, lbm/lbm mol
n	number of moles, lbm mol
n_p	number of moles produced in a pressure step (during CVD experiment), lbm mol
N_p	stock-tank condensate production, STB
OGIP	original gas in place
p	pressure, psia
p_d	dew point pressure, psia
R_s	solution gas−oil ratio, scf/STB
R_v	vaporized oil−gas ratio, STB/scf
y_i	mole fraction of the components in the gas, fraction
V_{eq}	vapor equivalent of primary separator liquid, scf/STB
z	z-factor at pressure, dimensionless
z_d	z-factor at dew point pressure, dimensionless
γ_o	specific gravity of stock-tank liquid, water = 1.0
γ_{oi}	specific gravity of component "i" as a liquid at standard conditions, water = 1.0
ρ_{oi}	density of component "i" as a liquid at standard conditions, lbm/ft^3

REFERENCES

Ahmadi, M.A., Elsharkawy, A., 2017. Robust correlation to predict dew point pressure of gas condensate reservoirs. Petroleum 3 (3). Available from: https://doi.org/10.1016/j.petlm.2016.05.001.

Ahmed, T., 2016. Equations of State and PVT Analysis, second ed Gulf Professional Publishing, United States. ISBN 9780128017524.

Al-Dhamen, M., and Al-Marhoun, M. 2011. New correlations for dew-point pressure for gas condensate. In: Paper SPE 155410 Presented at the SPE Saudi Arabia section Young Professionals Technical Symposium. 14−16 March, Dhahran, Saudi Arabia.

Alzahabi, A., El-Banbi, A.H., Trindade, A.A., Soliman, M., 2017. A regression model for estimation of dew point pressure from down-hole fluid analyzer data. J. Pet. Explor. Prod. Technol. Available from: https://doi.org/10.1007/s13202-016-0308-9.

Arukhe, I.N., Mason, W.E. 2012. The use of two phase compressibility factors in predicting gas condensate performance. In: Paper SPE 159080 Presented at the SPE Annual Technical Conference and Exhibition. 8—10 October, San Antonio, TX, United States.

Craft, B., Hawkins, M., Terry, R., 1991. Applied Petroleum Reservoir Engineering, second ed Prentice Hall PTR, Englewood Cliffs, NJ, United States.

Dindoruk, B. 2012. Development of a correlation for the estimation of condensate to gas ratio (CGR) and other key gas properties from density data. In: Paper SPE 160170 Presented at the SPE Annual Technical Conference and Exhibition, 8—10 October, San Antonio, Texas, United States.

El-Banbi, A.H., Abdel Fattah, K.A. and Sayyouh, M.H. 2006. New modified black oil correlations for gas condensate and volatile oil fluids. In: Paper SPE 102240 Presented at the SPE Annual Technical Conference and Exhibition, 24—27 September, San Antonio, TX, United States.

El-Banbi, A.H., Forrest, J.K., Fan, L., McCain, W.D., Jr. 2000a. Producing rich-gas-condensate reservoirs—case history and comparison between compositional and modified black-oil approaches. In: Paper SPE 58988 Presented at the SPE Fourth International Petroleum Conference and Exhibition, 1—3 February, Villahermosa, Mexico.

El-Banbi, A.H., McCain, W.D., Jr. and Semmelbeck, M.E. 2000b. Investigation of well productivity in gas-condensate reservoirs. In: Paper SPE 59773 Presented at the 2000 SPE/CERI Gas Technology Symposium, 3—5 April, Calgary, Canada.

Elsharkawy, A.M. 2001. Characterization of the plus fraction and prediction of the dew point pressure for gas condensate reservoirs. In: Paper SPE 68666 Presented at the SPE Western Regional Meeting, 26-30 March, Bakersfield, California.

Elsharkawy, A.M., 2011. Predicting the dew-point pressure for gas condensate reservoirs: empirical models and equations of state. Fluid Phase Equilibria vol 193 (issue 1-2), Elsevier Science B.V.

England, W.A., 2002. Empirical correlations to predict gas/gas condensate phase behavior in sedimentary basins. Org. Geochem. J. 33 (6), 665—673.

Fattah, K.A., El-Banbi, A.H., Sayyouh, M.H., 2006. Study compares PVT calculation methods for nonblack oil fluids. Oil Gas J. 104, 35—39. March 27.

Fevang, O., Singh, K., and Whitson, C.H. 2000. Guidelines for choosing compositional and black-oil models for volatile oil and gas-condensate reservoirs. In: Paper SPE 63087 Presented at the 2000 SPE Annual Technical Conference and Exhibition, 1—4 October, Dallas, TX, United States.

Godwin, O.N. 2012. A new analytical method for predicting dew-point pressure for gas condensate reservoirs. In: Paper SPE 162985 Presented at the 2012 SPE Nigerian Annual International Conference and Exhibition, 1—4, October, Nigeria.

Gold, D.K., McCain Jr., W.D., Jennings, J.W., 1989. An improved method for the determination of the reservoir-gas specific gravity for retrograde gases. J. Pet. Tech. 41, 747—752. Available from: https://doi.org/10.2118/17310-PA.

Humoud, A.A. and Al-Marhoun, M.A. 2001. A new correlation for gas-condensate dewpoint pressure prediction. In: Paper SPE 68230 Presented at the SPE Middle East Oil Show, 17—20 March, Manama, Bahrain. Available from: https://doi.org/10.2118/68230-MS.

Jones, J.R., Vo, D.T., Raghavan, R., 1989. Interoperation of pressure buildup responses in gas condensate wells. SPE Res. Eval. Eng 4 (1), 93—104.

Kamaria, A., Sattaria, M., Amir, H., Mohammad, A.H., Ramjugernatha, D., 2016. Rapid model for the estimation of dew point pressures in gas condensate systems. J. Taiwan Inst. Chem. Eng. 60, 258–266.

Marruffo, I., Maita, J., Him, J. and Rojas, G. 2001. Statistical forecast models to determine retrograde dew-point pressure and C7 + percentage of gas condensates on basis of production test data of eastern Venezuelan reservoirs. In: Paper SPE 69393 Presented at the SPE Latin American and Caribbean Petroleum Engineering Conference, 25–28 March, Buenos Aires, Argentina. Available from: https://doi.org/10.2118/69393-MS.

McCain Jr., W.D., 1994. Heavy components control reservoir fluid behavior. J. Pet. Technol. 46, 746–750. Technology Today Series; Trans., AIME, 297.

Nassar, I.S., El-Banbi, A.H. and Sayyouh, M.H. 2013. Modified black oil PVT properties correlations for volatile oil and gas condensate reservoirs. In: Paper SPE 164712 Presented at the 2013 North Africa Technical Conference and Exhibition. April 15–17, Cairo, Egypt.

Nemeth, L.K., Kennedy, H.T., 1967. A correlation of dewpoint pressure with fluid composition and temperature. SPEJ 7, 99–104. Trans., AIME, 240.

Olds, R.H., Sage, B.H., Lacy, W.N., 1945. Volumetric and phase behavior of oil and gas from Paloma field. Trans. AIME 160. Available from: https://doi.org/10.2118/945077-G.

Organick, E.I., Golding, B.H., 1952. Prediction of saturation pressures for condensate-gas and volatile-oil mixtures. Trans., AIME 195, 135–148.

Ovalle, A.P., Lenn, C.P., McCain Jr., W.D., 2007. Tools to manage gas/condensate reservoirs; novel fluid-property correlations on the basis of commonly available field data. SPE Res. Eval. Eng 10 (06). Available from: https://doi.org/10.2118/112977-PA.

Rayes, D.G., Piper, L.D., McCain Jr, W.D., Poston, S.W., 1992. Two-phase compressibility factors for retrograde gases. SPE Formation Eval. J. vol 7, 87–92. Available from: https://doi.org/10.2118/20055-PA.

Shokir, E.M., 2008. Dewpoint pressure model for gas condensate reservoirs based on genetic programming. Energy Fuels 22 (5), 3194–3200. Available from: https://doi.org/10.1021/ef800225b.

Whitson, C.H., Torp, S.B., 1983. Evaluating constant-volume depletion data. J. Pet. Tech. 35, 610–620.

Whitson, C.H., Brule, M.R., 2000. *Phase Behavior.* SPE Monograph 20 SPE Richardson, TX, United States.

PROBLEMS

5.1 You are given a sample of gas condensate fluid with the following composition. The sample comes from a newly discovered gas reservoir owned by the Good Petroleum Company. The dew point pressure was measured in the PVT laboratory and found to be 3115 psia. In the absence of laboratory data, dew point estimation models (correlations) are usually available to estimate the dew point with varying accuracy. These models are based on either knowledge of fluid composition or knowledge of some surface fluid properties data (e.g., GCR, API of stock-tank oil, and reservoir temperature). The data available for the sample are tabulated below.

Components	mol%	wt.%
H_2S	0	
CO_2	2.01	1.96
N_2	5.62	3.49
C-1	46.79	16.66
C-2	12.65	8.44
C-3	5.87	5.74
i-C4	6.04	7.79
n-C4	0	0
n-C5	0	0
i-C5	3.92	6.27
C6	4.78	9.14
C3−C5		19.8
C6 +		49.6
C7 + mol%	12.32	40.5
P_{sep}, psia	624.7	
T_{sep}, °F	100	
API	60	
Reservoir Gas Sp. Gr.	1.35	
Gas Sp. Gr. Sep1	0.762	
Dew Point Pressure	3115	
C7 + MW	148	
C7 + Sp. Gr.	0.804	
Field GOR, scf/STB	2265	
Reservoir T, °F	325	

Calculate the dew point pressure using the following correlations. Compare the calculated dew point pressure with the laboratory-measured value.

Approach 1 (From Composition)
Marruffo et al. (2001)
Nemeth and Kennedy (1967)
El-Sharkawy (2001)
Shokir (2008)
Olds et al. (1945)
Godwin (2012)
Alzahabi et al. (2017)

Approach 2 (From Field Data)
Humoud and Al-Marhoun (2001)
Ovalle et al. (2007)
Al-Dhamen and Al-Marhoun (2011)

5.2 Calculate dew point pressure for the following two samples using Marruffo et al. (2001), Humoud and Al-Marhoun (2001), Ovalle et al. (2007), and Al-Dhamen and Al-Marhoun (2011) correlation. Compare the calculated dew point pressures with the laboratory-measured values given in the following table. Also, calculate the reservoir gas specific gravity for the three gases.

Components	Sample 1	Sample 2
C7 + mol%	7.88	11.7
P_{sep}, psia	389.7	495
T_{sep}, °F	106	70
T, °F	301	238
Stock tank oil gravity, °API	61.72	42.4
Reservoir gas Sp. Gr.	1.095	1.25
Separator gas Sp. Gr.	0.777	0.702
Dew point pressure, psia	5724.7	4830
C7 + MW	164	169
C7 + Sp. Gr.	0.8197	0.813
Field GOR, scf/STB	5390	3503
CGR, STB/MMscf	185.5	285.4

5.3 The following data come from downhole fluid analyzer. The composition is given in weight percent. Calculate the dew point pressure using Alzahabi et al. (2017) correlation and compare with the laboratory-measured value of 4415 psia.

Components	wt.%
H_2S	0.03
CO_2	0.67
N_2	0
C-1	20.82
C-2	9.07
C-3	7.47
i-C4	5.21
n-C4	0
n-C5	0
i-C5	3.87
C6	3.64
C7 + %	49.1
P_{sep}, psia	202.7
T_{sep}, °F	70
API	49.9
Reservoir gas Sp. Gr.	1.18
Separator gas Sp. Gr.	0.77
Dew point pressure	4415
C7 + MW	193
C7 + Sp. Gr.	0.82
Field GOR, scf/STB	3410
Reservoir T, °F	190

5.4 The following gas composition is for the primary separator gas produced from a rich gas condensate reservoir in the Gulf of Mexico. Calculate the plant products using both lean oil plant and the more advanced refrigeration plant technologies. Provide the NGL products in GPM and in bbl/MMscf units.

Component	mol%
N_2	3.30
CO_2	6.82
C1	78.60
C2	6.41
C3	2.08
i-C4	0.44
n-C4	0.48
i-C5	0.18
n-C5	0.17
C6	0.16
C7 +	1.36
C7 + Molecular weight	139.00
C7 + Specific gravity	0.77

Chapter 6

Volatile Oils

Volatile oils (similar to gas condensates) are usually found in deeper reservoirs and higher temperature reservoirs. When volatile oil reservoirs are found in under-saturated state (initial reservoir pressure is higher than bubble point pressure), the fluid is essentially oil under reservoir conditions. As with black oils, when a volatile oil reservoir fluid is produced to surface, gas (which was originally dissolved in the oil) comes out of solution and becomes a free gas. There are several noticeable differences between black oils and volatile oils (McCain, 1993). Fig. 6.1 represents a typical volatile oil phase diagram.

The differences between black oils and volatile oils are as follows:

- In both volatile oil and black oil fluids, the reservoir temperature is less than the critical temperature. However, the reservoir temperature is closer to the critical temperature in volatile oil reservoirs than it is in black oil reservoirs.
- The volatile oil reservoirs produce initially with higher producing gas−oil ratio (GOR) than black oil reservoirs.
- The stock-tank oil is usually colored in volatile oil fluids, and darker in black oil fluids.
- Iso-volume lines in volatile oil fluids are shifted upward, while in black oil fluids they may be evenly spaced.
- Gas associated with volatile oil is gas condensate, while gas associated with black oil can usually be approximated with the dry gas concept (or the wet gas concept, when the amount of solution gas is relatively high).

The proximity of iso-volume lines to each other, in volatile oils, implies that a small pressure drop in the vicinity of the well will release large amounts of gas around the wellbore. El-Banbi and McCain (2001), through detailed compositional simulation, showed that the pressure drop can result in large compositional and saturation changes around the wellbore. The large amount of gas released around the wellbore can also reduce the oil relative permeability and may lead to choking the well with excessive gas.

PVT Property Correlations. DOI: https://doi.org/10.1016/B978-0-12-812572-4.00006-0

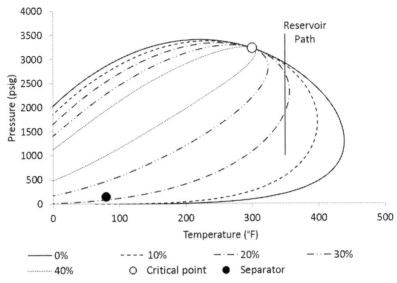

FIGURE 6.1   Typical phase diagram for volatile oil fluid.

The gas condensate produced with volatile oil contains liquids that will condense on surface. Jacoby and Berry (1957) and Cordell and Ebert (1965) highlighted the effect of the gas condensate associated with volatile oil on material balance calculations. They concluded that the material balance equation developed for typical black oil reservoirs cannot be used to handle volatile oil reservoirs. To handle material balance calculations for volatile oil, either the compositional material balance or the modified black-oil (MBO) approach must be used. Walsh (1994) and Walsh et al. (1994) presented the use of MBO material balance equations for volatile oil reservoirs.

PVT PROPERTIES FOR VOLATILE OILS

The same PVT properties defined for black-oil (specific gravity, bubble point pressure, oil formation volume factor, solution GOR, oil density, oil viscosity, and isothermal compressibility of oil) are used for engineering volatile oil reservoirs. The definitions and the behavior of these properties are explained in the Black Oil Chapter. An additional PVT property is required, which is the vaporized oil—gas ratio.

Vaporized Oil—Gas Ratio

Vaporized oil—gas ratio is defined as the amount of oil (condensate) that will be produced from the surface gas. In the petroleum literature, two symbols are used for vaporized oil—gas ratio (R_v and r_s). The vaporized oil—gas

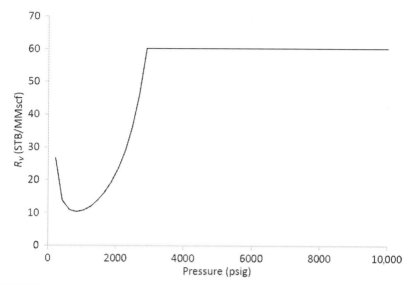

FIGURE 6.2 Vaporized oil–gas ratio calculated from EOS using Whitson and Torp (1983) procedure. *EOS*, equation of state.

ratio is a function of the composition of the fluid, reservoir temperature, and pressure. When the oil is above the bubble point pressure, R_v is constant. At these conditions, the gas produced on surface will drop a constant amount of liquid. When the reservoir pressure declines below the bubble point pressure, the gas produced on surface will contain less liquid than the initial gas, hence the reduction in R_v value with depleting pressure. Fig. 6.2 shows typical vaporized oil–gas ratio behavior for a volatile oil. The R_v in this figure is calculated by use of Whitson and Torp (1983) procedure and an equation-of-state (EOS) program.

Modified Black-Oil Approach for Volatile Oil

As with gas condensates, MBO approach can be used to model the behavior of volatile oil (Walsh, 1994; Walsh et al., 1994). Four PVT functions are required for MBO approach (oil formation volume factor, gas formation volume factor, solution GOR, and vaporized oil–gas ratio). It is preferable if the MBO PVT properties are generated from a tuned EOS model that matches laboratory observations of the volatile oil. Several techniques are available to derive the MBO properties from an EOS (Fattah et al., 2006). In absence of an EOS model, few correlations are available to derive these properties for volatile oils (El-Banbi et al., 2006; Nassar et al., 2013). As in the case of gas condensate, these correlations carry a great degree of uncertainty and should be used only when a representative EOS model is unavailable. The correlations for volatile oil are given in Appendix A, Oil Correlations Formulae.

HANDLING PVT PROPERTIES FOR VOLATILE OILS

Most volatile oil PVT properties can be determined from a reservoir fluid study. While five PVT experiments are usually performed for black oil fluids [composition measurement, constant composition expansion (CCE), differential liberation (DL), viscosity measurement, and separator experiment(s)], many recent PVT reports for volatile oil fluids include the results of an additional experiment [constant volume depletion (CVD)]. The PVT experiments are explained in other texts (e.g., Ahmed, 2016; Whitson and Brule, 2000). Unlike in black-oil PVT experiments, DL experiment alone is believed to be inadequate to explain the volatile oil behavior in the reservoir. CVD is added in modern volatile oil PVT reports to give more experimental observations that can be used to build reliable EOS models. The only PVT property that requires additional calculations is vaporized oil−gas ratio (R_v). Walsh and Towler (1994) presented a direct method to calculate the volatile oil PVT properties from the CVD experiment data. Other methods are also available, but require the use of a tuned EOS model for the fluid. Fattah et al. (2006) compared the most commonly used methods for calculating MBO properties and concluded that Whitson and Torp (1983) yields slightly better results than other methods.

If PVT reports and EOS models are not available, correlations can be used to estimate PVT properties for volatile oil. In general, the correlations reviewed in Chapter 7, Black Oils, can be used for volatile oils. These correlations include those for bubble point pressure, solution GOR, oil formation volume factor, oil density, oil viscosity, and isothermal compressibility of oil. The correlations for the free gas released in the reservoir when the pressure declines below the bubble point pressure can also be used to handle the free gas properties below the bubble point pressure. In addition to these correlations, the two R_v correlations (El-Banbi et al., 2006; Nassar et al., 2013) can be used for vaporized oil−gas ratio calculation.

Input data for correlations is the same for volatile oil and black oil fluids. However, it is more difficult sometimes to prepare volatile oil input data for correlations.

While it is relatively easy to estimate bubble point pressure from field data for black oils (when enough static reservoir pressure points are available), it is usually difficult to estimate bubble point pressure from field data in volatile oil reservoirs. This difficulty is due to the fact that reservoir pressure declines gently above and below the bubble point in volatile oil reservoirs. In black oil reservoirs, two distinct lines with significantly different slopes can usually be observed when static reservoir pressure is plotted against time or against cumulative oil production.

Estimation of initial producing GOR is also more difficult in volatile oils. The initial producing GOR is sensitive to separator conditions. Ibrahim et al. (2011) reports field cases that show significant variation in producing GOR when separator conditions change. These cases were reported for wells producing above the bubble point pressure (i.e., the producing GOR is not affected by free gas production from the reservoir). The total initial solution GOR is needed as input to the correlations. In case the stock-tank GOR is not measured, it needs to be estimated from the correlations presented in the Black Oil Chapter. Fig. 6.3 shows the GOR for the stock-tank and low pressure separator for many oil samples (covering the range of volatile oils, black oils, and low GOR oils). The total surface GOR is higher for more volatile oil (low C7 + mole percent oil). The amount of gas separated in low pressure separator and stock-tank can be considerable in volatile oil. When estimating the total solution GOR in volatile oils, care should be taken to use the oil correlations. The low pressure separator and stock-tank GOR can change significantly and are strong functions of separator conditions.

An alternative approach for volatile oil PVT prediction is the MBO approach. This approach has the advantage of estimating the vaporized oil−gas ratio. The MBO correlations for volatile oil can be applied to calculate the four MBO PVT functions (oil and gas formation volume factors, solution GOR, and vaporized oil−gas ratio). The MBO correlations for volatile oil are listed in Table 6.1, and the forms of the correlations are given in Appendix A, Oil Correlations Formulae.

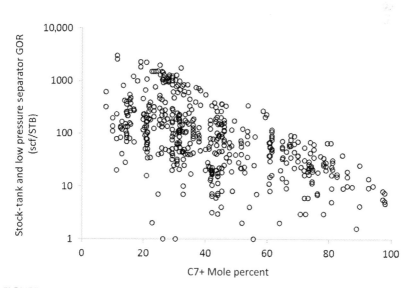

FIGURE 6.3   Stock-tank and low pressure separator GOR. *GOR*, gas−oil ratio.

TABLE 6.1 For Oil MBO PVT Properties Correlations

Oil Formation Volume Factor	Gas Formation Volume Factor	Solution Gas–Oil Ratio	Vaporized Oil–Gas Ratio
B_o	B_g	R_s	R_v
El-Banbi et al. (2006)	El-Banbi et al. (2006)	El-Banbi et al. (2006)	El-Banbi et al. (2006)
Nassar et al. (2013)	Nassar et al. (2013)	Nassar et al. (2013)	Nassar et al. (2013)

Sequence of Calculations

For estimation of volatile oil PVT properties, the same flow chart of Chapter 7, Black Oils, can be used. If vaporized oil–gas ratio is required, it must be calculated from any of the R_v correlations of Appendix A, Oil Correlations Formulae. If the MBO approach is to be used, the flow chart of Fig. 6.4 is applied. In this sequence of calculations, pressure, temperature, initial solution GOR, and separator conditions are specified. The bubble point pressure can be either specified by the user if known, or estimated from bubble point correlations (refer to the Black Oils Chapter) if not known. All these inputs (which also include API gravity of stock-tank oil and specific gravity of primary separator gas) will be used to calculate the four functions of the MBO model.

ENGINEERING VOLATILE OIL RESERVOIRS

Volatile oil reservoirs require the same PVT properties as black oil reservoirs (in addition to the vaporized oil–gas ratio) for engineering calculations. In general, liquid recovery from surface separators in volatile oil is more sensitive to separator conditions than in black oils. Figs. 6.5–6.8 show the effect of separator conditions on MBO PVT properties for a typical volatile oil. These properties were calculated at two different separator conditions using an EOS and Whitson and Torp (1983) procedure. In this example, all MBO PVT properties are affected by separator conditions except gas formation volume factor.

Problem 1—Calculation of Volatile Oil PVT Properties for Different Two-Stage Separator Conditions

The following represents the results of three two-stage separator experiments for a volatile oil. Use the total solution GOR (with other field measured

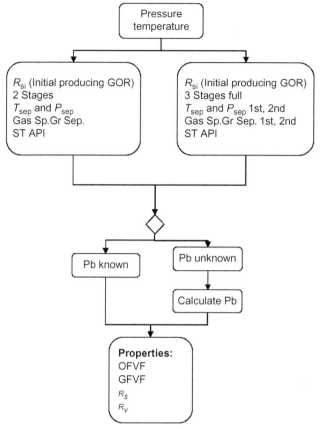

FIGURE 6.4   Flow chart for MBO PVT properties calculations for volatile oils. *MBO*, modified black oil.

values) and calculate the bubble point pressure, formation volume factor, solution GOR, oil density and oil viscosity from correlations (use Glaso, 1980; Standing, 1947; Lasater, 1958 correlations). Plot the results of your calculations to compare the values of PVT properties calculated from the information in the three separator tests. Compare your calculated bubble point pressure with the laboratory-measured bubble point pressure. Separator test data is given in Table 6.2.

Table 6.3 gives the uncorrected (for separator conditions) PVT laboratory measurements for all properties obtained from the CCE and DL experiments. Dead oil viscosity is 0.824 cp at the reservoir temperature of 232°F. Single stage flash of the fluid results in GOR of 3223 scf/STB.

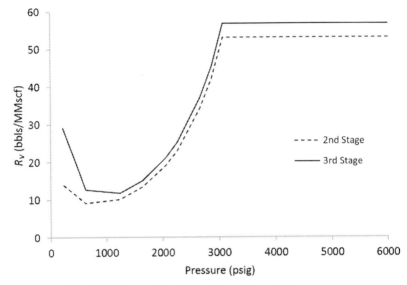

FIGURE 6.5   Vaporized oil−gas ratio at different separator conditions.

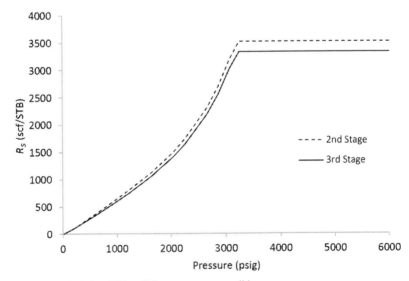

FIGURE 6.6   Solution GOR at different separator conditions.

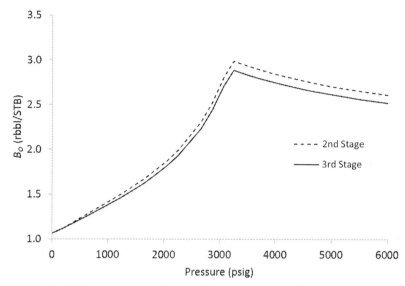

FIGURE 6.7   Oil formation volume factor at different separator conditions.

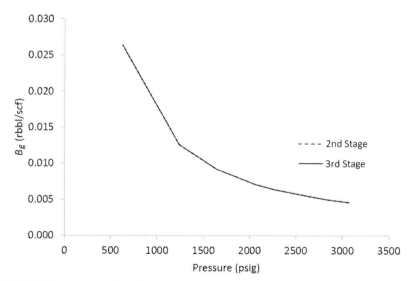

FIGURE 6.8   Gas formation volume factor at different separator conditions.

TABLE 6.2 Separator Test Data for the Fluid of Problem 1

Stage	Pressure, psia	T, °F	GOR, scf/STB	B_o, rbbl/STB	Gas Sp.Gr.	Stock-Tank Oil Density, lbm/cu.ft.	Oil Density at Saturation Pressure, lbm/cu.ft.	B_o at Saturation Pressure, rbbl/STB
Separator Test 1								
1	87.0	75	2867	1.08	0.87		32.96	2.59
2	14.7	60	102	1.00	1.26	49.95		
Total			2968					
Separator Test 2								
1	232.0	167	2551	1.17	0.83		32.96	2.52
2	14.7	140	282	1.00	1.20	49.82		
Total			2833					
Separator Test 3								
1	377.0	167	2319	1.23	0.80		32.90	2.48
2	14.7	140	435	1.00	1.18	49.81		
Total			2754					

GOR, gas–oil ratio.

TABLE 6.3 PVT Laboratory Data for the Fluid of Problem 1

Pressure, psia	B_o, rbbl/STB	R_s, scf/STB	z Factor	B_g cu.ft./scf	Gas Viscosity, cp	Gas Sp. Gr.	Oil Density, lbm/cu.ft.	Oil Viscosity, cp
9991	2.66	3612					37.7	0.248
9498	2.67	3612					37.5	0.244
9005	2.69	3612					37.2	0.241
8497	2.71	3612					37.0	0.237
8004	2.73	3612					36.6	0.234
7497	2.76	3612					36.3	0.230
7004	2.79	3612					36.0	0.227
6496	2.82	3612					35.5	0.223
6003	2.86	3612					35.1	0.220
5496	2.90	3612					34.6	0.215
5003	2.94	3612					34.1	0.212
4495	2.99	3612					33.5	0.209
4118	3.04	3612					33.0	0.206
3915	2.77	3059	0.823	0.004	0.042	1.114	33.2	0.214
3625	2.49	2534	0.789	0.004	0.035	1.053	34.0	0.222

(Continued)

TABLE 6.3 (Continued)

Pressure, psia	B_o, rbbl/STB	R_s, scf/STB	z Factor	B_g, cu.ft./scf	Gas Viscosity, cp	Gas Sp. Gr.	Oil Density, lbm/cu.ft.	Oil Viscosity, cp
3045	2.07	1783	0.765	0.005	0.027	0.984	36.0	0.238
2320	1.76	1213	0.783	0.006	0.020	0.900	38.3	0.258
1595	1.53	801	0.817	0.010	0.016	0.894	40.9	0.293
943	1.40	514	0.857	0.017	0.014	0.924	42.1	0.351
290	1.24	220	0.925	0.060	0.012	1.092	44.0	0.465

TABLE 6.4 Comparison Between Actual and Calculated Bubble Point Pressure From Different Correlations

Bubble Point Pressure, psia

Laboratory			4118	
	Flash	Test 1	Test 2	Test 3
Glaso (1980)	7143	6659	6618	6642
Standing (1947)	7661	6835	6759	6788
Lasater (1958)	4873	4822	4922	5031
Error Percent (%)				
	Flash	Test 1	Test 2	Test 3
Glaso (1980)	73	62	61	61
Standing (1947)	86	66	64	65
Lasater (1958)	18	17	20	22

Solution of Problem 1

Bubble Point Pressure

The bubble point pressure is calculated from the three correlations. Each time, the total solution GOR is used as an input to the correlation. The total solution GOR comes from the three separator experiments in addition to the flash solution GOR. The results are summarized in Table 6.4. The error between the calculated bubble point pressure (using the different input total solution GOR and different correlations) is also given in the results table.

Gas in Solution

Table 6.5 shows the corrected solution gas–oil ratio using different separator conditions. Fig. 6.9 shows the graphical comparison between the different solution GORs. The comparison is made with the uncorrected solution GOR obtained from the DL experiment. The correction of PVT properties to separator conditions is covered in Whitson and Brule (2000).

The separator test data of test 1 was used to calculate solution GOR from the three correlations (Glaso, 1980; Standing, 1947; Lasater, 1958). Fig. 6.10 shows comparison between the calculated solution GOR from the three correlations and the R_s data corrected using separator test 1. The plot shows that Lasater (1958) correlation is the closest to the laboratory-measured data for this volatile oil sample.

TABLE 6.5 Solution Gas–Oil Ratio Corrected to Different Separator Conditions

R_s, scf/STB

Pressure, psia	Laboratory	Separator Test 1	Separator Test 2	Separator Test 3
9991	3612	2968	2833	2754
9498	3612	2968	2833	2754
9005	3612	2968	2833	2754
8497	3612	2968	2833	2754
8004	3612	2968	2833	2754
7497	3612	2968	2833	2754
7004	3612	2968	2833	2754
6496	3612	2968	2833	2754
6003	3612	2968	2833	2754
5496	3612	2968	2833	2754
5003	3612	2968	2833	2754
4495	3612	2968	2833	2754
4118	3612	2968	2833	2754
3915	3059	2497	2375	2302
3625	2534	2049	1940	1873
3045	1783	1410	1318	1259
2320	1213	924	846	793
1595	801	572	505	456
943	514	327	266	221
290	220	77	23	0

Fig. 6.11 shows Lasater (1958) calculated using the solution GOR from flash, and the three separator tests. The comparison plot shows that different separator conditions can affect the solution GOR. This implies that we need to use the proper total solution GOR when calculating PVT properties for engineering applications.

The plot also shows the importance of predicting bubble point pressure as accurately as possible. The large difference observed between Glaso (1980)

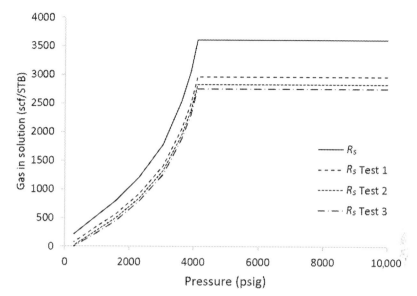

FIGURE 6.9   Solution GOR corrected to different separator conditions.

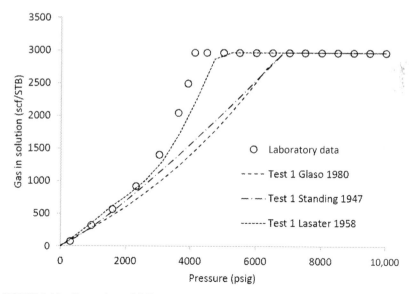

FIGURE 6.10   Comparison of different correlations results using separator test 1 information.

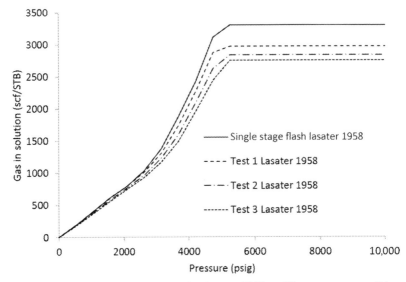

FIGURE 6.11 Solution GOR calculated using Lasater (1958) at different separator conditions.

and Standing (1947) correlation prediction and the laboratory-measured data is due to the correlations inadequate bubble point pressure prediction.

Oil Formation Volume Factor

Table 6.6 and Fig. 6.12 show the laboratory values of oil formation volume factor (uncorrected) with the calculated values at the three separator tests.

Fig. 6.13 shows comparison results between corrected laboratory data (using separator test 1) and formation volume factors calculated using Glaso (1980), Standing (1947), and Lasater (1958) correlations. The total solution gas—oil ratio that was used as input to all correlations was that of separator test 1.

The effect of total solution GOR on the calculated oil formation volume factor is shown by Fig. 6.14, which gives the calculated oil formation volume factor using Glaso (1980) correlation with four different input solution GORs. The input solution GORs are those of the three separator tests in addition to the flash GOR. The figure shows that oil formation volume factor can change significantly with separator conditions.

Oil Density

Table 6.7 and Fig. 6.15 show the oil density computed by the three different correlations. The input solution GOR to the correlations is that of separator test 1.

TABLE 6.6 Corrected Oil Formation Volume Factor

B_o, rbbl/STB

Pressure, psi	Laboratory	Separator Test 1	Separator Test 2	Separator Test 3
9991	2.66	2.26	2.20	2.17
9498	2.67	2.28	2.21	2.19
9005	2.69	2.29	2.23	2.20
8497	2.71	2.31	2.25	2.22
8004	2.73	2.33	2.27	2.24
7497	2.76	2.35	2.29	2.26
7004	2.79	2.38	2.31	2.28
6496	2.82	2.40	2.34	2.31
6003	2.86	2.43	2.37	2.33
5496	2.90	2.47	2.40	2.37
5003	2.94	2.51	2.44	2.40
4495	2.99	2.55	2.48	2.45
4118	3.04	2.59	2.52	2.48
3915	2.77	2.36	2.29	2.26
3625	2.49	2.12	2.06	2.03
3045	2.07	1.77	1.72	1.69
2320	1.76	1.50	1.46	1.44
1595	1.53	1.30	1.27	1.25
943	1.40	1.19	1.16	1.15
290	1.24	1.06	1.03	1.02

Oil Viscosity

Table 6.8 shows the oil viscosity calculated from the three correlations. Fig. 6.16 shows the computed oil viscosity using the different corrections (using total solution GOR of separator test 1 as input) compared with the laboratory-measured oil viscosity.

The figure shows low values of oil viscosity, which are expected for volatile oil fluids.

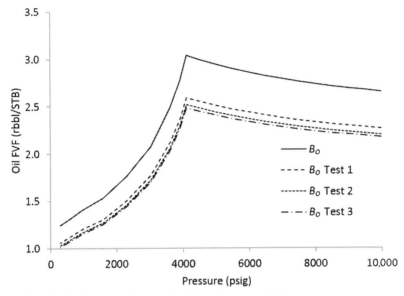

FIGURE 6.12 Laboratory and corrected oil formation volume factor.

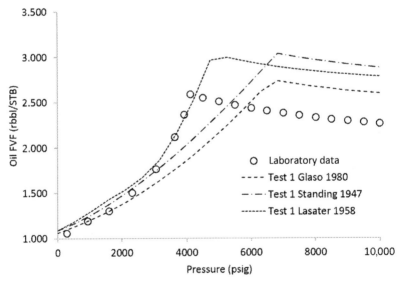

FIGURE 6.13 Laboratory-measured oil formation volume factor compared with calculated oil formation volume factor from three correlations.

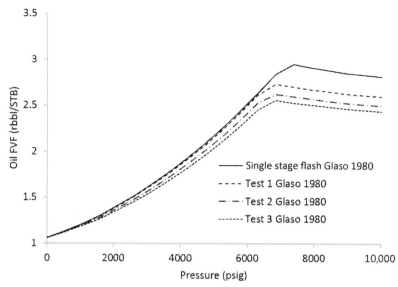

FIGURE 6.14 Oil formation volume factor calculated by Glaso (1980) correlation at different separator conditions.

TABLE 6.7 Oil Density Using the Different Correlations

Pressure, psia	Oil Density, lbm/cu.ft.		
	Glaso (1980)	Standing (1947)	Lasater (1958)
0	48.1	46.6	46.7
526	46.7	45.3	44.8
1053	45.2	43.6	42.8
1579	43.5	41.8	40.8
2105	41.9	40.0	39.3
2632	40.2	38.4	37.5
3158	38.7	36.8	35.3
3684	37.3	35.3	32.7
4211	36.0	34.0	30.4
4737	34.9	32.8	28.2
5263	33.8	31.7	28.0
5789	32.9	30.6	28.4
6316	32.2	29.7	28.8
6842	31.6	28.8	29.1
7368	31.4	28.0	29.4
7895	31.8	27.8	29.7
8421	32.1	28.1	29.9
8947	32.4	28.4	30.1
9474	32.7	28.6	30.2
10,000	32.9	28.9	30.4

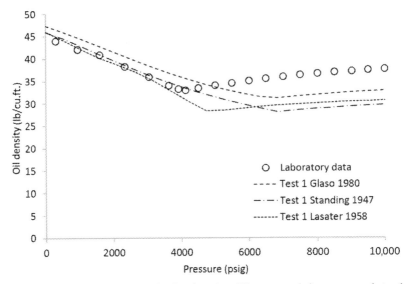

FIGURE 6.15 Oil density calculated using the different correlations compared to the laboratory-measured oil density.

Problem 2—Calculation of Modified Black-Oil PVT Properties for Volatile Oil

A volatile oil fluid at reservoir temperature of 232°F has stock-tank oil specific gravity of 0.83 and separator gas specific gravity of 0.828. Initial producing GOR is 2833 scf/STB for a three-stage separation system. Primary separator pressure is 232 psi and separator temperature is 110°F, while secondary separator pressure and temperature are 70 psi and 70°F, respectively. The stock-tank is at standard pressure and temperature. Use Nassar et al. (2013) volatile oil correlations to calculate bubble point pressure, oil and gas formation volume factors, solution GOR, and vaporized oil–gas ratio as function of pressure. Plot the calculated MBO PVT properties.

Solution of Problem 2

Using Nassar et al. (2013) volatile oil saturation pressure correlation (bubble point pressure):

$$P_{sat} = [A_0 \times R_{si}^{A1}] \times [(X+Y)^{A2}] \times [STO^{A3}] \times [Tr^{A4}]$$
$$X = \frac{SG1}{P_{sep1}}$$
$$Y = \frac{SG2}{P_{sep2}}$$

(6.1)

TABLE 6.8 Oil Viscosity Calculated Using Different Correlations

Pressure, psig	Oil Viscosity, cp		
	Glaso (1980)	Standing (1947)	Lasater (1958)
0	0.464	0.391	0.620
526	0.392	0.345	0.458
1053	0.344	0.300	0.351
1579	0.304	0.262	0.280
2105	0.271	0.231	0.231
2632	0.241	0.206	0.195
3158	0.216	0.185	0.168
3684	0.193	0.167	0.147
4211	0.173	0.152	0.130
4737	0.155	0.139	0.116
5263	0.140	0.127	0.114
5789	0.126	0.117	0.119
6316	0.113	0.109	0.125
6674	0.106	0.101	0.132
6842	0.107	0.101	0.138
7368	0.113	0.106	0.146
7895	0.119	0.112	0.153
8421	0.125	0.118	0.161
8947	0.131	0.124	0.169
9474	0.138	0.130	0.178
10,000	0.145	0.137	0.187

Where, $A0 = 762$, $A1 = -0.06$, $A2 = -0.017$, $A3 = 0.96$, and $A4 = 0.38$.

The bubble point pressure is calculated to be 6717 psia. The four MBO functions are calculated using Nassar et al. (2013) correlations of Appendix A, Oil Correlations Formulae, and the results are given in Table 6.9. The results at the bubble point pressure are highlighted in bold.

Figs. 6.17−6.20 show the calculated MBO PVT properties for solution GOR, vaporized oil−gas ratio, oil formation volume factor, and gas formation volume factor, respectively.

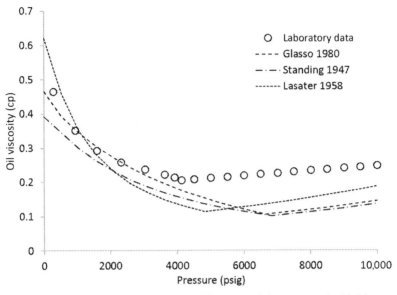

FIGURE 6.16 Oil viscosity calculated using different correlations compared with laboratory-measured values.

TABLE 6.9 Calculated MBO PVT Properties for Problem 2

P, psia	R_s, Mscf/STB	R_v, STB/Mscf	B_o, rbbl/STB	B_g, rbbl/Mscf
717	0.031	0.037	1.18	3.16
1017	0.044	0.033	1.22	2.38
1317	0.059	0.031	1.28	1.93
1617	0.076	0.030	1.34	1.64
1917	0.096	0.031	1.40	1.43
2217	0.118	0.035	1.48	1.27
2517	0.143	0.040	1.56	1.15
2817	0.170	0.047	1.66	1.05
3117	0.199	0.057	1.75	0.97
3417	0.231	0.068	1.86	0.90
3717	0.265	0.081	1.98	0.84
4017	0.302	0.096	2.10	0.79
4317	0.341	0.113	2.23	0.74

(Continued)

TABLE 6.9 (Continued)

P, psia	R_s, Mscf/STB	R_v, STB/Mscf	B_o, rbbl/STB	B_g, rbbl/Mscf
4617	0.383	0.132	2.37	0.70
4917	0.427	0.153	2.51	0.67
5217	0.473	0.176	2.66	0.64
5517	0.522	0.201	2.82	0.61
5817	0.573	0.227	2.99	0.58
6117	0.627	0.256	3.17	0.56
6417	0.683	0.287	3.35	0.54
6717	**0.741**	**0.319**	**3.54**	**0.52**
7217	0.741	0.319	3.52	0.49
7717	0.741	0.319	3.49	0.46
8217	0.741	0.319	3.45	0.44
8717	0.741	0.319	3.42	0.42
9217	0.741	0.319	3.38	0.40
9717	0.741	0.319	3.35	0.39
10,000	0.741	0.319	3.33	0.38

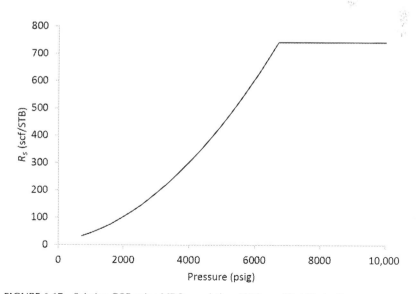

FIGURE 6.17 Solution GOR using MBO correlation. *MBO*, modified black oil.

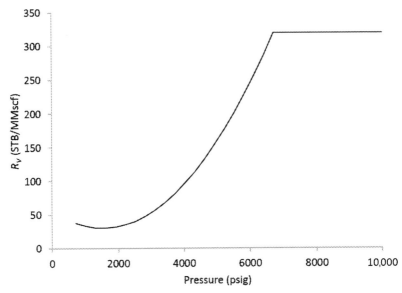

FIGURE 6.18 Vaporized oil–gas ratio using MBO correlation. *MBO*, modified black oil.

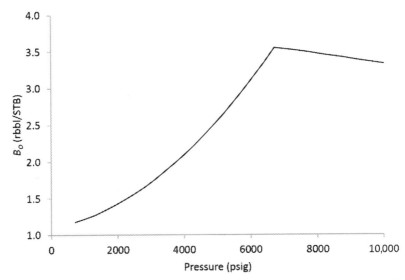

FIGURE 6.19 Oil formation volume factor using MBO correlation. *MBO*, modified black oil.

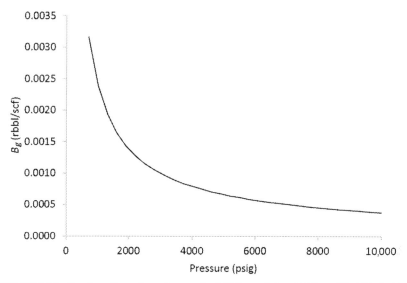

FIGURE 6.20   Gas formation volume factor using MBO correlation. *MBO*, modified black oil.

NOMENCLATURE

B_g gas formation volume factor, rcf/scf
B_o oil formation volume factor, rbbl/STB
CCE   constant composition expansion
CVD   constant volume depletion
DL   differential liberation
EOS   equation of state
GOR   gas–oil ratio, scf/STB
P_{sat} saturation pressure, psia
P_{sep1} primary separator pressure, psia
P_{sep2} secondary separator pressure, psia
r_s vaporized oil–gas ratio, STB/MMscf
R_s solution gas–oil ratio, scf/STB
R_{si} initial solution gas–oil ratio, scf/STB
R_v vaporized oil–gas ratio, STB/MMscf
SG_1 primary separator gas specific gravity, dimensionless
SG_2 secondary separator gas specific gravity, dimensionless
STO   stock-tank oil gravity, API
T_r reservoir temperature, °F

REFERENCES

Ahmed, T., 2016. Equations of State and PVT Analysis, 2nd ed Gulf Professional Publishing, USA, ISBN: 9780128017524.
Cordell, J.C., Ebert, C.K., 1965. A case history-comparison of predicted and actual performance of a reservoir producing volatile crude oil. J. Pet. Tech. 17, 1291–1293.

El-Banbi, A.H., McCain, W.D., Jr. 2001. Sampling volatile oil wells. In: Paper SPE 67232 Presented at the 2001 SPE Production and Operations Symposium. March 24—26, Oklahoma City, OK.

El-Banbi, A.H., Abdel Fattah, K.A., Sayyouh, M.H. 2006. New modified black oil correlations for gas condensate and volatile oil fluids. In: Paper SPE 102240 Presented at the SPE Annual Technical Conference and Exhibition. September 24—27, San Antonio, TX.

Fattah, K.A., El-Banbi, A.H., Sayyouh, M.H., 2006. Study compares PVT calculation methods for nonblack oil fluids. Oil Gas J 104 (12), 35—39.

Glaso, O., 1980. Generalized pressure—volume—temperature correlations. J. Pet. Tech 32, 785—795.

Ibrahim, M., El-Banbi, A.H., El-Tayeb, S., Sayyouh, H., 2011. Changing separator conditions during black-oil and modified black-oil simulation runs. In: Paper SPE 142462 Presented at the SPE Middle East Oil and Gas Show and Conference, March 6—9, Manama, Bahrain.

Jacoby, R.H., Berry Jr., V.J., 1957. A method for predicting depletion performance of a reservoir producing volatile crude oil. Trans. AIME 210, 27—33.

Lasater, J.A., 1958. Bubble point pressure correlations. J. Pet. Technol. 10 (5), 65—67. SPE-957-G.

McCain Jr., W.D., 1993. Black oils and volatile oils — what's the difference? Pet. Eng. Intl 65, 24—27.

Nassar, I.S., El-Banbi, A.H., Sayyouh, M.H. 2013. Modified black oil PVT properties correlations for volatile oil and gas condensate reservoirs. In: Paper SPE 164712 Presented at the 2013 North Africa Technical Conference and Exhibition. April 15—17, Cairo, Egypt.

Standing, M.B., 1947. A pressure-volume-temperature correlation for mixtures of California oils and gases. Drilling and Production Practice. American Petroleum Institute, New York, USA.

Walsh, M.P., 1994. A generalized approach to reservoir material balance calculations. J. Can. Pet. Technol 34, 1.

Walsh, M.P., Towler, B.F., 1994. Method computes PVT properties for gas condensate. Oil Gas J 93, 83—86.

Walsh, M.P., Ansah, J., and Raghavan, R. 1994. The new generalized material balance as an equation of a straight-line: Part 1—Application to undersaturated and volumetric reservoirs. In: SPE 27684, Presented at the 1994 SPE Permian Basin Oil and Gas Recovery Conference, March 16—18, Midland, TX.

Whitson, C.H., Brule, M.R., 2000. Phase Behavior, SPE Monograph, 20. SPE Richardson, TX, USA.

Whitson, C.H., Torp, S.B., 1983. Evaluating constant-volume depletion data. J. Pet. Tech. 35 (3), 610—620.

PROBLEMS

6.1 Two-stage separator test data for a volatile oil sample are given below. Calculate the oil PVT properties using both Lasater (1958) and Standing (1947) correlations. Plot the calculated solution GOR using the three separator tests and compare the results. Also, make similar comparison plots for oil formation volume factor using Standing (1947) correlation.

Stage	Pressure, psia	T, °F	GOR, scf/ STB	B_o, rbbl/ STB	Gas Sp. Gr.	Stock-Tank Oil Density, lbm/cu.ft.	Oil Density at Saturation Pressure, lbm/cu.ft.	B_o at Saturation Pressure, rbbl/STB
Separator Test 1								
1	100.0	75	3500	1.10	0.88		31.96	2.79
2	14.7	60	90	1.00	1.27	48.95		
	Total		3590					
Separator Test 2								
1	250.0	167	2800	1.13	0.85		31.96	2.62
2	14.7	140	289	1.00	1.21	48.82		
	Total		3089					
Separator Test 3								
1	400.0	167	2550	1.33	0.81		31.90	2.58
2	14.7	140	380	1.00	1.19	48.81		
	Total		2930					

6.2 Calculate the four MBO PVT properties for the volatile oil given in the above problem. Compare the calculated properties from the three separator initial solution GOR. Assume that reservoir temperature is 320°F.

Chapter 7

Black Oils

Black oils constitute the majority of the oils we produce. They are present in nearly every basin. The majority of black-oil reservoirs are initially discovered as undersaturated reservoirs (initial pressure is higher than the bubble-point pressure). Black-oil fluids consist of a large number of hydrocarbon components from methane to sometimes very high carbon number molecules. They may also include impurities of variable amounts of nonhydrocarbons.

A typical phase diagram of black oils is shown in Fig. 7.1. In black-oil phase diagrams, the reservoir temperature is usually far lower than the critical temperature. The iso-volume lines are spaced approximately evenly around the reservoir temperature. In addition, the gas that is liberated in the reservoir at pressures below the bubble point can usually be considered dry gas for engineering purposes.

According to McCain, oils that produce initially with gas—oil ratio (GOR) less than 1750 scf/STB or have more than 20% C7 + mole percent are considered black oils. Oils that produce initially with GOR higher than that level, or contain less C7 + than 20%, are considered volatile oils. The distinction between black oils and volatile oils is not very sharp. For practical purposes, we divided black oils into two groups (low GOR oils and moderate GOR oils) in Chapter 2, Reservoir-Fluid Classification. The low GOR oils have distinct characteristics that distinguish them from black oils with higher GOR.

PVT PROPERTIES FOR BLACK OILS

PVT properties for black oils are required in almost all reservoir, production and surface facilities calculations to manage and forecast oil reservoir behavior. They also affect calculations that are sensitive to the economic value of oil reservoirs (e.g., reserves and production forecast).

The engineering of black-oil fluids requires knowledge of multiple PVT properties including bubble-point pressure, oil formation volume factor, solution GOR, oil density, oil viscosity, and isothermal compressibility of oil. The following sections define these properties and comment on their expected range of values.

PVT Property Correlations. DOI: https://doi.org/10.1016/B978-0-12-812572-4.00007-2
147

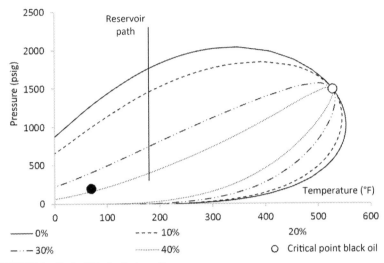

FIGURE 7.1 Typical black-oil phase diagram.

Specific Gravity of Oil

The oil specific gravity is defined as the ratio of the oil density to water density, both measured at the same pressure and temperature. The water density (reference fluid) is nearly always taken at the highest density point (4°C or 39.2°F). In the petroleum industry, specific gravity of both oil and water are usually measured at 60°F and atmospheric pressure. Another term that is commonly used for specific gravity of oil (API gravity) is related to oil specific gravity by the following equation:

$$\text{API} = \frac{141.5}{\gamma_o} - 131.5 \qquad (7.1)$$

The API gravity is commonly measured for stock-tank oil and is usually less than 45°API for black oils. API gravity of oil is used to indicate the quality of the oil and is also commonly used in oil pricing.

Bubble-Point Pressure

Bubble-point pressure is defined as the pressure at which the first bubble of gas appears at a specific temperature. The phase diagram of typical black oils shows that the bubble-point pressure could be different at different temperatures. In the petroleum industry, if bubble-point pressure value is mentioned without reference to a particular temperature, the temperature is implicitly assumed to be the reservoir temperature.

When the reservoir is depleted and its pressure falls below the bubble-point pressure, free gas starts to form in the reservoir. Since gas has higher mobility than oil, the producing GOR is expected to increase when the reservoir pressure decreases below the bubble-point pressure. Other PVT properties also undergo significant changes when the reservoir pressure passes through the bubble-point pressure, as will be shown in the next sections.

Formation Volume Factor of Oil

Oil formation volume factor is defined as the volume of oil (and dissolved gas) at reservoir pressure and temperature required to produce one stock tank barrel of oil at the surface. The oil formation volume factor is expressed in units of reservoir volume over standard volume (usually rbbl/STB). At surface, the oil loses the dissolved gas it contained under high pressure and temperature in the reservoir. Three things happen to the reservoir oil (with dissolved gas) when it moves from reservoir conditions to surface:

1. The oil loses mass due to the loss of dissolved gas on the oil trip from the reservoir to the surface.
2. The oil shrinks slightly due to the reduction in temperature.
3. The oil expands slightly due to the reduction in pressure.

The volume reduction due to temperature and the volume increase due to pressure usually have minimal effect on the oil and tend to cancel each other. The mass loss due to liberation of dissolved gas represents the major effect and is primarily responsible for the shape of oil formation volume factor below the bubble point. Fig. 7.2 shows that oil formation volume factor increases with reduction in pressure until the oil reaches the bubble-point pressure. The volume increase at pressures above the bubble point is due to the expansion of oil (with its dissolved gas). Below the bubble point, and with the continued reduction in pressure, the oil formation volume factor is reduced primarily due to mass loss with the additional release of dissolved gas. For black oils, the highest value for formation volume factor is usually less than 2 rbbl/STB (McCain, 1993).

Solution Gas—Oil Ratio

Solution GOR (also called gas in solution) is defined as the gas dissolved in oil at any pressure and temperature. At standard pressure and temperature, dissolved gas is completely released from the oil, and therefore the oil contains no gas. The gas liberated from the oil will be released in the form of free gas. Oil at any pressure and temperature has a certain capacity to dissolve gas. In general, the more volatile the oil (due to the presence of higher amounts of lighter hydrocarbons), the greater the capacity of oil to dissolve gas. Therefore, volatile oils often have a higher solution GOR initially and at any pressure, than do black oils. The bubble-point pressure at a particular temperature is defined by the amount of gas available to go into solution (and the capacity of the oil to dissolve the gas). Above the bubble-point pressure, the gas in solution is constant. Fig. 7.3 represents the solution GOR. The unit that is often used for solution GOR is scf/STB. At surface (theoretically, at the standard pressure and temperature), all gas comes out of solution, and therefore, the solution GOR is equal to zero.

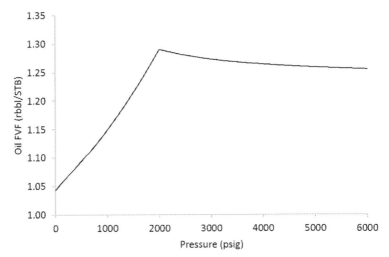

FIGURE 7.2 Oil formation volume factor.

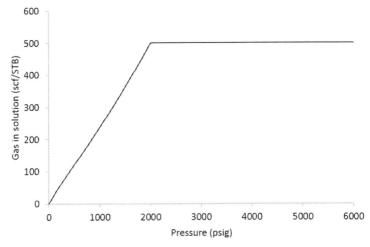

FIGURE 7.3 Solution gas−oil ratio.

Oil Density

The density of oil is defined as the mass per unit volume. In oil field units, density is expressed in units of lbm/ft^3. Oil density is usually measured at the reservoir temperature and different pressures. Fig. 7.4 is an example oil density plot at a particular temperature. In typical oil reservoirs, the oil density decreases with depletion of pressure until it reaches a minimum value at the bubble point. The higher oil density at higher pressures is a reflection of the packing of more molecules of oil in the same unit volume. At pressures below the bubble-point pressure and with reduction in pressure, gas comes out of solution and forms free gas. The released gas will contain some of the intermediate components of oil, leaving the heavier components in the reservoir oil. This release of intermediate components accounts for the reversal in the density versus pressure trend below the bubble-point pressure.

A common representation of oil density is the oil gradient (expressed in psi/ft.). The importance of oil gradient is connected with the common use of measurement of fluid pressure versus depth (or RFT measurement) to reveal the type of reservoir fluid according to the value of the fluid gradient. Fig. 7.5 shows an example of RFT measurements in a sandstone reservoir. The figure shows a clear difference between oil and water gradients and contributes to determination of the reservoir fluid type and the location of contacts. For this reservoir, the OWC is estimated at 9198′ TVD (true vertical depth). The line connecting the pressure points in the oil zone gives an oil gradient of 0.401 psi/ft, while the water gradient is 0.449 psi/ft. A downhole oil sample was collected from the reservoir at a depth of 9160′ TVD and a water sample was collected from depth 9300′ TVD. The fluid samples confirmed the interpretation of the pressure gradients for this well.

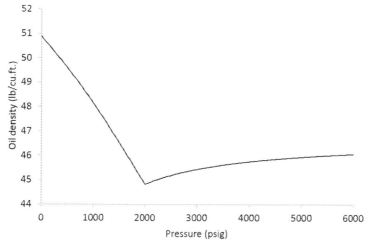

FIGURE 7.4 Oil density.

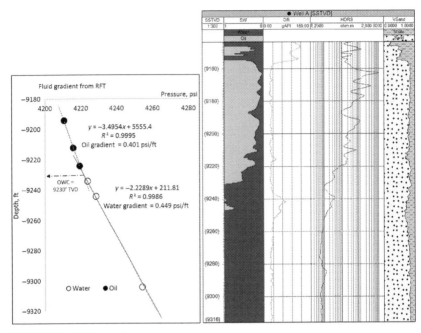

FIGURE 7.5 RFT example.

Oil Viscosity

Dynamic or absolute viscosity is defined as resistance to flow exerted by a fluid (usually measured in centipoise, cp). Viscosity of oil is highly affected by the oil composition in addition to amount of solution gas and temperature. It is a measure of the magnitude of internal friction between fluid layers as they move relative to each other. High viscosity is usually associated with low GOR oils; lower viscosity is usually seen with high GOR oils. We generally expect wells with lower viscosity to have higher flow rates, and reservoirs with lower viscosity oils to recover more reserves. Static (or kinematic) viscosity is obtained by dividing the dynamic viscosity by the density of the fluid.

Fig. 7.6 is a typical oil viscosity plot versus pressure (measured at reservoir temperature). The behavior of viscosity is similar to that of density. The figure shows that oil viscosity decreases with reduction in pressure, due to reduced friction between fluid layers with the reduction in pressure. The viscosity of oil records a minimum value at the bubble point. With the continued depletion of reservoir pressure and the release of free gas, the reservoir oil becomes heavier (i.e., more viscous) due to loss of lighter components to the free gas. Viscosity values of black oils can differ significantly from one oil to another.

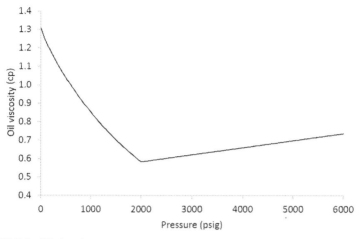

FIGURE 7.6   Oil viscosity.

Compressibility of Oil

Oil compressibility (also called isothermal oil compressibility) is defined as the change in fluid volume with respect to the change in pressure at isothermal conditions. Oil compressibility is expressed in psi^{-1}. Another unit (microsip) is also used. A microsip is equal to $1 \times 10^{-6} psi^{-1}$. Oil compressibility is determined during the laboratory procedure known as CCE (constant composition expansion). In that procedure, the oil in the PVT cell is allowed to expand by reduction of pressure, and both the increase in oil volume and the decrease in oil pressure are measured. When the oil passes through the bubble-point pressure, free gas accumulates in the PVT cell and is not released out of the cell.

The following equation is used as the definition for oil compressibility above the bubble-point pressure:

$$c_o = -\frac{1}{V}\left(\frac{\partial V}{\partial p}\right)_T = -\frac{1}{B_o}\left(\frac{\partial B_o}{\partial p}\right)_T \qquad (7.2)$$

Below the bubble-point pressure, the oil compressibility definition includes the compressibility of the free gas that is accumulated in the PVT cell. Oil compressibility is therefore given by the following equation:

$$c_o = -\frac{1}{B_o}\left[\left(\frac{\partial B_o}{\partial p}\right)_T - B_g\left(\frac{\partial R_s}{\partial p}\right)_T\right] \qquad (7.3)$$

Due to the significant difference in free gas and oil compressibility, the oil compressibility below the bubble-point pressure is dominated by the free gas compressibility. A point of sudden increase in compressibility therefore exists at the bubble-point pressure. Fig. 7.7 is a typical oil compressibility

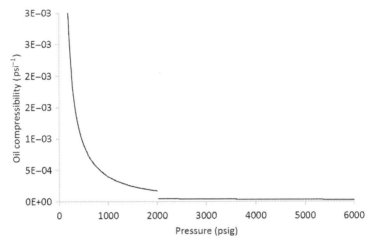

FIGURE 7.7 Isothermal compressibility of oil.

plot, and it shows the sudden increase in oil compressibility between the values above and below the bubble-point pressure.

Compressibility of oil above the bubble point is of particular importance because it affects the material balance calculations for volumetric oil reservoirs. It is also the value used in pressure transient analysis for single phase oil wells. Oil compressibility is also used in vertical lift performance calculations and others. The value of oil compressibility can range from as low as $3 \times 10^{-6}\,psi^{-1}$ for low GOR undersaturated oils to approximately $150 \times 10^{-6}\,psi^{-1}$ for saturated high GOR oils.

HANDLING PVT PROPERTIES FOR BLACK OILS

PVT properties for black oils can be determined in the PVT laboratory with the reservoir fluid study. For black oils, the reservoir fluid study usually consists of five experiments [composition, CCE, differential liberation (DL), separator test(s), and viscosity measurement]. When a representative fluid sample or PVT laboratory report is not available, we rely on correlations to calculate the PVT properties for oils. Other texts have discussed the use of PVT laboratory reports in calculation of PVT properties for black oils (e.g., McCain, 1990; Dandekar, 2006; Ahmed, 2010).

Over the years, numerous correlations to calculate PVT properties for oils have been developed. The correlations are usually simple expressions (equations) that allow engineers to calculate the PVT properties from readily available inputs. The correlations inputs should not rely on measured values that can be arrived at only in the PVT laboratory.

Correlations are usually developed from large datasets of measured PVT properties. Many correlations rely on data collected from oil samples in a particular region or basin. Others use oil samples of worldwide origin. The correlation is usually developed by fitting a predetermined form of an equation (or more) with the measured data and calculating the equation(s) parameters by use of multivariable regression techniques. It is important to note the range of applicability of each correlation, as the performance of any correlation prediction is not guaranteed outside its range of applicability. Tables 7.1−7.12 summarize the majority of the known correlations for oil PVT properties.

The equations of many of these correlations are listed in Appendix A, Oil Correlations Formulae. Applicability ranges for these correlations are given in Appendix C, Oil Correlations Range of Applicability.

In the majority of the referenced literature that discusses the development of new PVT correlations, the authors compare their new correlation(s)

TABLE 7.1 Bubble-Point Correlations

Bubble-Point Pressure		Al-Marhoun (1988)	Al-Mehaideb (1997)
Al-Najjar et al. (1988)	Al-Shammasi (1999)	Asgarpour et al. (1988)	Bolondarzadeh et al. (2006)
De Ghetto et al. (1995)	Dindoruk and Christman (2001)	Doklah and Osman (1992)	Elam (1957)
El-Banbi et al. (2006)	Elmabrouk et al. (2010)	Elsharkawy and Alikhan (1997)	Farshad et al. (1996)
Glasso (1980)	Hanafy et al. (1997)	Hassan (2011)	Hemmati and Kharrat (2007)
Ikiensikimama (2008)	Ikiensikimama and Ogboja (2009)	Karimnezhad et al. (2014)	Kartoatmdjo and Schmidt (1991)
Kartoatmdjo and Schmidt (1994)	Khairy et al. (1998)	Khamehchi and Ebrahimian (2009)	Khazam et al. (2016)
	Labedi (1982)	Labedi (1990)	Lasater (1958)
Levitan and Murtha (1999)	Macary and El-Batanony (1992)	Mazandarani and Asghari (2007)	McCain (1991)
Mehran et al. (2006)	Moradi et al. (2010)	Movagharnejad and Fasih (1999)	Obomanu and Okpobiri (1987)
Okoduwa and Ikiensikimama (2010)	Omar and Todd (1993)	Ostermann and Owolabi (1983)	Owolabi (1984)
Petrosky (1990)	Petrosky and Farshad (1993)	Petrosky and Farshad (1998)	Standing (1947)
Standing (1981)	Steve (1993)	Vasquez and Beggs (1980)	Velarde et al. (1997)

TABLE 7.2 Gas in Solution Correlations

Gas in Solution	Al-Marhoun (1988)	Al-Shammasi (1999)	Asgarpour et al. (1988)
Casey and Cronquist (1992)	De Ghetto et al. (1995)	Dindrouk and Christman (2001)	Doklah and Osman (1992)
Elsharkawy and Alikhan (1997)	Farshad et al. (1996)	Glasso (1980)	Hanafy et al. (1997)
Hassan (2011)	Hemmati and Kharrat (2007)	Ikiensikimama (2008)	Kartoatmdjo and Schmidt (1991)
Kartoatmdjo and Schmidt (1994)	Khamehchi and Ebrahimian (2009)	Khazam et al. (2016)	Knopp and Ramsey (1960)
	Labedi (1982)	Labedi (1990)	Lasater (1958)
Macary and El-Batanony (1992)	Mazandarani and Asghari (2007)	McCain (1991)	Petrosky (1990)
Petrosky and Farshad (1993)	Petrosky and Farshad (1998)	Standing (1947)	Standing (1981)
Vasquez and Beggs (1980)	Velarde et al. (1997)		

TABLE 7.3 Saturated Oil Compressibility Correlations

Saturated Oil Compressibility	Ahmed (1989)	Al-Marhoun (1992)	Al-Mehaideb (1997)
Calhoun (1947)	Chew and Connally (1959)	De Ghetto et al. (1995)	Dindoruk and Christman (2004)
Elsharkawy and Alikhan (1997)	Farshad et al. (1996)	Ikiensikimama (2008)	Kartoatmdjo and Schmidt (1991)
Kartoatmdjo and Schmidt (1994)	Khazam et al. (2016)	Labedi (1982)	McCain (1991)
McCain et al. (1988)	Petrosky (1990)	Petrosky and Farshad (1993)	Petrosky and Farshad (1998)
Standing (1947)	Standing (1981)	Steve (1993)	Vasquez and Beggs (1980)

against existing correlations to show superior prediction results over the dataset used in development of the new correlation(s). Examples of such comparison studies include Sutton and Farshad (1990), Hanafy et al. (1997), and Al-Marhoun (2004). Additional, more elaborate comparison studies using larger databases are also available (McCain et al., 2011).

TABLE 7.4 Saturated Oil Formation Volume Factor Correlations

Saturated Oil Formation Volume Factor	Abdul-Majeed and Salman (1988)	Ahmed (1989)	
Al-Marhoun (1988)	Al-Marhoun (1992)	Al-Marhoun (2003)	Al-Mehaideb (1997)
Al-Najjar et al. (1988)	Al-Shammasi (1999)	Arps (1962)	Asgarpour et al. (1989)
Bolondarzadeh et al. (2006)	Casey and Cronquist (1992)	Dindrouk and Christman (2001)	Doklah and Osman (1992)
Elam (1957)	El-Banbi et al. (2006)	Elmabrouk et al. (2010)	Elsharkawy and Alikhan (1997)
Farshad et al. (1996)	Galsso (1980)	Hanafy et al. (1997)	Hemmati and Kharrat (2007)
Ikiensikimama and Ogboja (2009)	Karimnezhad et al. (2014)	Kartoatmdjo and Schmidt (1991)	Kartoatmdjo and Schmidt (1994)
Khairy et al. (1998)	Khazam et al. (2016)	Knopp and Ramsey (1960)	Labedi (1982)
Levitan and Murtha (1999)	Macary and El-Batanony (1992)	Mazandarani and Asghari (2007)	McCain (1991)
Mehran et al. (2006)	Moradi et al. (2013)	Obomanu and Okpobiri (1987)	Omar and Todd (1993)
Ostermann and Owolabi (1983)	Owolabi (1984)	Petrosky (1990)	Petrosky and Farshad (1993)
Petrosky and Farshad (1998)	Standing (1947)	Standing (1981)	Sulaimon et al. (2014)
Vasquez and Beggs (1980)	Velarde et al. (1997)		

TABLE 7.5 Saturated Oil Density Correlations

Saturated Oil Density	Ahmed (1989)	Hanafy et al. (1997)	McCain (1991)
Standing (1947)	Standing (1981)		

Investigating the summary tables reveals that correlations are developed either as individual correlations for a particular PVT property, or as groups (families) of correlations developed from the same dataset for several PVT properties (e.g., Al-Marhoun, Hanafy et al., Petrosky and Farshad, Standing,

TABLE 7.6 Saturated Oil Viscosity Correlations

Saturated Viscosity	Abdul-Majeed et al. (1990)	Abu-Khamsim and Al-Marhoun (1991)	Al-Khafaji (1987)
Al-Mehaideb (1997)	Beggs and Robinson (1975)	Bergman (2004)	Bergman (2007)
Chew and Connally (1959)	De Ghetto et al. (1995)	Dindrouk and Christman (2001)	Elsharkawy and Alikhan (1997)
Glasso (1980)	Hanafy et al. (1997)	Hossain et al. (2005)	Ikiensikimama (2008)
Kartoatmdjo and Schmidt (1991)	Kartoatmdjo and Schmidt (1994)	Khamehchi and Ebrahimian (2009)	Khan et al. (1987)
Khazam et al. (2016)	Labedi (1982)	Labedi (1992)	McCain (1991)
Naseri et al. (2005)	Osorio (1990)	Petrosky (1990)	Petrosky and Farshad (1993)
Petrosky and Farshad (1995)	Petrosky and Farshad (1998)	Standing (1947)	Standing (1981)
Steve (1993)			

TABLE 7.7 Undersaturated Oil Compressibility Correlations

Undersaturated Oil Compressibility	Ahmed (1989)	Al-Marhoun (1992)	Al-Marhoun (2003)
Al-Marhoun (2006)	Al-Mehaideb (1997)	De Ghetto et al. (1995)	Dindoruk and Christman (2004)
Elsharkawy and Alikhan (1997)	Farshad et al. (1996)	Hanafy et al. (1997)	Ikiensikimama (2008)
Kartoatmdjo and Schmidt (1991)	Kartoatmdjo and Schmidt (1994)	Labedi (1982)	McCain (1991)
McCain et al. (1988)	Petrosky (1990)	Petrosky and Farshad (1993)	Petrosky and Farshad (1998)
Spivey et al. (2007)	Standing (1947)	Standing (1981)	Vasquez and Beggs (1980)

and McCain). In the absence of measured PVT data, the choice of correlation (or families of correlations) to be used for a particular fluid is typically a matter of personal preference or past experience.

TABLE 7.8 Stock-Tank Gas–Oil Ratio Correlations

Tank GOR	Elmabrouk and Shirif (2003)	Okeke and Sylvester (2016)	Rollins et al. (1990)
Valko and McCain (2003)			

TABLE 7.9 Undersaturated Oil Formation Volume Factor Correlations

Undersaturated Oil Formation Volume Factor	Ahmed (1992)	Al-Marhoun (1992)	General Equation

TABLE 7.10 Undersaturated Oil Density Correlations

Undersaturated Oil Density	General Equation		

TABLE 7.11 Undersaturated Oil Viscosity Correlations

Undersaturated Oil Viscosity	Abdul-Majeed et al. (1990)	Al-Khafaji et al. (1987)	Al-Mehaideb (1997)
Beal (1946)	Beggs and Robinson (1975)	Bergman (2004)	Bergman and Sutton (2006)
De Ghetto et al. (1995)	Dindoruk and Christman (2001)	Elsharkawy and Alikhan (1999)	Glasso (1980)
Hossain et al. (2005)	Ikiensikimama (2008)	Kartoatmdjo and Schmidt (1991)	Kartoatmdjo and Schmidt (1994)
Khan et al. (1987)	Khazam et al. (2016)	Kouzel et al. (1965)	
Labedi (1982)	Labedi (1992)	McCain (1991)	Naseri et al. (2005)
Petrosky (1990)	Petrosky and Farshad (1993)	Petrosky and Farshad (1995)	Petrosky and Farshad (1998)
Standing (1947)	Standing (1981)	Twu (1985)	Vasquez and Beggs (1980)

TABLE 7.12 Undersaturated Dead Oil Viscosity Correlations

Dead Oil Viscosity	Al-Khafaji et al. (1987)	Ansrade (1930)	Beal (1946)
Beggs and Robinson (1975)	Bennison (1998)	Bergman (2004)	Bergman and Sutton (2006)
De Ghetto et al. (1995)	Dindrouk and Christman (2001)	Elsharkawy and Alikhan (1999)	Elsharkawy and Gharbi (2001)
Fitzgerlad (1997)	Glasso (1980)	Hossain et al. (2005)	Ikiensikimama (2008)
Kartoatmdjo and Schmidt (1991)	Kartoatmdjo and Schmidt (1994)	Kaye (1985)	Khan et al. (1987)
Khazam et al. (2016)	Labedi (1982)	Labedi (1992)	McCain (1991)
Naseri et al. (2005)	Naseri et al. (2012)	Ng and Egbogah (1983)	Oyedeko and Ulaeto (2011)
Petrosky (1990)	Petrosky and Farshad (1993)	Petrosky and Farshad (1995)	Petrosky and Farshad (1998)
Sattarina et al. (2007)	Standing (1947)	Standing (1981)	Ulaeto and Oyedeko (2014)
Whitson and Brule (2000)			

Field Data for Oil PVT Correlations

Calculating PVT properties for oils requires four inputs, which can be obtained from readily available field information. These inputs include: (1) total solution GOR at the bubble point; (2) API gravity of stock-tank oil; (3) specific gravity of separator gas; and (4) temperature.

Several oil PVT properties require knowledge of bubble-point pressure. Bubble-point pressure (at the required temperature) can be estimated from any of the bubble-point pressure correlations. If the bubble-point pressure is known from external sources, it can be directly used in calculating other oil PVT properties from their respective correlations. Prior knowledge of bubble-point pressure improves the accuracy of most correlations.

Estimating Solution Gas−Oil Ratio

Total solution GOR is obtained from oil and gas sales data. In theory, the solution GOR is constant as long as the reservoir pressure is above the bubble-point pressure; in practice, however, some wells produce below the bubble-point pressure and therefore have higher producing GOR than solution GOR. It is usually advisable to make a plot of producing GOR

versus time or cumulative oil production (the latter is preferred) to help
determine a good value to use as initial solution GOR. An example is given
in Fig. 7.8. The figure shows both the instantaneous (producing) GOR and
the cumulative GOR. Cumulative GOR is defined according to the following
equation. The same data is plotted against time in Fig. 7.9.

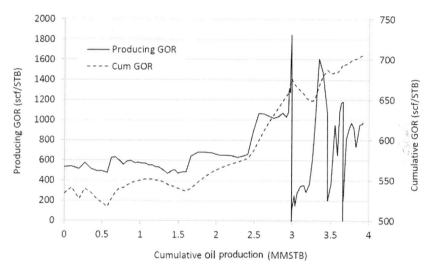

FIGURE 7.8 Example producing GOR and cumulative GOR versus cumulative oil production.
GOR, gas–oil ratio.

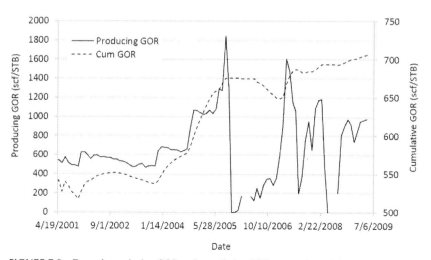

FIGURE 7.9 Example producing GOR and cumulative GOR versus time. *GOR*, gas–oil ratio.

$$R_p = \frac{G_p}{N_p} \qquad (7.4)$$

Fig. 7.8 shows that the field producing GOR is a constant value (around 580 scf/STB) for some time until cumulative oil production has reached 0.5 MMSTB. GOR starts to increase and reaches 1000 scf/STB. The GOR data generally show fluctuations. In this example, considerable variation in GOR data occurred in 2005 and after, due to changes in wells completion and poor maintenance of gas meters.

Consideration of the entire field GOR rather than individual wells GOR is important for selecting an appropriate value for use in PVT correlations. In addition, taking a 3 or 4-month moving average for the data reduces the scattering in GOR data and obtains a more representative average initial GOR. The correct PVT correlations input is the total GOR (and not the separator GOR). In the majority of oil field operations, the gas produced from the stock-tank (and sometimes the low-pressure separator) is vented or sent to the flare. Therefore, the value of the stock-tank GOR is not usually available. Correlations are used to estimate the stock-tank GOR. These correlations are dependent on primary separator conditions in addition to the usual correlations input parameters.

Estimating Stock-Tank Gas—Oil Ratio

Investigating the summary tables for oil correlations shows that four correlations exist for calculating stock-tank GOR. Further information on their applicability ranges is given in Appendix C, Oil Correlations Range of Applicability. These correlations require primary separator conditions in addition to separator GOR and stock-tank oil gravity to calculate an estimate for stock-tank GOR. Once the stock-tank GOR is calculated, it is added to the separator GOR to estimate the total GOR required for oil PVT correlations.

Estimating Stock-Tank Oil Gravity

Stock-tank oil gravity is usually available from the oil sales data. Theoretically, stock-tank oil gravity is constant as long as the oil is above the bubble point. Stock-tank oil gravity slightly decreases in black oils, while it slightly increases for volatile oils when the reservoir pressure falls below the bubble point. The change in API gravity value occurs due to changes in oil composition when gas is released from the oil in the reservoir. Occasionally, these changes in API are not recorded properly. The API value that is required as input in oil correlations is the initial value (when the oil is still undersaturated).

Estimating Separator Gas Gravity

Separator gas gravity is usually measured by taking a sample of the separator gas (sales gas) and measuring its composition. The separator gas is usually sold, and therefore it is necessary to calculate its heat content. The heat content calculation requires that the gas composition be known. The standard procedure for calculating gas specific gravity from composition (see Chapter 3: Dry Gases) is used.

Estimating Temperature

If oil PVT properties are required for a reservoir oil sample, the PVT properties will be calculated at the reservoir temperature. The reservoir temperature is usually obtained from logging tools and is usually available in the log header. If reservoir temperature data is unavailable, the following equation can be used to estimate an approximate reservoir temperature (assuming normal temperature gradient). The temperature unit in this equation is °F and the depth is recorded in feet.

$$T = T_{\text{surface}} + 15 \left(\frac{\text{Depth}}{1000} \right) \tag{7.5}$$

The temperature gradient varies from one location to another, and the local temperature gradient is usually available.

Knowledge of Bubble-Point Pressure

If bubble-point pressure is not known for the reservoir fluid, we first calculate the bubble-point pressure from a correlation, then use it in the following calculations for other PVT properties. In some cases, bubble-point pressure is known or can be estimated from field data. Fig. 7.10 is a plot of static reservoir pressure (measured from several wells in the field) versus cumulative oil production from the reservoir. Also shown in the plot is the cumulative GOR versus cumulative oil production. A weighted moving averaging method was used to estimate the reservoir pressure curve within the measured points. At cumulative oil production of approximately 0.5 MMSTB, it is observed that the static pressure curve flattens and changes in slope to a gentler slope. This point indicates that the reservoir has reached the bubble-point pressure. The use of this value of bubble point (2700 psia) as input to the oil PVT correlations to compute other PVT properties will result in more accurate PVT properties than will the use of an estimated bubble point from a correlation. For this particular example, a bottom-hole fluid sample was available. The sample indicated that the bubble point pressure is 2760 psia which is close to the value estimated from the static pressure versus cumulative oil production plot.

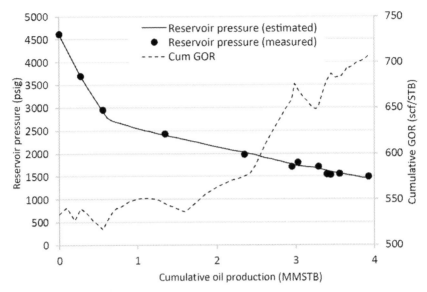

FIGURE 7.10 Example static reservoir pressure versus cumulative oil production plot.

Utilizing the Known Bubble-Point Pressure to Improve Accuracy of Oil PVT Correlations

Knowledge of bubble-point pressure can be used to improve the accuracy in prediction of other PVT properties, as the bubble point will not be calculated by a correlation that carries some error (McCain, 1990). The known bubble-point value can also be used to improve the accuracy of the PVT correlations prediction even further. McCain (1990) outlines a simple procedure to shift the calculated solution GOR and formation volume factor. In this procedure, McCain calculates a "delta pressure" between the calculated bubble-point pressure from a specified correlation and the actual bubble-point pressure. He then shifts the pressure values horizontally so the calculated solution GOR and oil formation volume factor at the bubble point coincide with the correct bubble-point pressure. McCain's procedure can be applied to limited properties and is considered to be a one-dimensional shift in properties. Kanu and Ikiensikimama (2014) proposed a technique that uses a multiplier and a shift parameter to move the calculated PVT property from any correlation to match the actual PVT data when available. When the bubble-point pressure is the only known property, their procedure cannot be used.

The technique presented here is based on two-dimensional shift of the correlations' calculated properties. Fig. 7.11 illustrates the 2D shift technique in the solution GOR property curve. In this graph, the calculated bubble-point pressure value from a particular correlation is 3432 psig. If the correct bubble-point pressure is 3000 psig, a backward shift is needed. If the known bubble-point pressure is 4000 psig, a forward shift is to be made.

Fig. 7.12 shows how a point on the correlation-calculated R_s curve is moved to the new position (in 2D) on a modified R_s curve. This 2D shift in R_s values improves the accuracy of the calculated R_s from any correlation (if

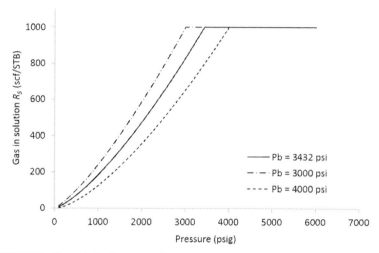

FIGURE 7.11   Example for R_s curve shift.

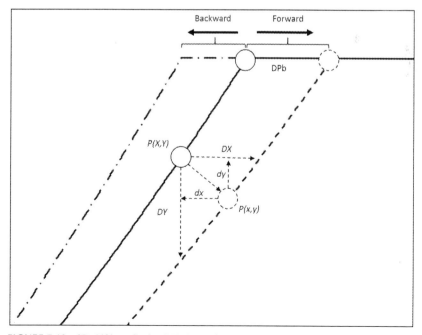

FIGURE 7.12   2D shifting of point $P(X,Y)$ to point $P(x,y)$ on the R_s curve.

the correct bubble-point pressure is known). The modified R_s curve is to be used in further calculations of the remaining PVT properties.

Sequence of Calculations

The following sequence of calculations for black-oil PVT properties from correlations are in fact general steps for oil correlations and can be applied to volatile oils as well.

The calculations first determine whether or not the user input for initial solution GOR is the total solution GOR. If the user inputs the GOR based on measuring gas rates at the separator and oil rates at the stock-tank (the usual situation), it is necessary to use a correlation to estimate the stock-tank GOR. Then the stock-tank GOR is added to separator GOR to obtain the total solution GOR by the following equation:

$$tGOR = sGOR + STGOR \tag{7.6}$$

The second check is for bubble point. If the user inputs the bubble-point pressure at the specified temperature, then the input bubble point is used in further calculations. If bubble-point pressure is not available, it should be calculated from any of the bubble-point correlations.

After the bubble-point pressure has been determined, dead oil viscosity is calculated from any of the dead oil viscosity correlations. Dead oil viscosity is the viscosity of the oil assuming no gas in solution. Other PVT properties for oil (solution GOR, oil formation volume factor, oil compressibility, oil density, and oil viscosity) are computed for all saturated pressure points (pressures below the bubble-point pressure). The PVT property values are also calculated at the bubble-point pressure. The following step is to calculate the same PVT properties for undersaturated pressure points (pressures above the bubble-point). The sequence of calculations is presented in the flow diagram of Fig. 7.13.

APPLICATIONS OF OIL PVT CORRELATIONS

In the absence of representative PVT samples, engineers still must use some estimates of oil PVT properties. This section explains several critical points to be taken into consideration for calculation of good estimates of PVT properties. The procedures to furnish the correct values to be used as input parameters in oil PVT correlations are illustrated with example calculations.

Problem 1—Estimating Input Data for Oil Correlations

An oil reservoir was discovered recently and has been producing with initial reservoir pressure of 4100 psig and GOR of 600 scf/STB (GOR is measured based on separator data). Primary separator conditions are 200 psig and 75°F. A sample from sales gas (separator gas) was taken to measure the

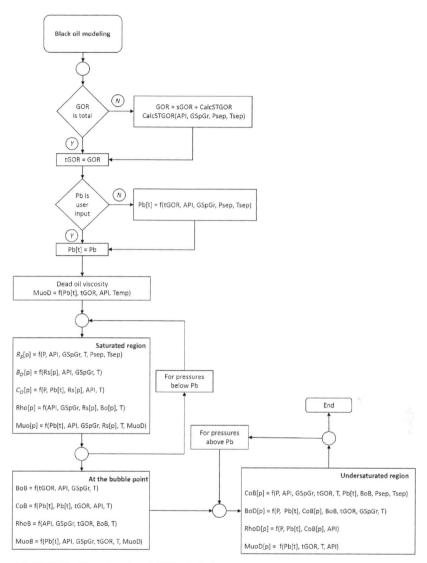

FIGURE 7.13   Flow chart for oil PVT calculations.

composition and its heat content. The specific gravity of separator gas is 0.82. The stock-tank gas is vented. Oil gravity was measured for sales oil to be 40.4°API. The log reading of temperature at the pay zone is 217°F. The preliminary volumetric calculation resulted in OOIP of 25 MMSTB while the reservoir permeability was 312 md and the initial water cut was 20%. Estimate the input data required for calculating oil PVT properties from correlations for this reservoir fluid.

Solution of Problem 1

The correlation inputs are

GOR = 600 scf/STB as Separator GOR.

For calculating the Stock-Tank GOR

Separator pressure = 200 psig
Separator temperature = 75°F
Stock-tank API = 40.4
Separator gas Sp. Gr. = 0.82.

Reservoir temperature = 217°F

Note: Stock-tank API gravity and separator gas gravity are used for calculation of both the stock-tank GOR and other black-oil fluid properties.

Stock-tank GOR = 147 scf/STB (from Valko and McCain, 2003 correlation)
Total GOR = 747 scf/STB.

Problem 2—Calculation of PVT Properties for a Black Oil

A PVT report on a representative fluid sample is available for the fluid in Problem 1. Use the correlation input data of Problem 1 to estimate bubble-point pressure from selected correlations of the ones given in Appendix A, Oil Correlations Formulae (Standing, 1947; Al-Marhoun, 1988; and McCain, 1991). Compare the calculated bubble-point pressure to the measured value of 2635 psia.

Also calculate oil formation volume factor, solution GOR, oil density, oil viscosity, and oil compressibility using Standing (1947), Al-Marhoun (1988), and latest McCain correlations (when applicable) at the bubble-point pressure, and compare the calculated values to the actual measured values from the PVT report:

Measured values from the PVT report are given below.

- Oil formation volume factor = 1.48 rbbl/STB
- Solution GOR = 780 scf/STB
- Oil density = 41.0 lb/ft^3
- Oil viscosity = 0.373 cp
- Oil compressibility = $18.8E^{-6}$ psi^{-1}

Solution of Problem 2

Separator GOR = 600 scf/STB. At the given separator conditions, the stock-tank GOR is estimated to be 127 scf/STB from Valko and McCain (2003) correlation (Appendix A: Oil Correlations Formulae).

Table 7.13 summarizes the calculated values at the bubble-point pressure from different correlations compared with the laboratory-measured values.

Table 7.14 summarizes the error for the fluid properties compared with actual values for the different correlations. The error is calculated using the absolute average percent error.

TABLE 7.13 Problem 2 Solution Results

Fluid Property	Laboratory	Standing (1947)	Al-Marhoun (1988)	McCain (1991)
Bubble-point pressure, psia	2635	2531	2674	3962
Oil formation volume factor, rbbl/STB	1.48	1.53	1.45	1.43
Solution GOR, scf/STB	780	727	727	727
Oil density, lbm/ft^3	40.9	40.5	40.6	40.6
Oil viscosity, cp	0.37	0.29	0.3	0.25
Oil compressibility, psi^{-1}	18.8E − 6	8.42E − 06	8.39E − 06	8.48E − 06

TABLE 7.14 Problem 2 Solution Result Error

Fluid Property	Standing (1947) (Error %)	Al-Marhoun (1988) (Error %)	McCain (1991) (Error %)
Bubble point pressure, psia	− 4	1	50
Oil formation volume factor, rbbl/STB	3	− 2	− 3
Solution GOR, scf/STB	− 7	− 7	− 7
Oil density, lbm/ft^3	− 1	− 1	− 1
Oil viscosity, cp	− 21	− 20	− 32
Oil compressibility, psi^{-1}	− 55	− 55	− 55

Problem 3—Effect of Preknowledge of Bubble-Point Pressure

In Problem 2, assume that the correct value of the bubble-point pressure (2635 psia) is known. Use this value and compute the solution GOR, formation volume factor, oil density, and oil viscosity for Standing (1947), Al-Marhoun (1988), and McCain (1991) correlations at pressure of 1100 psia. Compare the computed values with actual values.

Solution of Problem 3

Table 7.15 shows the calculated properties from the three correlations. The laboratory-measured values are also given in the table. Table 7.16 shows the absolute average percent error for the properties at pressure of 1100 psia calculated from the three correlations when compared with the laboratory-measured values.

TABLE 7.15 Problem 3 Solution Results

Fluid Property	Laboratory	Standing	AlMarhoun	McCain
Pressure, psia	1100	1100	1100	1100
Oil formation volume factor, rbbl/STB	1.28	1.48	1.21	1.26
Solution GOR, scf/STB	374	258	219	358
Oil density, lbm/cu.ft	44.0	45.0	45.4	44.0
Oil viscosity, cp	0.54	0.44	0.46	0.33

TABLE 7.16 Problem 3 Solution Result Error

Fluid Property	Standing (%)	Al-Marhoun (%)	McCain (%)
Pressure, psia	0	0	0
Oil formation volume factor, rbbl/STB	16	−5	−1
Solution GOR scf/STB	−31	−41	−4
Oil density, lbm/ft³	2	3	0
Oil viscosity, cp	−18	−15	−40

Problem 4—Shifting Calculated PVT Properties

In Problem 2 above, use the known bubble-point pressure value (2635 psia) and shift R_s curve for one of the correlations (McCain, 1991). Compute the remaining PVT properties using McCain (1991) correlations and apply the 2D shift on the calculated properties. Compare the shifted curves with the actual values.

Solution of Problem 4

The basic input data was used to compute all PVT properties for pressures between 100 and 5000 psia. The 2D shifting procedure explained in this chapter was also used to compute modified values of PVT properties. Both the un-shifted and shifted values are given in the following plots.

Fig. 7.14 shows the results calculated by McCain (1991) correlation family with no R_s data shift and after R_s data shift to the correct bubble-point pressure.

Figs. 7.15−7.18 show the remaining fluid properties and comparison between the calculated PVT properties (using McCain, 1991 correlations) before and after application of the 2D shift in R_s data.

Comparison of the calculated PVT properties before and after 2D shift reveals that in some cases, the procedure for shifting the data improves the accuracy of the predicted data significantly. Knowledge of bubble-point pressure for the reservoir fluid can therefore improve the accuracy of PVT correlations.

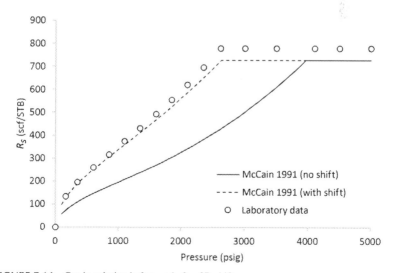

FIGURE 7.14   Gas in solution before and after 2D shift.

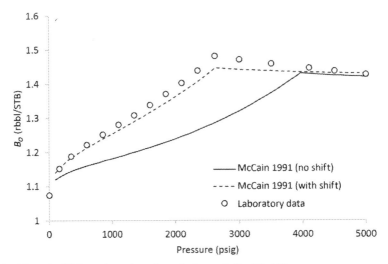

FIGURE 7.15 Oil formation volume factor before and after 2D shift.

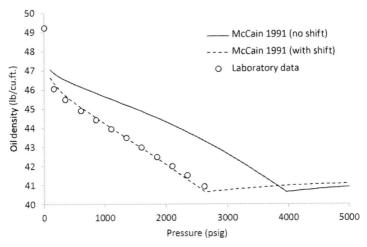

FIGURE 7.16 Oil density before and after 2D shift.

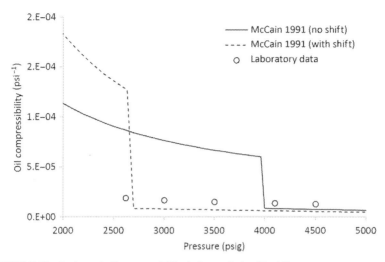

FIGURE 7.17 Isothermal oil compressibility before and after 2D shift.

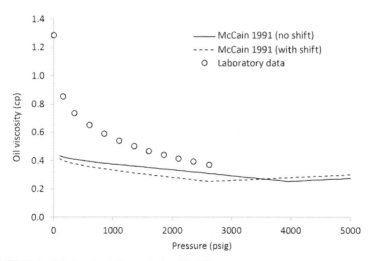

FIGURE 7.18 Oil viscosity before and after 2D shift.

NOMENCLATURE

API	density of stock-tank liquid, °API
B_g	gas formation volume factor, rcf/scf
B_o	oil formation volume factor, rbbl/STB
c_o	isothermal oil compressibility, psi^{-1}
CCE	constant composition expansion
DL	differential liberation
GOR	gas−oil ratio, scf/STB
G_p	cumulative gas production, scf
N_p	cumulative oil production, STB
R_p	cumulative gas−oil ratio, scf/STB
R_s	solution gas−oil ratio, scf/STB
RFT	repeat formation tester (measurement of fluid pressure versus depth)
p_b	bubble-point pressure, psia
sGOR	separator GOR, scf/STB
STGOR	stock-tank GOR, scf/STB
tGOR	total GOR, scf/STB
T	temperature, °F
T_{surf}	surface temperature, °F

REFERENCES

Abdul-Majeed, G.H., Salman, N.H., 1988. An empirical correlation for oil FVF prediction. JCPT 27, 6.

Abdul-Majeed, G.H., Kattan, R.R., Salman, N.H., 1990. New correlation for estimating the viscosity of undersaturated crude oils. JCPT 29 (3), 80−85. SPE-90-03-10.

Ahmed, T., 1989. *Hydrocarbon Phase Behavior*. Vol. 7. Gulf Publishing Company, Houston, United States.

Ahmed, T., 2010. Reservoir Engineering Handbook, fourth. ed. Gulf Publishing Company, Boston, MA, United States.

Al-Khafaji, A.H., Abdul-Majeed, G.H., Hassoon, S.F., 1987. Viscosity correlation for dead, live and undersaturated crude oil. J. Petrol. Res. 6, 1−16.

Al-Marhoun, M.A., 1988. Pressure-volume-temperature correlations for Saudi crude oils. JPT 40 (May). Available from: https://doi.org/10.2118/13718-PA.

Al-Marhoun, M.A., 1992. New correlations for formation volume factors of oil and gas mixtures. JCPT 31, 3. Available from: https://doi.org/10.2118/92-03-02.

Al-Marhoun, M.A. 2003. The coefficient of isothermal compressibility of black oils. In: Paper SPE-81432-MS Presented at the 2003 Middle East Oil Show, June 9−12, Manama, Bahrain, https://doi.org/10.2118/81432-MS.

Al-Marhoun, M.A., 2004. Evaluation of empirically derived PVT properties for middle east crude oils. J. Pet. Sci. Eng. 42, 209−221.

Al-Marhoun, M.A., 2006. A new correlation for undersaturated isothermal oil compressibility, In: Paper SPE-81432-SUM, Society of Petroleum Engineers.

Al-Mehaideb, R.A. 1997. Improved PVT correlations for UAE crude oils. In: Paper SPE-37691-MS Presented at the Middle East Oil Show and Conference, March 15−18, Manama, Bahrain. https://doi.org/10.2118/37691-MS.

Al-Najjar, H.S., Al-Soof, N.B.A., Al-Khalisy, K.M., 1988. Correlations for bubble-point pressures, gas oil ratios and formation volume factors for Iraqi crude oils. J. Petrol. Res. 7 (June), 13−35.

Abu-Khamsin, S.A., Al-Marhoun, M.A., 1991. Development of a new correlation for bubble point oil viscosity. Arab. J. Sci. Eng. 16 (April), 2A.

Al-Shammasi, A.A. 1999. Bubble point pressure and oil formation volume factor correlations. In: Paper SPE-53185-MS Presented at the Middle East Oil Show and Conference, February 20−23, Manama, Bahrain. https://doi.org/10.2118/53185-MS

Arps, J.J., 1962. Estimation of primary oil and gas reserve. In: Frick, T.C. (Ed.), *Petroleum Production Handbook*. Volume II. SPE, Dallas, Chapter 37.

Asgarpour, S., McLauchlin, L., Wong, D., Cheung, V., 1989. Pressure-volume-temperature correlations for western Canadian gases and oils. JCPT 28 (4), 103−111. SPE-89-04-08.

Beal, C., 1946. Viscosity of air, water, natural gas, crude oil and its associated gases at oil field temperature and pressures. Trans. AIME 165, 94−115.

Beggs, H.D., Robinson, J.R., 1975. Estimating the viscosity of crude oil systems. JPT 27, 1140−1141.

Bennison, T. 1998. Prediction of heavy oil viscosity. In: Paper Presented at the IBC Heavy Oil Field Development Conference, December 2−4, London, United Kingdom.

Bergman, D.F., 2004. Don't forget viscosity. In: Paper Presented at the Petroleum Technology Transfer Council 2nd Annual Reservoir Engineering Symposium, July 28, Lafayette, LA, United States.

Bergman, D.F., and Sutton, R.P. 2006. Undersaturated oil viscosity correlation for adverse conditions. In: Paper SPE 103144 Presented at 2006 SPE Annual Technical Conference and Exhibition. September 24−27, San Antonio, TX, United States.

Bergman, D.F. and Sutton, R.P. 2007. An update to viscosity correlations for gas saturated crude oils. In: Paper SPE 110195 Presented at 2007 SPE Annual Technical Conference and Exhibition, November 11−14, CA, United States.

Bolondarzadeh, A., et al., 2006. The new PVT generated correlations of Iranian oil properties. 4th Iranian Petroleum Engineering Student conference, Tehran, Iran.

Calhoun Jr., J.C., 1947. Fundamentals of Reservoir Engineering. University of Oklahoma Press, Norman, OK, United States, p. 35.

Casey, J.M., Cronquist, C., 1992. Estimate GOR and FVF using dimensionless PVT analysis. World Oil 213, 83−87.

Chew, J., Connally Jr., C.A., 1959. A viscosity correlation for gas-saturated crude oils. Transactions of the American Institute of Mining, Metallurgical, and Petroleum Engineers, Vol. 216. Society of Petroleum Engineers of AIME, Dallas, TX, United States, p. 23.

Dandekar, A.Y., 2013. Petroleum reservoir rock and fluid properties, second ed CRC Press, Boca Raton, FL, United States.

De Ghetto, G. and Villa, M. 1994. Reliability analysis on PVT correlations. In: Paper SPE 28904 Presented at the European Petroleum Conference, October 25−27, London, United Kingdom.

De Ghetto, G., Paone, F., and Villa, M. 1995. Pressure-volume-temperature correlations for heavy and extra heavy oils. In: Paper SPE 30316 Presented at the SPE International Heavy Oil Symposium, June 19−21, Calgary, Canada.

Dindoruk, B. and Christman, P.G. 2001. PVT properties and viscosity correlations for Gulf of Mexico oils. In: Paper SPE 71633 Presented at the 2001 SPE Annual Technical Conference and Exhibition, September 3−October 3, New Orleans, LA, United States.

Dindoruk, B., Christman, P.G., 2004. PVT properties and viscosity correlations for Gulf of Mexico oils. SPE Res. Eval. Eng. 28 (4), 427−437.

Doklah, M., Osman, M., 1991. Correlation of PVT properties for UAE crudes. SPE Form. Eval. 7 (1). Available from: https://doi.org/10.2118/20989-PA.

Elmabrouk, S. Zekri, A., and Shirif, E. 2010. Prediction of bubblepoint pressure and bubblepoint oil formation volume factor in the absence of PVT analysis. In: Paper SPE 137368 Presented at the SPE Latin American & Caribbean Petroleum Engineering Conference, Lima, Peru, December 1−3. https://doi.org/10.2118/137368-MS.

Elam, F.M. 1957. Prediction of Bubble Point Pressures and Formation Volume Factors From Field Data (MS thesis). University of Texas at Austin, Austin, TX, United States.

El-Banbi, A.H., Abdel Fattah, K.A., Sayyouh, M.H. 2006. New modified black oil correlations for gas condensate and volatile oil fluids. In: Paper SPE 102240 Presented at the SPE Annual Technical Conference and Exhibition, 24−27 September, San Antonio, TX, United States.

Elsharkawy, A.M., Alikhan, A.A., 1997. Correlations for predicting solution gas/oil ratio, oil formation volume factor, and undersaturated oil compressibility. J. Pet. Sci. Eng. 17, 291−302.

Elsharkawy, A.M., Alikhan, A.A., 1999. Models for predicting the viscosity of Middle East crude oils. Fuel 78 (8), 891−903.

Elsharkawy, A.M., Gharbi, R.B.C., 2001. Comparing classical and neural regression techniques in modelling crude oil viscosity. Adv. Eng. Softw. 32 (3), 215−224.

Farshad, F.F., Leblance, J.L., Garber, J.D. and Osorio, J.G. 1996. Empirical PVT correlations for Colombian crude oils. In: Paper SPE 36105 Presented at SPE Latin America/Caribbean Petroleum Engineering Conference, April 23−26, Port of Spain, Trinidad and Tobago.

Fitzgerald, D.J., 1994. A Predictive Method for Estimating the Viscosity of Undefined Hydrocarbon Liquid Mixtures. Pennsylvania State University, State College, PA, United States (MS thesis).

Glaso, O., 1980. Generalized pressure-volume-temperature correlations. JPT 32, 785−795.

Hanafy, H.H., Macary, S.M., ElNady, Y.M., Bayomi, A.A., and El Batanony, M.H. 1997. Empirical PVT correlations applied to Egyptian crude oils exemplify significance of using regional correlations. In: Paper SPE 102240 Presented at 1997 SPE International Symposium on Oilfield Chemistry, 18-21 February, Houston, TX, United States.

Hassan, O.F., 2011. Correlation for solution gas-oil ratio of Iraqi oils at pressures below the bubble point pressure. Iraqi J. Chem. Pet. Eng. 12 (2), 1−8. June 2011.

Hemmati, M.N., Kharrat, R. 2007. A correlation approach for prediction of crude oil PVT properties. In: Paper SPE 104543 Presented at the 15th SPE Middle East Oil and Gas Show, March 11−14, Bahrain.

Hossain, M.S., Sarica, C., Zhang, H.Q., Rhyne, L., and Greenhill, K.L. 2005. Assessment and development of heavy oil viscosity correlations. In: Paper SPE 97907 Presented at SPE International Thermal Operations and Heavy Oil Symposium, November 1−3, Calgary, Alberta, Canada.

Ikiensikimama, S. and Ogboja, O. 2009. New bubble point pressure empirical PVT correlation. In: Paper SPE 128893 Presented at Nigeria Annual International Conference and Exhibition, 1-3 November, Abuja, Nigeria.

Ikiensikimama, S., 2008. Oil and Gas PVT Characterization. University of Lagos, Akoka, Nigeria (Ph.D. thesis).

Ikiensikimama S.S. and Azubuike, I.I. 2012. Modeling approach for Niger-delta oil formation volume factor prediction using artificial neural network. In: SPE-162987 Presented at the Nigeria Annual International Conference and Exhibition, August 6−8, Lagos, Nigeria. https://doi.org/10.2118/162987-MS.

Kanu, A.U. and Ikiensikimama, S.S. 2014. Globalization of black oil PVT correlations. In: Paper SPE 172494 Presented at SPE Nigeria Annual International Conference and Exhibition, August 18−21, Lagos, Nigeria.

Karimnezhad, M., Heidarian, M., Kamari, M., Jalalifar, H., 2014. A new empirical correlation for estimating bubble point oil formation volume factor. J. Nat. Gas Sci. Eng. 18, 329−335.

Kartoatmodjo, T.R S. and Schmidt, Z. 1991. New correlations for crude oil physical properties. In: Paper SPE 23556-MS. Society of Petroleum Engineers, Richardson, TX, unsolicited paper.

Kartoatmodjo, T.R.S., Schmidt, Z., 1994. Large data bank improves crude physical property correlation. Oil Gas J. 4, 51−55.

Kaye, S.E. 1985. Offshore California viscosity correlations. Technical Report, No. TS85000940, Chevron Oil Field Research Co. (COFRC), La Habra, CA (August).

Khairy, M., El-Tayeb, S., Hamdallah, M., 1998. PVT correlations developed for Egyptian crudes. Oil and Gas J. 96, 114−116.

Khamehchi, R., Ebrahimian, R., 2009. Novel empirical correlations for estimation of bubble point pressure, saturated viscosity and gas solubility of crude oils. Pet. Sci. 6, 86−91. Available from: https://doi.org/10.1007/s12182-009-0016-x.

Khan, S.A., Al-Marhoun, M.A., Duffuaa, S.O., and Abu-Khamsin, S.A. 1987. Viscosity correlations for Saudi Arabian crude oils. In: Paper SPE 15720 Presented at the Fifth SPE Middle East Oil Show, March 7−10, Manama, Bahrain.

Khazam, M., Shlak, M., Alkhaboli, M., 2016. New PVT correlations based on Libyan crudes for predicting fluid physical properties. Oil Gas Res. 2, 122. Available from: https://doi.org/10.4172/2472-0518.1000122.

Knopp, C.R., Ramsey, L.A., 1960. Correlation of oil formation volume factor and solution gas-oil ratio. JPT 12, 27−29. Available from: https://doi.org/10.2118/1433-G.

Kouzel, B., 1965. How pressure affects liquid viscosity. Hydrocarb. Process. Pet. Refiner 44 (3), 120.

Labedi, R.M., 1982. PVT Correlations of the African Crudes. Colorado School of Mines, Leadville, CO, United States (Ph.D. thesis).

Labedi, R.M. 1990. Use of production data to estimate the saturation pressure solution GOR and chemical composition of reservoir fluids. In: Paper SPE 21164 Presented at the SPE Latin America Petroleum, October 14−19, Rio de Janeiro, Brazil.

Labedi, R.M., 1992. Improved correlations for predicting the viscosity of light crudes. J. Pet. Sci Eng 8, 221−234.

Lasater, J.A., 1958. Bubble point pressure correlations. J. Pet. Technol 10 (5), 65−67. SPE-957-G.

Levitan, L.L., Murtha, M., 1999. New correlations estimate Pb, FVF. Oil Gas J. 97 (10)).

Macary, S.M. and El-Batanoney, M.H. 1992. Derivation of PVT correlations for the Gulf of Suez crude oils. In: Proc., 11th EGPC Petroleum Exploration and Production Conference, Cairo, Egypt, Vol 36, (1993)no. 6, p. 472-478.

Mazandarani, M.T., Asghari, S.M., 2007. Correlations for predicting solution gas-oil ratio, bubble point pressure and oil formation volume factor at bubble-point of Iran crude oils. European Congress of Chemical Engineering (ECCE-6), Copenhagen, September 16−20.

McCain Jr., W.D., Rollins, J.B., Lanzi, A.J.V., 1988. The coefficient of isothermal compressibility of black oils at pressures below the bubblepoint. SPE Form. Eval. 3, 659−662. Available from: https://doi.org/10.2118/15664-PA.

McCain Jr., W.D., 1990. The Properties of Petroleum Fluids, second ed. PennWell Books, Tulsa, OK, United States.

McCain Jr., W.D., 1991. Reservoir-fluid property correlations-state of the art (includes associated papers 23583 and 23594). SPE Res. Eng. 6 (2). Available from: https://doi.org/10.2118/18571-PA.

McCain Jr., W.D., 1993. Chemical composition determines behavior of reservoir fluids. Pet. Eng. Int. J. 65, 10.

McCain Jr., W.D., Spivey, J.P., Lenn, C.P., 2011. Petroleum Reservoir Fluid Property Correlations. PennWell Publishing Co., Tulsa, OK, United States.

Mehran, F. Movagharnejad, K. and Didanloo, A. 2006. New correlation for estimation of formation volume factor and bubble point pressure for Iranian oil fields. In: Paper Presented at the First Iranian Petroleum Engineering Congress, Tehran, Iran.

Moradi, B., Malekzadeh, E., Amani, M., Boukadi, F.H., and Kharrat, R., 2010. Bubble Point Pressure Empirical Correlation. In: Paper SPE-132756 -MS Presented at the Trinidad and Tobago Energy Resources Conference, Port of Spain, Trinidad.

Moradi, B., Malekzadrh, E., Kharrat, R., 2013. New oil formation volume factor empirical correlation for middle east crude oils. Academic Research Online Publisher. Available from: http://www.aropub.org/journals/international-journal-of-petroleum-and-geoscience-engineering-ijpge/ijpge-archive/htmlnew-oil-formation-volume-factor-empirical-correlation-for-middle-east-crude-oils.

Movagharnejad, and Fasih. 1999. The new correlation for prediction bubble point pressure and oil formation volume factor for Iranian reservoirs. Research Institute of Petroleum Industry, National Iranian Oil Company, Iran.

Naseri, A., Nikazar, M., Mousavi-dehghani, S.A., 2005. A correlation approach for prediction of crude oil viscosity. J. Pet. Sci. Eng. 47, 163−174.

Naseri, A., Yousefi, S.H., Sanaei, A., Gharesheikhlou, A.A., 2012. A neural network model and an updated correlation for estimation of dead crude oil viscosity. Braz. J. Pet. Gas 6 (1), 31−41.

Ng, J.T.H. and Egbogah, E.O. 1983. An improved temperature-viscosity correlation for crude oil systems. In: Paper Presented at the Annual Technical Meeting, Banff, Canada, May 10−13. PETSOC-83-34-32.

Obomanu, D.A., Okpobiri, G.A., 1987. Correlating the PVT properties of Nigerian crudes. J. Energy Resour. Technol. 109 (4), 214−217.

Okeke, H., and Sylvester, O. 2016. Improved correlation for predicting stock tank gas-oil ratio in Niger Delta. In: Paper SPE 184374 Presented at SPE Nigeria Annual International Conference and Exhibition, August 2−4, Lagos, Nigeria.

Okoduwa, I.G. and Ikiensikimama, S.S. 2010. Bubble point pressure correlations for Niger Delta crude oils. In: Paper SPE 136968 presented at the 34th Annual SPE International Conference and Exhibition, July 1−August 7, Tinapa, Calabar, Nigeria.

Omar, F., 2011. Correlation for solution gas -oil ratio of Iraqi oils at pressures below the bubble point pressure. Iraqi J. Chem. Pet. Eng. 12 (2 − 8), ISSN: 1997-4884.

Omar, M.I. and Todd, A.C. 1993. Development of new modified black oil correlations for Malaysian crudes. In: Paper SPE 25338 Presented at the 34th Annual SPE International Conference and Exhibition, February 8−10, Singapore.

Osorio, J.G., 1990. Empirical PVT Correlations for Colombian Crude Oils. University of Southwestern Louisiana, Lafayette, LA, United States (MS thesis).

Ostermann, R.D. and Owolabi, O.O. 1983. Correlations for the reservoir fluid properties of Alaskan crudes. In: Paper SPE 11703 Presented at SPE California Regional Meeting, March 23−25, Ventura, CA, United States. https://doi.org/10.2118/11703-MS.

Owolabi, O.O., 1984. Reservoir Fluid Properties of Alaskan Crudes. University of Alaska, Fairbanks, AK, United States (MS thesis).

Oyedeko, K.F., Ulaeto, U.W., 2011. Predicting the dead oil viscosity of reservoir fluids: a case study of the Niger delta. J. Energy Technol. Policy 3 (13), 2013.

Petrosky, G.E., 1990. PVT Correlations for Gulf of Mexico Crude Oils. University of Southwestern Louisiana, Lafayette, LA, United States (MS thesis).

Petrosky, G.E. and Farshad, F.F. 1993. Pressure-volume-temperature correlations for Gulf of Mexico crude oils. In: Paper SPE 26644 Presented at SPE Annual Technical Conference and Exhibition, October 3−6, Houston, TX, United States.

Petrosky, G.E. and Farshad, F.F. 1995. Viscosity correlations for Gulf of Mexico crude oils. In: Paper SPE 29468 Presented at SPE Production Operations Symposium, April 2–4, Oklahoma City, OK, United States.

Petrosky, G.E., Farshad, F., 1998. Pressure-volume-temperature correlations for Gulf of Mexico crude oils. SPE Res. Eval. Eng. J. 1 (05), 416–420. Available from: https://doi.org/10.2118/51395-PA.

Rollins, J.B., McCain Jr., W.D., Creeger, J.T., 1990. Estimation of solution GOR of black oils. J. Pet. Tech. 42, 92–94. January. Transactions, AIME. Vol. 289.

Sattarina, M., Modarresi, H., Bayata, M., Teymoria, M., 2007. New viscosity correlations for dead crude oils. Pet. Coal 49 (2), 33–39. ISSN 1335-3055.

Spivey, J.P., Valko, P.P., McCain Jr., W.D., 2007. Applications of the coefficient of isothermal compressibility to various reservoir situations with new correlations for each situation. SPE Res. Eval. Eng. J. 10, 43–49.

Standing, M.B., 1947. A pressure-volume-temperature correlation for mixtures of California oils and gases, Drilling and Production Practice. American Petroleum Institute, New York, NY, United States, January 1.

Standing, M.B., 1981. Volumetric and Phase Behavior of Oil Field Hydrocarbon Systems, ninth ed. Society of Petroleum Engineers of AIME, Richardson, TX, United States.

Steve, S.K. Sim, 1993. Pressure-Volume-Temperature Correlations for Crude Oils from the Illinois Basin. Illinois State Geological Survey 140.

Sulaimon, A.A., Ramli, N., Adeyemi, B.J. and Saaid, I.M. 2014. New correlation for oil formation volume factor. In: Paper SPE 172396 Presented at the SPE Nigeria Annual International Conference and Exhibition, August 5–7, Lagos, Nigeria.

Sutton, R.P., Farshad, F., 1990. Evaluation of empirically derived PVT properties for Gulf of Mexico crude oils. SPE Res. Eng. J. 5 (1). Available from: https://doi.org/10.2118/13172-PA.

Twu, C.H., 1985. Internally consistent correlation for predicting liquid viscosities of petroleum fractions. Ind. Eng. Chem. Process Des. Dev. 34 (4), 1287–1293.

Ubong, U., Oyedeko, K., 2014. Improved dead oil viscosity model. J. Energy Technol. Policy 4 (7), ISSN (Online) 2225-0573.

Valko, P.P., McCain Jr., W.D., 2003. Reservoir oil bubblepoint pressures revisited; solution gas-oil ratios and surface gas specific gravities. J. Pet. Sci. Eng. 37, 153–169. Available from: https://doi.org/10.1016/S0920-4105(02)00319-4.

Vasquez, M., Beggs, H.D., 1980. Correlation for fluid physical property predictions. JPT 32, 968–970.

Velarde, J., Blasingame, T.A., McCain Jr., W.D., 1997. Correlation of black oil properties at pressures below bubble point pressure - a new approach. JCPT 38 (13). Available from: https://doi.org/10.2118/97-93.

Whitson, C.H., Brule, M.R., 2000. Phase Behavior. SPE Monograph, Vol. 20. SPE Richardson, TX, United States.

FURTHER READING

Craft, B.C., Hawkins, M.F., 1959. Applied Petroleum Reservoir Engineering. Prentice-Hall, Englewood Cliffs, NJ.

McCain Jr., W.D., 1994. Heavy components control reservoir fluid behavior. J. Pet. Technol., pp. 746–750. Technology Today Series; Trans., AIME, 297.

PROBLEMS

7.1 A bottom-hole fluid sample from a black-oil reservoir at depth of 6126 ft., initial reservoir pressure of 2338 psia, reservoir temperature of 194°F, and separator conditions of 100°F and 500 psia has the data given in the below table. Answer the following questions.

Pressure, psig	Oil Density, g/cc	Oil FVF, rbbl/STB	Gas FVF, cu. ft./scf	R_s, scf/ STB	Oil Viscosity, cp
3000	0.599	1.39		861	0.31
2750	0.596	1.40		861	0.30
2338	0.593	1.40		861	0.28
2200	0.591	1.41		861	0.28
2000	0.588	1.42		861	0.27
1800	0.586	1.42		861	0.27
1600	0.584	1.43		861	0.26
1550	0.584	1.43		861	0.26
1500	0.589	1.42	0.008	839	0.26
1400	0.599	1.39	0.008	791	0.27
1200	0.621	1.34	0.010	687	0.28
1000	0.645	1.29	0.012	580	0.30
800	0.671	1.24	0.016	473	0.32
600	0.699	1.19	0.023	358	0.36
400	0.729	1.14	0.038	238	0.41
200	0.763	1.09	0.080	138	0.47
60	0.788	1.06	0.240	65	0.55
0	0.799	1.04	1.244	0	0.69

(a) Plot oil formation volume factor. Determine bubble-point pressure from the plot and the data given in the table.

(b) Construct density, solution gas–oil ratio and oil viscosity curves versus pressure.

(c) For the pressure range given in the table, calculate the isothermal compressibility of oil from the following correlations.

- Ahmed (1989)
- Al-Marhoun (2006)
- Al-Mehaidab (1997)
- De Ghetto et al. (1995)
- Dindoruk and Chrisman (2004)
- ElSharkawy and Alikhan (1997)
- Farshad et al. (1996)
- Hanafy et al. (1997)
- Kartoatmdjo and Schmidt (1991)
- McCain (1991)
- McCain et al. (1988)
- Petrosky and Farshad (1998)
- Spivey et al. (2007)
- Standing (1947)
- Vasquez and Beggs (1980)

(d) Compare the calculated undersaturated oil compressibility from the different correlations with the following laboratory-measured compressibility values.

Pressure, psig	Oil Compressibility x 10E6, psi^{-1}
3000	
2750	12.909
2339	14.034
2200	15.159
2000	16.284
1800	17.184
1600	18.084
1550	18.309

7.2 An oil reservoir was discovered in 2015 and has been producing with initial reservoir pressure of 2283 psi and GOR of 420 scf/STB (GOR is measured based on separator data). Primary separator conditions are 150 psig and 100°F. A sample from sales gas (separator gas) was taken to measure the gas composition and its heat content. The specific gravity of separator gas is 0.92. The stock-tank gas is vented. Oil gravity was measured for sales oil to be 49°API. The log reading of temperature at the pay zone is 175°F. The preliminary volumetric calculations showed OOIP of 25 MMSTB while the average reservoir permeability is 322 md and the initial water cut is 22%. Estimate the input data required for calculating oil PVT properties from correlations for this reservoir fluid.

7.3 A PVT report for a representative reservoir fluid sample is available for the fluid in Problem 7.2. Use the correlation input data of Problem 7.2 to estimate bubble-point pressure from correlations given in Appendix A, Oil Correlations Formulae (Standing, 1947; Al-Marhoun, 1988; and McCain, 1991). Compare the calculated bubble-point pressure to the measured value of 1430 psia. Calculate oil formation volume factor, solution GOR, oil density, oil viscosity, and oil compressibility using Standing (1947), Al-Marhoun (1988), and latest McCain (1991) correlations (when applicable) at the bubble-point pressure. Compare the calculated values to the laboratory-measured values from the PVT report:

Measured values from the PVT report at the bubble-point pressure are:

- Oil formation volume factor = 1.0810 rbbl/STB
- Solution GOR = 411 scf/STB
- Oil density = 48.0 lbm/ft3
- Oil viscosity = 1.230 cp
- Oil compressibility = $8.3526E - 6 \text{ psi}^{-1}$

Chapter 8

Low Gas—Oil Ratio Oils

Low gas—oil ratio (GOR) oils are black oils that contain a low amount of solution gas. They have low bubble point pressures that may never be reached in the reservoir under practical operating conditions. These oils produce with low GOR that can be considered constant (since the reservoir pressure remains above the bubble point pressure for nearly the entire life of the reservoir). Low GOR oils are not necessarily heavy oils as they sometimes have reasonably low viscosity and can flow easily under normal conditions of temperature. Despite the low GOR, these oils may have high stock-tank oil API gravity. Fig. 8.1 shows a plot of bubble point pressure versus total producing GOR for oil samples collected from worldwide origin and representing different types of oils (each point represents an oil sample with full PVT report).

Fig. 8.1 shows that the bubble point pressure is proportional to the initial solution GOR. It also shows that the bubble point pressure is generally below 1000 psia for oils with solution GOR less than 150 scf/STB. At these levels of low GOR, the field measurements of gas production may not be available.

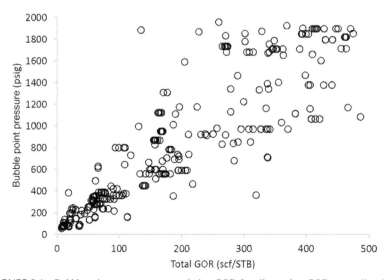

FIGURE 8.1   Bubble point pressure versus solution GOR for oil samples. GOR, gas—oil ratio.

PVT Property Correlations. DOI: https://doi.org/10.1016/B978-0-12-812572-4.00008-4

The little gas production is usually flared at these quantities for economic reasons. Additionally, at these low levels of bubble point pressure, the reservoir pressure may not ever reach the bubble point pressure (the reservoir may be abandoned before reaching the bubble point pressure). In some cases (e.g., shallow reservoirs that are developed with the high density of production wells), little free gas is generated in the reservoir when the pressure declines below the bubble point pressure. Even in those cases, gas production will not increase significantly, and the concept of low GOR oils can be justified.

Low GOR oils usually exist at shallow and medium depths. One can argue that oil trapped near the surface may not contain large amounts of gas as the volatile components of oil may have escaped the oil and found their way to the surface over geologic ages. Low GOR oils are usually discovered at low initial reservoir pressures.

As discussed in Chapter 2, Reservoir-Fluid Classification, many classifications for reservoir fluid types have been suggested. The most widely used classification is that of McCain (1994). This classification divides oils into two types: black and volatile. The low GOR oil is included in the black-oil classification of McCain (1994). However, low GOR oils, as a class of fluids on their own, differ from typical black oils as follows:

1. Low GOR oils produce very little gas, which is often ignored in field measurements.
2. Production remains above the bubble point for the entire life of the reservoir.
3. Calculation of PVT properties is reduced to the undersaturated portion of the PVT properties curve.
4. Solution GOR can be neglected for reservoir engineering calculations. However, it can be assumed to be a constant value for production engineering calculations (e.g., vertical-lift performance calculations of production wells).

A typical phase diagram of low GOR oil is shown in Fig. 8.2. This figure shows that at reservoir temperature, bubble point pressure is low and may not be reached during reservoir depletion. The oil, therefore, remains in the single-phase liquid region for the entire life of the reservoir. The phase diagram of low GOR oil also shows that even if the reservoir pressure declines below the bubble point pressure, little gas will be formed in the reservoir. Efficient recovery of the oil from the reservoirs will require a significant number of wells and/or water injection or enhanced oil recovery.

Fig. 8.3 represents three different phase diagrams plotted using an EOS program for three different low GOR oil samples taken from the same region. The three samples represent oils that are produced from stacked reservoirs. The depth of the three samples ranges from 5000 to 7000 ft TVDss.

The composition of the three samples is given in Table 8.1. Although the three samples differ in composition (values of C7 + mole percent range

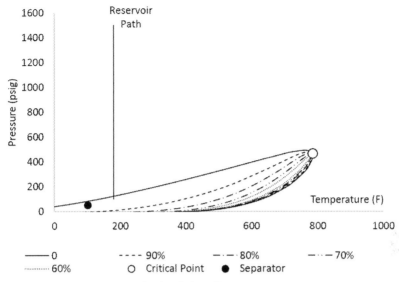

FIGURE 8.2   Typical low gas–oil ratio oil phase diagram.

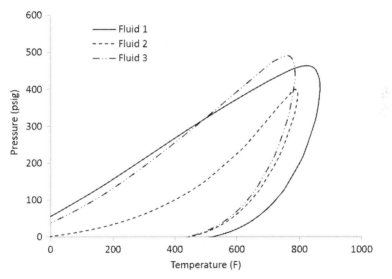

FIGURE 8.3   Phase diagrams for three different low GOR oil samples from the same region. *GOR*, gas–oil ratio.

between 76.6% and 82.7%), their phase behavior is close. All three samples have bubble point pressures less than 100 psia and total solution GOR of less than 90 scf/STB. The reservoirs in this region and in this range of depth produce very little gas, such that the gas production is not even routinely measured. Due to the low bubble point pressure, it is believed that these reservoirs will always produce above their bubble point pressure.

TABLE 8.1 Composition and Properties of Plus Fraction of Three Low GOR Oil Samples

	N_2	CO_2	H_2S	C1	C2	C3	i-C4	n-C4	i-C5	n-C5	C6	C7+	C7+ MW	C7+ SpGr
Fluid 1	0.10	0.12	0.00	3.13	0.84	1.36	2.02	2.84	2.06	3.67	7.10	76.78	193.00	0.83
Fluid 2	0.05	0.03	0.01	0.56	0.59	0.95	0.10	1.37	1.45	2.03	9.14	82.73	188.01	0.85
Fluid 3	0.10	0.13	0.00	2.03	1.92	3.52	1.78	2.17	2.22	2.72	6.82	76.60	175.74	0.87

As shown by these and many other samples, it is evident that low GOR oils contain low concentrations of C1. In general, C1 mole percent is usually less than 15% for these oils and C7 + mole percent is higher than 75%.

PVT PROPERTIES FOR LOW GAS–OIL RATIO OILS

The PVT properties required to engineer low GOR oils are the same as those required for black oils. They also have the same definitions as PVT properties for black oils. The most challenging PVT property to predict in low GOR oils is viscosity. The following discussion addresses the issue of oil viscosity prediction in low GOR oils.

Low Gas–Oil Ratio Oil Viscosity

In general, estimation of oil viscosity from correlations poses a challenge with low GOR oils. Viscosity correlations depend on simple surface data (initial producing GOR, API gravity of stock-tank oil, specific gravity of separator gas, and temperature). For many oils, viscosity is strongly correlated with temperature and API gravity of stock-tank oil. However, for low GOR oils, viscosity correlations prove inaccurate in prediction of the actual values. For low GOR oils, it seems that temperature, solution GOR, and API gravity of stock-tank oil are inadequate to predict the viscosity. Knowledge of oil composition requires performance of measurements in a PVT laboratory. The problem of viscosity prediction by correlation in low GOR oils is illustrated below.

Table 8.2 shows the field values required for PVT correlation calculations for two low GOR oil samples with very close field values. Table 8.3 shows the measured values of viscosity for each sample. Comparison of the data reveals that although the two samples have very similar correlation input values, their actual viscosity values differ significantly (one fluid is twice as viscous as the other). The correlations would predict nearly the same viscosity value for each sample.

TABLE 8.2 Surface Data for Two Low GOR Oil Samples Used as Input to Viscosity Correlations

	Sample 1	Sample 2
Reservoir temperature, °F	165	163
Total producing gas–oil ratio, scf/STB	38	36
Stock tank, °API	24	25
Separator gas specific gravity	0.91	0.87

TABLE 8.3 Laboratory-Measured Viscosity Values for Two Low GOR Oil Samples

Sample 1		Sample 2	
Pressure, psia	Oil Viscosity, cp	Pressure, psia	Oil Viscosity, cp
2175	5.14	5000	4.84
2100	5.12	4000	4.38
2030	5.09	3000	3.91
1885	5.05	2000	3.44
1740	5.00	1479	3.20
1595	4.96	1000	2.98
1450	4.91	500	2.64
1305	4.87	233	2.60
1160	4.82	150	2.61
1015	4.78	100	2.65
870	4.74	50	2.71
725	4.69	15	6.50
580	4.65		
218	4.53		
145	4.51		
73	4.49		
22	5.98		
15	6.33		

Another complication is shown in Fig. 8.4. In this figure, the laboratory-measured viscosity values for the two fluid samples are plotted with the viscosity calculated from three commonly used viscosity correlations (Beal, 1946; Beggs and Robinson, 1975; Petrosky and Farshad, 1995). All three correlations overestimated viscosity values for the two samples.

Dead and Live Oil Viscosity

Dead oil viscosity is defined as the viscosity of oil after removal of solution gas. In many oil field operations, oil samples are collected from the stock-tank oil and viscosity is measured in the field laboratory. Oil viscosity may also be measured at different temperatures (but at atmospheric pressure) to yield data for dead oil viscosity. Live oil viscosity is the viscosity of the oil

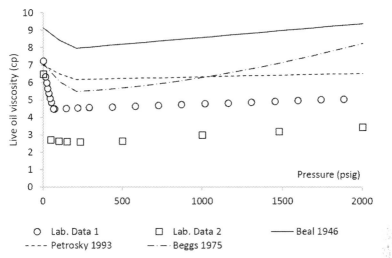

FIGURE 8.4 Comparison of actual viscosity values for two similar samples to calculated viscosity.

before removal of solution gas. It is usually measured at reservoir temperature for a range of pressures. Live oil viscosity is often measured with a rolling ball viscometer in the PVT laboratory.

In absence of a complete PVT laboratory report, engineers should look for any available dead oil viscosity values. Different dead oil viscosity correlations can be compared to determine which correlation calculates the dead oil viscosity values that come closest to the measured values. With the exception of the dead oil correlation by Dindoruk and Christman (2004), which requires the additional inputs of solution GOR and specific gravity of solution gas, dead oil viscosity correlations (Appendix A: Oil Correlations Formulae) use API gravity and temperature as inputs. Fig. 8.5 shows dead oil viscosity for a low GOR oil (available from the field laboratory) plotted with dead oil viscosity results from three different correlations.

The above plot shows that Beggs and Robinson (1975) dead oil viscosity correlation provides the closest match, among the three correlations, to the actual values. Based on this result, the Beggs and Robinson (1975) correlation should be used to calculate the live oil viscosity at reservoir temperature. Fig. 8.6 shows the comparison of the same three correlations with the actual values of live oil viscosity for the oil sample. Although the three calculated live oil viscosity curves do not match the laboratory-measured live oil viscosity, the closest accuracy is achieved by the Beggs and Robinson (1975) correlation.

HANDLING PVT PROPERTIES FOR LOW GAS–OIL RATIO OILS

In absence of representative fluid samples and PVT laboratory reports, oil PVT correlations can be used to generate PVT properties for low GOR oils.

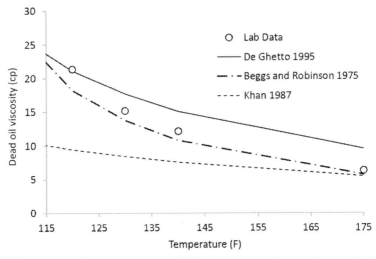

FIGURE 8.5 Dead oil viscosity for a low GOR oil compared with viscosity calculated from different correlations. *GOR*, gas−oil ratio.

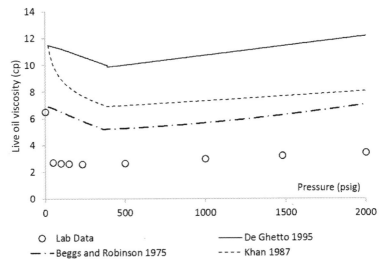

FIGURE 8.6 Live oil viscosity for a low GOR oil compared with viscosity calculated from different correlations. *GOR*, gas−oil ratio.

The main distinction, however, is that only undersaturated correlations are usually necessary. Bubble-point pressure correlations can be used to ensure that the predicted bubble point pressure is well below the reservoir pressures that will be encountered during reservoir depletion.

Appendix A, Oil Correlations Formulae, shows a large number of available oil PVT correlations. Appendix C, Oil Correlations Range of

Applicability, lists the ranges of applicability of each correlation. In general, the variation of most PVT properties with pressure (at reservoir temperature) is not significant for low GOR oils. This fact is true for oil formation volume factor, oil density, and oil compressibility. Many correlations can predict the PVT properties (except for oil viscosity, as discussed earlier) with sufficient accuracy. Oil formation volume factor can also pose a challenge in prediction for some correlations, as will be illustrated in Problem 1.

The inputs to PVT correlations for low GOR oils include initial producing GOR, API gravity of stock-tank oil, specific gravity of separator gas, and reservoir temperature. The initial producing GOR is a low value in the case of low GOR oils and is sometimes unavailable in field production data. In this case, it can be assumed at a low value. The same applies to specific gravity of separator gas. API gravity of stock-tank oil and reservoir temperature are usually readily available.

Sequence of Calculations

The sequence of calculations for PVT properties of low GOR oils is the same as that of black-oils. The only difference is that the saturated part of the curve is not usually calculated. After data input, bubble point pressure is either input (e.g., estimated from offset wells), or calculated using any of bubble-point pressure correlations. The oil formation volume factor is then calculated with the selected correlation(s). Following that calculation, oil compressibility, formation volume factor of oil, and oil density are calculated for the undersaturated portion of the curve. Dead oil viscosity is then calculated, followed by live oil viscosity using the undersaturated viscosity correlations. The sequence of calculations is provided in the flow chart of Fig. 8.7.

ENGINEERING LOW GAS–OIL RATIO OILS

Problem 1—Calculating PVT Properties for Low Gas–Oil Ratio Oil Reservoirs

You are given the following basic field information for three different low GOR fluids. Use Standing (1947) and De Ghetto et al. (1995) correlations to calculate the bubble point pressure for the three fluids and compare with laboratory-measured values of bubble point pressure. Use Standing (1947) correlations family to compute solution GOR, oil formation volume factor, oil density, oil compressibility, and oil viscosity for Fluid 1. Then, use the two correlations to calculate oil formation volume factor, oil compressibility, oil density, and oil viscosity for Fluid 3. Compare the calculated properties with laboratory-measured values. The basic fluid parameters and laboratory-measured values at the bubble point pressure for the three fluids are given in Tables 8.4 and 8.5.

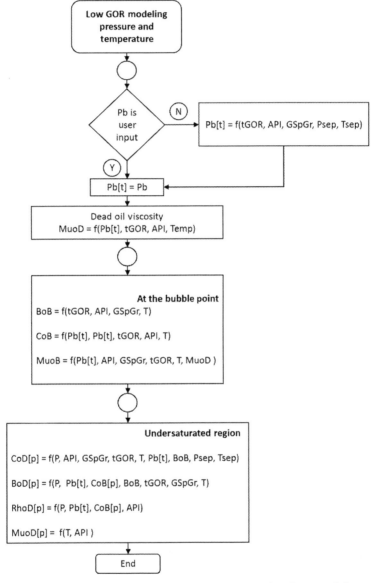

FIGURE 8.7 Flowchart for low GOR oil PVT properties calculations from correlations. *GOR*, gas—oil ratio.

Solution of Problem 1

Standing (1947) and De Ghetto et al. (1995) family of correlations are used to calculate the bubble point pressure for the three fluids. Table 8.6 shows the comparison between calculated values and laboratory-measured bubble

TABLE 8.4 Basic Field Information for the Three Fluids

	Fluid 1	Fluid 2	Fluid 3
Tr, °F	152	167	177
Stock tank, °API	24	24	24
Surface producing GOR, scf/STB	73	10	37
Separator gas specific gravity	0.83	0.83	0.83

GOR, gas–oil ratio

TABLE 8.5 Laboratory-Measured Values at the Bubble Point Pressure for the Three Fluids

	Fluid 1	Fluid 2	Fluid 3
Bubble point pressure, psia	200	75	109
Oil formation volume factor at the Pb, rbbl/STB	1.07	1.04	1.06
Oil viscosity at the Pb, cp	1.63E + 00	4.18E + 00	1.67E + 00
Oil density at the Pb, lbm/ft^3	49.53	53.64	50.84

TABLE 8.6 Bubble-Point Pressure Calculation With Different Correlations

Bubble Point Pressure, psi

	Fluid 1	Fluid 2	Fluid 3
Laboratory data	199.7	74.7	108.7
Standing (1947)	489.9	76.7	283.5
De Ghetto et al. (1995)	706.4	138.5	370.4

point pressure values. Table 8.7 shows Standing (1947) correlations results for all PVT properties for Fluid 1.

Tables 8.8 and 8.9 show the results of Standing (1947) and De Ghetto et al. (1995) correlations for all PVT properties for Fluid 3, respectively.

TABLE 8.7 Standing (1947) Correlations Calculations for Fluid 1

Pressure, psia	Gas in Solution, scf/STB	Oil Formation Volume Factor, rbbl/STB	Oil Density, lb/cu. ft.	Oil Compressibility, psi^{-1}	Oil Viscosity, cp
14.7	0	1.421	53.79	—	6.09
278	36	1.426	53.39	1.6E − 03	5.39
283	37	1.426	53.38	1.6E − 03	5.37
556	37	1.424	53.46	5.3E − 06	5.51
833	37	1.422	53.52	4.9E − 06	5.66
1111	37	1.421	53.58	4.6E − 06	5.81
1389	37	1.419	53.63	4.2E − 06	5.97
1667	37	1.418	53.67	3.9E − 06	6.13
1944	37	1.418	53.70	3.6E − 06	6.30
2222	37	1.417	53.72	3.3E − 06	6.47
2500	37	1.416	53.74	3.1E − 06	6.64

TABLE 8.8 Standing (1947) Correlations Results for Fluid 3

Pressure, psia	Gas in Solution, scf/STB	Oil Formation Volume Factor, rbbl/STB	Oil Density, lb/cu.ft.	Oil Compressibility, psi^{-1}	Oil Viscosity, cp
14.7	0	1.070	53.05	1.4E − 02	6.17
200	18	1.077	52.87	1.5E − 03	5.34
350	35	1.084	52.69	9.3E − 04	4.71
370	37	1.085	52.67	5.4E − 05	4.63
490	37	1.079	52.96	4.1E − 05	4.68
560	37	1.076	53.10	3.6E − 05	4.72
830	37	1.067	53.52	2.4E − 05	4.90
1010	37	1.063	53.73	2.0E − 05	5.06
1390	37	1.057	54.08	1.4E − 05	5.45
1670	37	1.053	54.28	1.2E − 05	5.80
2000	37	1.049	54.47	1.0E − 05	6.28

TABLE 8.9 De Ghetto et al. (1995) Correlations Results for Fluid 3

Pressure, psia	Gas in Solution, scf/STB	Oil Formation Volume Factor, rbbl, STB	Oil Density, lb/ft^3	Oil Compressibility, psi^{-1}	Oil Viscosity, cp
14.7	0	1.070	53.05	1.4E − 02	6.17
200	18	1.077	52.87	1.5E − 03	5.34
350	35	1.084	52.69	9.3E − 04	4.71
370	37	1.085	52.67	5.4E − 05	4.63
490	37	1.079	52.96	4.1E − 05	4.68
560	37	1.076	53.10	3.6E − 05	4.72
830	37	1.067	53.52	2.4E − 05	4.90
1010	37	1.063	53.73	2.0E − 05	5.06
1390	37	1.057	54.08	1.4E − 05	5.45
1670	37	1.053	54.28	1.2E − 05	5.80
2000	37	1.049	54.47	1.0E − 05	6.28

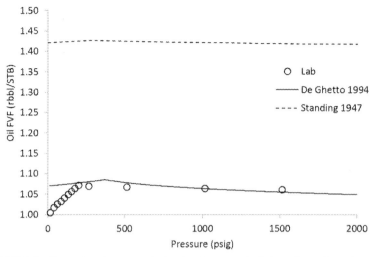

FIGURE 8.8   Comparison between calculated and measured oil formation volume factor for Fluid 3.

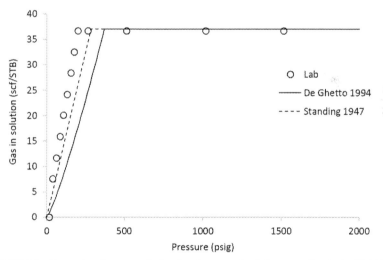

FIGURE 8.9   Comparison between calculated and measured solution gas−oil ratio for Fluid 3.

Figs. 8.8−8.12 show the comparison plots for different fluid PVT properties for Fluid 3. The comparison plots show laboratory-measured values with results of Standing (1947) and De Ghetto et al. (1995) correlations.

Examination of the oil formation volume factor plots indicates that some correlations (e.g., Standing, 1947) overestimate values. A check of the

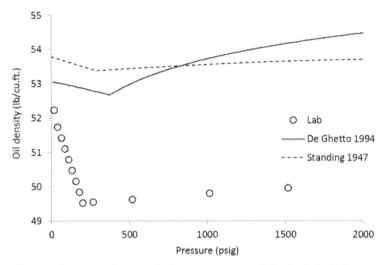

FIGURE 8.10 Comparison between calculated and measured oil density for Fluid 3.

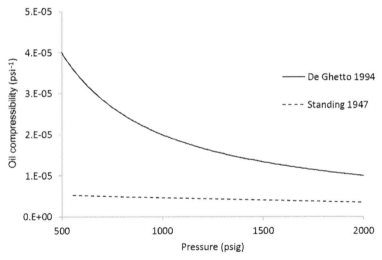

FIGURE 8.11 Calculated isothermal oil compressibility for Fluid 3.

correlation applicability range from Appendix C, Oil Correlations Range of Applicability, for oil formation volume factor shows that the data of Fluid 3 is within the data range used to develop the correlation. However, the correlation overestimates the formation volume factor, most likely due to the biases of the correlation to lighter oil data. Many other correlations predict an oil formation volume factor less than 1.0 rbbl/STB for low GOR oils. These values should be rejected.

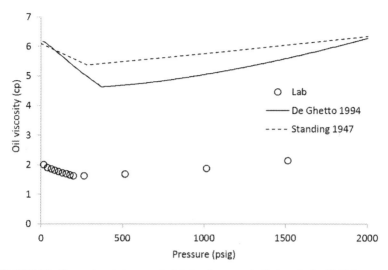

FIGURE 8.12 Comparison between calculated and measured oil viscosity for Fluid 3.

NOMENCLATURE

API	density of stock-tank liquid, °API
EOR	enhanced oil recovery
GOR	gas–oil ratio, scf/STB
p_b	bubble point pressure, psia
T_r	reservoir temperature, °F
TVDss	true vertical depth subsea

REFERENCES

Beal, C., 1946. Viscosity of air, water, natural gas, crude oil and its associated gases at oil field temperatures and pressures. Trans. AIME 165, 94–115.

Beggs, H.D., Robinson, J.R., 1975. Estimating the viscosity of crude oil systems. J. Pet. Tech 27, 1140–1141. Available from: https://doi.org/10.2118/5434-PA.

De Ghetto, G. and Villa, M. 1994. Reliability analysis on PVT correlations. In: Paper SPE-28904-MS Presented at the European Petroleum Conference, London, United Kingdom, 25–27 October.

De Ghetto, G., Paone, F., Villa, M. 1995. Pressure–volume–temperature correlations for heavy and extra heavy oils. In: Paper SPE-30316-MS Presented at the SPE International Heavy Oil Symposium. Calgary, 19–21 June.

Dindoruk, B. and Christman, P. G. 2004. PVT properties and viscosity correlations for Gulf of Mexico oils. In: Paper SPE-71633-MS Presented at the SPE Annual Technical Conference and Exhibition, 30 September–3 October, New Orleans, Louisiana, United States.

Khan, S.A., Al-Marhoun, M.A., Duffuaa, S.O. and Abu-Khamsin, S.A. 1987. Viscosity correlations for Saudi Arabian crude oils. In: Paper SPE-15720-MS Presented at the Middle East Oil Show, 7−10 March, Bahrain. Available from: https://doi.org/10.2118/15720-MS.

McCain Jr., W.D., 1994. Heavy components control reservoir fluid behavior. J. Pet. Tech. 46, 746−750. Technology Today Series; Trans., AIME, 297.

Petrosky, G.E. and Farshad, F.F. 1995. Viscosity correlations for Gulf of Mexico crude oils. In: Paper SPE-29468-MS Presented at the SPE Production Operations Symposium, 2−4 April, Oklahoma City, Oklahoma, United States. Available from: https://doi.org/10.2118/29468-MS.

Standing, M.B., 1947. A pressure-volume-temperature correlation for mixtures of California oils and gases, Drilling and Production Practice. American Petroleum Institute, New York, United States, 1st January.

FURTHER READING

Al-Marhoun, M.A., 1988. PVT correlations for Middle East crude oils. Soc. Pet. Eng . Available from: https://doi.org/10.2118/13718-PA.

Beal, C., 1946. Viscosity of air, water, natural gas, crude oil and its associated gases at oil field temperatures and pressures. Trans. AIME 165, 94−115.

De Ghetto, G., Paone, F., Villa, M., 1995. Pressure−volume−temperature correlations for heavy and extra heavy oils. In: Presented at the SPE International Heavy Oil Symposium. Calgary, 19−21 June. SPE-30316-MS.

Khan, S.A., Al-Marhoun, M.A., Duffuaa, S.O., Abu-Khamsin, S.A., 1987. Viscosity correlations for Saudi Arabian crude oils. Soc. Pet. Eng . Available from: https://doi.org/10.2118/15720-MS.

Petrosky, G.E., Farshad, F.F., 1995. Viscosity correlations for Gulf of Mexico crude oils. Soc. Pet. Eng . Available from: https://doi.org/10.2118/29468-MS.

PROBLEMS

8.1 You are given the following basic field information for three different low gas−oil ratio fluids. Use Standing (1947) and De Ghetto et al. (1995) correlations to calculate the bubble-point pressure for the three fluids. Also, use the two correlations to calculate oil formation volume factor, oil compressibility, oil density, and oil viscosity for Sample 3.

	Sample 1	Sample 2	Sample 3
Reservoir temperature, °F	180	177	190
Total producing gas−oil ratio, scf/STB	29	49	80
Stock tank, °API	18	19	20
Separator gas specific gravity	0.87	0.86	0.88

8.2 Use the correlation of Beal (1946), Petrosky and Farshad (1995)., Beggs and Robinson (1975), and De Ghetto et al. (1995) to calculate the dead oil viscosity at the temperatures of 75, 104, 120, 130, 140, and 175°F. Compare the dead oil viscosity values calculated from different correlations at different temperatures. The following surface data is available for the oil:

Total producing gas–oil ratio, scf/STB	50
Stock tank gravity, °API	18
Separator gas specific gravity	0.87

Chapter 9

Selection of PVT Correlations

Different PVT correlations sometimes yield significantly different results from the same input. Arbitrary selection of PVT correlations can lead to significant errors in engineering calculations.

Fig. 9.1 shows an example single-phase z-factor calculation compared against laboratory-measured data. The figure shows that different correlations produce different results. To use the same basis for comparison, the input to all these correlations was pseudo-reduced pressure and temperature calculated with the Piper et al. (1993) correlation. Although some correlations show a better match with actual data in a portion of the curve, the same correlation is not guaranteed to produce good results on all occasions. In addition, one correlation may work for a particular PVT property, but a sister correlation (from the same family of correlations) may not work for another PVT property for the same fluid.

In absence of actual PVT laboratory measurements, it is challenging to generate accurate values of PVT properties. Engineers are faced with the

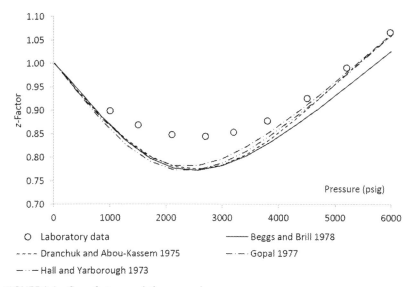

FIGURE 9.1 Gas z-factor correlation comparison.

PVT Property Correlations. DOI: https://doi.org/10.1016/B978-0-12-812572-4.00009-6

problem of estimating the best correlation(s) to use. Many practicing engineers follow a completely arbitrary process to select PVT correlations. Others may follow more structured approaches such as those discussed in the following section.

SELECTION OF PVT OIL PROPERTIES CORRELATIONS

Numerous PVT correlations exist for each oil (and gas) property. The number of oil correlations available exceeds that of gas correlations (Appendix A: Oil Correlations Formulae and Appendix B: Gas Correlations Formulae).

Oil and gas PVT correlations can usually be grouped into correlation families. Researchers of PVT correlations usually develop a separate correlation for each PVT property from their PVT laboratory-measured data. For selection of correlation(s) to use for a specific fluid, the following approaches are available: (1) use of geographical location, (2) use of the type or obvious characteristics of the fluid, (3) use of correlations based on application, (4) use of data ranges and ranking, and (5) use of an expert system or set of rules.

Correlation Selection Based on Geographical Location

Many oil PVT correlations were developed with data sets that belong to a specific region. Among the commonly known geographical correlations are Al-Marhoun (1988) correlations for Saudi Arabian crudes; Standing (1947) for California crudes; Petrosky (1990) for Gulf of Mexico crudes; Hanafy et al. (1997) for Egyptian crudes; and Omar and Todd (1993) for Malaysian crudes. Appendix C, Oil Correlations Range of Applicability, gives the application ranges of each correlation and identifies the geographical origin of each correlation, where applicable.

In absence of laboratory-measured PVT data, many engineers use specific correlation families for crude oils based on geographical origin. Although it is logical to assume that oil in different reservoirs in a particular geographical location is of the same origin, it is erroneous to assume that this approach is consistently reliable. The geographical PVT correlation families may have been developed from PVT data collected from relatively shallow reservoirs. Deeper reservoirs developed at a later stage usually contain more volatile oil and richer gas condensate fluids. New correlations are required for the fluids more recently discovered in these traditional regions when sufficient data becomes available. Therefore, caution is advisable in use of the geographical correlation approach.

In addition, Al-Shammasi (2001), in a large comparison study, showed that geographical location-based correlations are not guaranteed to work.

Correlation Selection Based on Type of Fluid

Several correlations that pertain to specific fluid type were developed for oils or gases. For example, the correlations of De Ghetto et al. (1995) were developed for heavy and extraheavy crude oils, and therefore, use of these correlations is recommended for heavy oils. Use of these correlations outside their range of applicability usually leads to errors that may be easily avoided.

Correlation Selection Based on Required PVT Model (Application)

In some applications, the required PVT model determines the correlations to use. If the black-oil model is to be used, many correlations are available. If the modified black-oil (MBO) model is needed for a certain application, fewer options are available. Although individual PVT properties are available for some MBO PVT properties, two correlation families are available for all four MBO PVT properties [solution gas−oil ratio (GOR); vaporized oil−gas ratio; oil formation volume factor; and gas formation volume factor]. These correlations and their applications were discussed in Chapter 5, Gas Condensates, and Chapter 6, Volatile Oil (El-Banbi et al., 2006; Nassar et al., 2013).

Correlation Selection Based on Data Range and Ranking

One successful practice in selection of PVT correlations based on input data entails identification of a suitable range for every correlation. Al-Shammasi (2001) reviewed multiple correlations and compared their performance against a large database of oil PVT data. Al-Shammasi (2001) performed a study for bubble point pressure and formation volume factor correlations only. The study concluded that greater accuracy in prediction from PVT correlations could be achieved if correlation selection is based on API gravity values input. According to Al-Shammasi's grouping idea, Glaso (1980) bubble point pressure correlation is most suitable for very heavy oil (API gravity less than 10°API); Macary and El-Batanoney (1992), for medium API gravity oil (between 10 and 20°API); and Al-Marhoun (1988), for higher quality oil (20−30°API).

Al-Zahaby et al. (2012) expanded Al-Shammasi's work and created clusters of input data. For each cluster, they found the best correlation(s) to use among a reasonable number of oil PVT correlations out of all oil PVT properties. They also collected the ranges of applicability of the correlations recommended in their work. Their approach is based on two steps: (1) the selection of a subset of correlations for every PVT property by use of the input data and applicability ranges for each correlation and (2) the ranking of

PVT correlations within the cluster by use of a PVT data set. A limited data set was used in this work.

The applicability ranges for gas and oil correlations can be used to reduce the number of applicable correlations for a particular fluid. These ranges are based on the database used to develop each correlation. The ranges are given in Appendix C, Oil Correlations Range of Applicability and Appendix D, Gas Correlations Range of Applicability, for oil and gas correlations, respectively.

Correlation Selection Based on Expert System or Set of Rules

Al-Gathe (2015) developed an expert system (a set of rules) for oil PVT correlation selection. The study checked the accuracy of correlations against a large database of PVT laboratory-measured data. From the input data for each correlation, the correlations are ranked according to their accuracy. In these calculations, intermediate PVT properties required as inputs to the following correlation calculations are assumed to be accurate. Fig. 9.2 shows the input data required to compute each PVT property. The usual four inputs are required in the selection process for all PVT properties. The workflow starts with selection of the bubble point pressure correlation. The value of the bubble point pressure is needed in the selection of the other PVT correlations such as solution GOR, saturated oil formation volume factor, saturated oil density, undersaturated oil compressibility, and saturated oil viscosity correlations. The value of dead oil viscosity is also required in selection of both saturated and undersaturated oil viscosity correlations. In absence of laboratory-measured PVT data, dead oil viscosity must be estimated from correlations. In ranking correlations for saturated and undersaturated oil

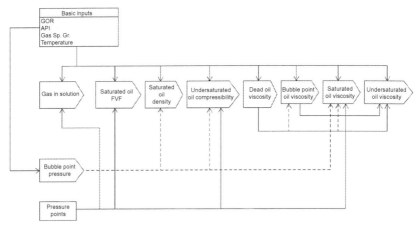

FIGURE 9.2 Flow chart showing inputs needed for each correlation in PVT correlation selection.

viscosity, Al-Gathe (2015) assumed that dead oil viscosity is known. Nevertheless, testing this approach on a large data set (which was not used in development of the expert system) demonstrated the accuracy of this approach in the majority of cases. This result means that the expert system is capable of selecting one (sometimes two) correlation(s) for every PVT property that provides the values closest to laboratory-measured values.

The rules developed by Al-Gathe (2015) are represented graphically in Appendix F, Worksheets for Oil PVT Correlations Selection. Although the correlations used in development of these selection rules were fewer than those presented in this book, the rules increase the chances of selecting correlations that provide lower error compared to measured values.

Selim (2018) developed an expert system similar to that of Al-Gathe (2015), but dependent on selection of correlation families (rather than individual correlations for each property). Selim (2018) identified the best correlation family to be used for any set of input data. In his study, the input data for every PVT property correlation are grouped in clusters. One family of correlations is selected to calculate all PVT properties (e.g., bubble point pressure, solution GOR, oil formation volume factor, oil compressibility, oil density, and oil viscosity). The correlation families considered include Al-Marhoun (1988), Standing (1947), Farshad et al. (1996), etc.

APPLICATIONS FOR PVT CORRELATION SELECTION

Problem 1—Showing the Variation of Calculated Dew Point From Different Correlations

The gas with the composition given in Table 9.1 is produced from a particular reservoir. Additional information for the gas is given in Table 9.2. Calculate the dew point pressure using several correlations for comparison. The laboratory-measured dew point pressure for this gas is 3260 psia.

Solution of Problem 1

The composition and separator data are used to calculate dew point pressure from four correlations that require basic input and four correlations that require gas composition. The correlations are given in Appendix B, Gas

TABLE 9.1 Composition of an Example Gas Condensate Sample

H_2S	CO_2	N_2	C1	C2	C3	n-C4	i-C4	n-C5	i-C5	C6	C7+
0.00	6.42	3.16	50.26	9.68	6.20	2.19	3.73	1.88	1.85	3.04	11.59

TABLE 9.2 Basic Field Information for the Gas Condensate Sample

Separator pressure	850	psig
Separator temperature	109	°F
Gas–oil ratio	3483	scf/STB

TABLE 9.3 Calculated Dew Point Pressure From Several Correlations for the Gas of Problem 1

Correlation	Correlation Base	Dew Point Pressure, psia	Error (%)
Laboratory		3260	0.0
Nassar et al. (2013)	Basic input	4360	33.8
Marruffo et al. (2001)	Basic input	3330	2.2
Al-Dhamen and Al-Marhoun (2011)	Basic input	3254	− 0.2
Ovalle et al. (2007)	Basic input	3251	− 0.3
Elsharkawy (2001)	Composition	3473	6.5
Nemeth and Kennedy (1967)	Composition	3320	1.8
Godwin (2012)	Composition	3380	3.7
Kamaria et al. (2016)	Composition	2440	− 25.1

Correlations Formulae. Table 9.3 lists the calculated dew point from the eight different correlations with the percent error.

Comments on Problem 1

The above comparison of the different correlations reveals that significant variation is possible in the results from different correlations. The comparison also shows that sometimes correlations requiring fewer basic inputs may outperform correlations requiring additional inputs (e.g., composition). In addition, the correlations that perform well in prediction of the dew point pressure in some samples may not perform well in other samples.

Problem 2—Selection of PVT Correlations Based on Rules

A shallow reservoir producing oil with API of 37.7 and initial producing GOR of 675 scf/STB is considered for further development. The well logs indicate that reservoir temperature is 190°F. Separator information shows that specific gravity of separator gas is 1.221. Use the worksheets of Appendix F, Worksheets for Oil PVT Correlations Selection, to select PVT correlations to calculate PVT properties for this oil. Calculate fluid properties with the selected correlation(s) and compare results with laboratory data using appropriate plots. Table 9.4 shows the laboratory-measured data for this oil sample.

Solution of Problem 2

The workflow in Fig. 9.2 and worksheets of Appendix F, Worksheets for Oil PVT Correlations Selection, are followed to select the best correlation for every PVT property. Details of the procedure are given in the following sections.

Bubble Point Pressure

Fig. 9.3 shows the four basic inputs intersecting the different correlations. Table 9.5 lists the calculated bubble point pressure and its related errors. According to the applicability limits of the four inputs, two correlations can be selected (Al-Shammasi, 2001; Lasater, 1958). The bubble point was estimated for this fluid using all eight correlations. Comparison between the results of all correlations reveals that Dindoruk and Christman (2004) have the lowest error for this fluid. The two selected correlations (Al-Shammasi, 2001; Lasater, 1958) have the second and third-lowest errors.

Gas in Solution

Fig. 9.4 shows that three of the four solution GOR correlations are selected. Fig. 9.5 shows the comparison between the different solution GOR correlation results with the laboratory-measured values. The comparison shows that all solution GOR correlations (including the selected three) produce values that reasonably match the laboratory data. The Farshad et al. (1996) correlation is selected for further work on the problem.

Oil Formation Volume Factor

Fig. 9.6 shows that the inputs for selection of oil formation volume factor correlation intersect only two correlations (Kartoatmodjo and Schmidt, 1994; Vasquez and Beggs, 1980). Fig. 9.7 shows the comparison plot for all correlations with the laboratory-measured data. The Vasquez and Beggs (1980) correlation is indeed the best correlation out of all.

Saturated Oil Density

The selection of the most appropriate correlation for saturated oil density is shown in Fig. 9.8. By use of the intersections of the inputs with the applicability ranges for all correlations, one correlation only is selected (Standing,

TABLE 9.4 Laboratory-Measured Data for the Oil of Problem 2

Laboratory Data

Pressure, psia	Gas in Solution, scf/STB	Oil FVF, rbbl/STB	Oil Density, lbm/cu.ft.	Oil Viscosity, cp	Oil Compressibility, psi^{-1}
2914.7	675	1.420	43.3	0.473	1.4E − 05
2769.7	675	1.423	43.2	0.469	1.4E − 05
2624.7	675	1.426	43.1	0.464	1.5E − 05
2479.7	675	1.429	43.0	0.460	1.5E − 05
2410.7	675	1.431	43.0	0.457	1.6E − 05
2334.7	675	1.432	42.9	0.455	1.6E − 05
2189.7	675	1.436	42.8	0.451	1.7E − 05
2044.7	675	1.439	42.7	0.446	1.8E − 05
1899.7	675	1.443	42.6	0.442	1.9E − 05
1754.7	675	1.447	42.5	0.437	2.0E − 05
1682.7	675	1.449	42.4	0.435	2.1E − 05
1656.7	675	1.449	42.4	0.434	2.0E − 05
1319.7	552	1.381	43.5	0.478	
1029.7	455	1.328	44.2	0.524	
739.7	359	1.276	45.0	0.584	
449.7	250	1.214	46.0	0.679	
159.7	99	1.096	47.8	0.921	
14.7	0	1.014	49.2	1.632	

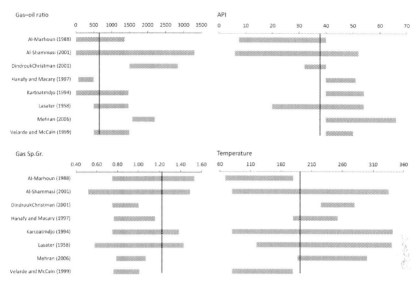

FIGURE 9.3 Selection of bubble point pressure correlation from the four basic inputs for Problem 2.

TABLE 9.5 Bubble Point Pressure Results From Different Correlations for Problem 2

Correction	Bubble Point Pressure (psia)	Error (%)
Laboratory	1657	0
Al-Marhoun (1988)	1198	− 28
Al-Shammasi (2001)	1754	6
Dindoruk and Christman (2004)	1609	− 3
Hanafy et al. (1997)	2616	58
Kartoatmodjo (1994)	2000	21
Lasater (1958)	1944	17
Velarde and McCain (1997)	2282	38

1947). Fig. 9.9 shows the comparison plot between the laboratory-measured data and the two correlations available for saturated oil density. The comparison shows that the Standing (1947) correlation impressively predicts the saturated oil density for the oil of Problem 2. The undersaturated oil density was calculated from the basic equation in the two correlations presented in the figure.

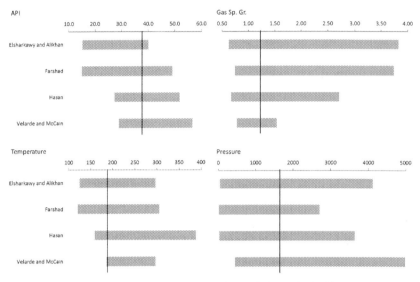

FIGURE 9.4 Selection of solution GOR from the four basic inputs for Problem 2.

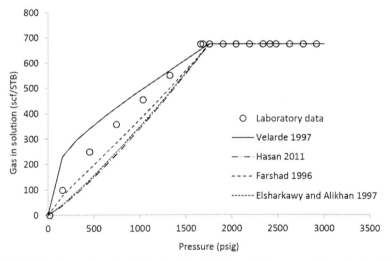

FIGURE 9.5 Comparison of calculated solution GOR from different correlations for the oil of Problem 2.

Undersaturated Oil Compressibility

Figs. 9.10 and 9.11 show the applicability ranges of six undersaturated oil compressibility correlations and seven input values for the correlation selection. Only one correlation is selected because all other correlations were

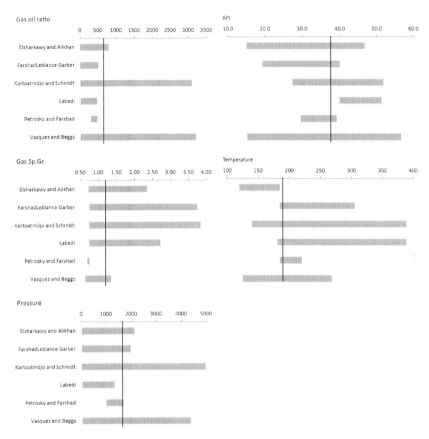

FIGURE 9.6   Selection of oil formation volume factor correlation from inputs for the oil of Problem 2.

eliminated by one or more of the input criteria. The selected correlation is Dindoruk and Christman (2004).

The available laboratory-measured values were compared with the results of all six correlations. These results are shown in Fig. 9.12. The comparison results show that the selected correlation is the closest to the laboratory-measured data.

Dead Oil Viscosity

Dead oil viscosity selection criteria depend on the four basic inputs for the oil PVT correlations. Correlations of dead oil viscosity calculate one value for the viscosity of gas-free oil at standard conditions. The four basic inputs for Problem 2 are applied in Fig. 9.13 to select the most appropriate dead oil viscosity correlation. The four basic inputs resulted in elimination of all

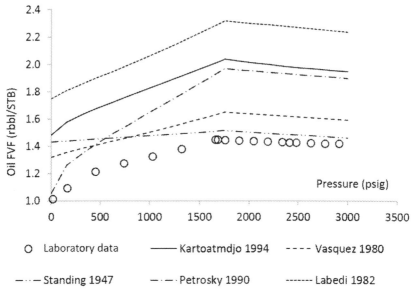

FIGURE 9.7   Comparison of oil formation volume factor calculated from different correlations for the oil of Problem 2.

dead oil viscosity correlations except for Labedi (1992). Table 9.6 lists the calculated dead oil viscosity from all seven correlations. This table shows that the selected correlation (Labedi, 1992) is the second-best correlation.

Oil Viscosity at the Bubble Point

The calculated dead oil viscosity is used to select the best correlation for oil viscosity at the bubble point pressure. Fig. 9.14 shows that the four basic inputs in addition to the dead oil viscosity value intersect only the Chew and Connally (1959) correlation. The other correlations are eliminated. The resultant value of oil viscosity at the bubble point pressure will be used in the selection of saturated and undersaturated oil viscosity.

Saturated Oil Viscosity

Fig. 9.15 shows the selection of the saturated oil viscosity correlations. Eight inputs are used in the selection process. The selection for the saturated oil viscosity shows that two correlations are possible (Elsharkawy and Alikhan, 1999; Naseri et al., 2005).

Fig. 9.16 shows the comparison between the saturated oil viscosity calculated from different correlations and the laboratory-measured values. The figure shows that the majority of the correlations (including the selected ones) are in agreement with the observed values.

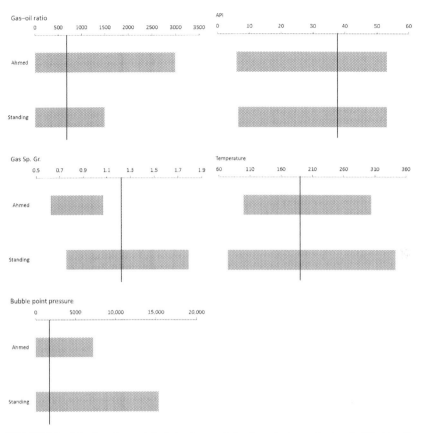

FIGURE 9.8   Selection of saturated oil density correlation from the inputs for the oil of Problem 2.

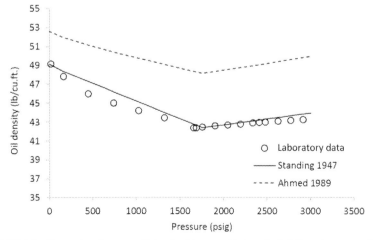

FIGURE 9.9   Comparison between oil density (above and below the bubble point pressure) cal-culated from different correlations with the laboratory-measured values for the oil of Problem 2.

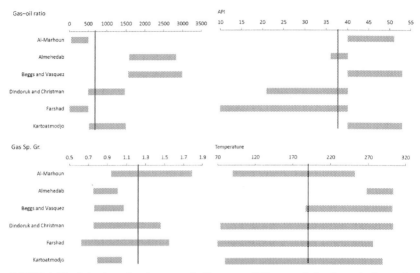

FIGURE 9.10 Selection of undersaturated oil compressibility correlation from the inputs for the oil of Problem 2 (Part 1).

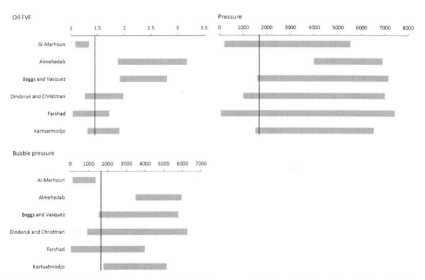

FIGURE 9.11 Selection of undersaturated oil compressibility correlation from the inputs for the oil of Problem 2 (Part 2).

Undersaturated Oil Viscosity

Selection for undersaturated oil viscosity depends on two basic parameters (solution GOR and API) in addition to bubble point pressure, dead oil viscosity, oil viscosity at the bubble point and the pressure steps. From the

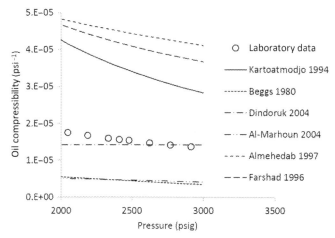

FIGURE 9.12   Comparison between calculated undersaturated oil compressibility from different correlations with the laboratory-measured values for the oil of Problem 2.

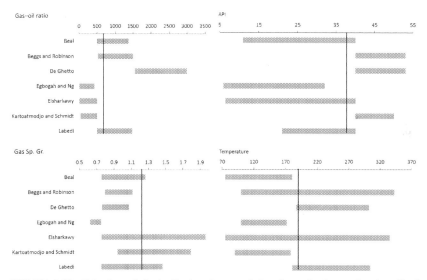

FIGURE 9.13   Selection of dead oil viscosity correlation from the inputs for the oil of Problem 2.

intersections of the inputs with the undersaturated oil viscosity correlations in Fig. 9.17, it is observed that only the Beal (1946) correlation is selected. Fig. 9.18 shows the comparison between the laboratory-measured data for undersaturated viscosity and the results of different correlations. The plot shows that Beal (1946) is indeed the correlation closest to the measured data.

TABLE 9.6 Dead Oil Viscosity Results From Different Correlations for the Oil of Problem 2

Correlation	Dead Oil Viscosity (cp)	Error (%)
Laboratory	1.6	0
Beal (1946)	4.3	164
Beggs and Robinson (1975)	13.3	714
De Ghetto et al. (1995)	6.8	317
Ng and Egbogah (1983)	0.6	−64
Elsharkawy and Alikhan (1999)	3.5	115
Kartoatmodjo and Schmidt (1991)	4.3	163
Labedi (1992)	2.9	75

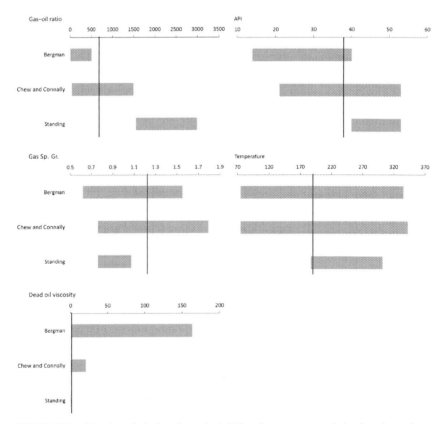

FIGURE 9.14 Selection of oil viscosity at the bubble point pressure correlation from inputs for the oil of Problem 2.

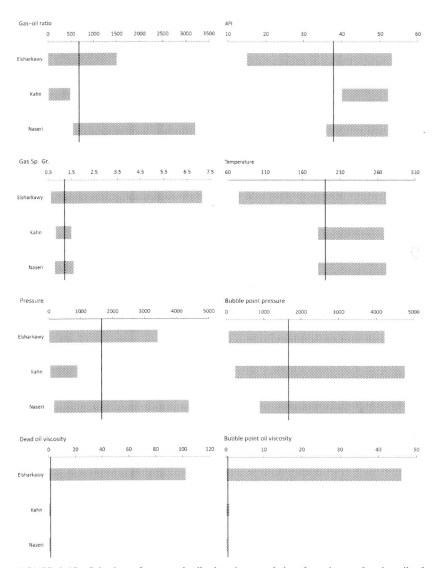

FIGURE 9.15   Selection of saturated oil viscosity correlation from inputs for the oil of Problem 2.

Table 9.7 lists the selected correlations for the oil of Problem 2.

Comments on Problem 2

Problem 2 shows that the use of rules for selection of PVT correlations can sometimes lead to selection of the most accurate correlations. Due to the considerable variability of PVT properties and the fact that few parameters

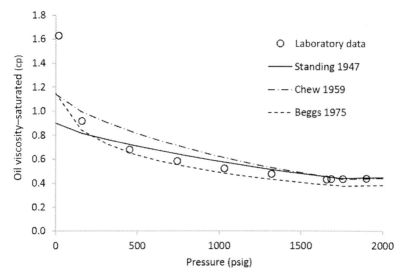

FIGURE 9.16 Comparison between calculated saturated oil viscosities from different correlations with laboratory-measured values for the oil of Problem 2.

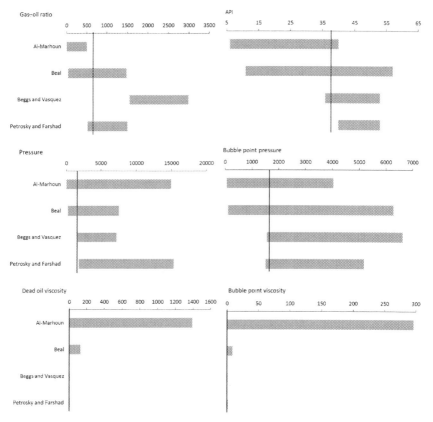

FIGURE 9.17 Selection of undersaturated oil viscosity correlation from inputs for the oil of Problem 2.

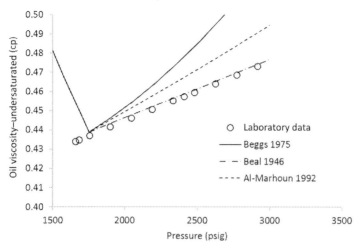

FIGURE 9.18 Comparison between calculated undersaturated oil viscosities from different correlations with laboratory-measured values for the oil of Problem 2.

TABLE 9.7 Summary of Selected Correlations for Problem 2

Selected Correlation	
Bubble point pressure	Al-Shammasi (2001)
Gas in solution	Farshad et al. (1996)
Oil formation volume factor	Vasquez and Beggs (1980)
Dead oil viscosity	Labedi (1992)
Saturated oil density	Standing (1947)
Saturated oil viscosity	Standing (1947)
Viscosity at the bubble point	Chew and Connally (1959)
Undersaturated oil compressibility	Dindoruk and Christman (2004)
Undersaturated oil viscosity	Beal (1946)

are routinely used to characterize the fluids, it is expected that sets of rules established for PVT correlation selection might yield inaccuracies in some situations. However, the use of such rules is still beneficial, as Problem 2 demonstrates. The use of rules can often result in selection of reasonably accurate correlations, even if the technique does not lead to selection of the most accurate correlation(s) for every fluid or for every PVT property.

REFERENCES

Al-Dhamen, M., Al-Marhoun, M., 2011. New correlations for dew-point pressure for gas condensate. In: Paper SPE 155410 Presented at SPE Saudi Arabia section Young Professionals Technical Symposium, 14-16 March, Dhahran, Saudi Arabia. Available from: https://doi.org/10.2118/155410-MS.

Al-Gathe, A.A., 2015. Development of Artificial Intelligence Systems for Black Oil PVT Properties Prediction (Ph.D. thesis). Cairo University, Cairo, Egypt.

Al-Marhoun, M.A., 1988. PVT correlations for middle east crude oils. JPT 30 (5), 650–666. Available from: https://doi.org/10.2118/13718-PA.

Al-Shammasi, A.A., 2001. A review of bubblepoint pressure and oil formation volume factor correlations. SPE Res. Eval. Eng. J. 4 (2). Available from: https://doi.org/10.2118/71302-PA.

Al-Zahaby, A.M., El-Banbi, A.H., Sayyouh, M.H., 2012. Guidelines for selecting appropriate PVT correlations for black oils. In: Paper SPE 150650 Presented at the North Africa Technical Conference and Exhibition, 20–22 February, Cairo, Egypt.

Beal, C., 1946. Viscosity of air, water, natural gas, crude oil and its associated gases at oil field temperature and pressures. Trans. AIME 165, 94–115.

Beggs, H.D., Robinson, J.R., 1975. Estimating the viscosity of crude oil systems. JPT 27, 1140–1141.

Chew, J., Connally Jr., C.A., 1959. A viscosity correlation for gas-saturated crude oils. Trans. AIME 216, 23.

De Ghetto, G., Paone, F., and Villa, M., 1995. Pressure-volume-temperature correlations for heavy and extra heavy oils. In: Paper SPE 30316 Presented at the SPE International Heavy Oil Symposium, 19-21 June, Calgary, Canada.

Dindoruk, B., Christman, P.G., 2004. PVT properties and viscosity correlations for Gulf of Mexico oils. SPE Res. Eval. Eng. J. 28 (4), 427–437.

El-Banbi, A.H., Abdel Fattah, K.A., Sayyouh, M.H., 2006. New modified black oil correlations for gas condensate and volatile oil fluids. In: Paper SPE 102240 Presented at the SPE Annual Technical Conference and Exhibition, 24–27 September, San Antonio, TX., United States.

Elsharkawy, A.M., Alikhan, A.A., 1997. Correlations for predicting solution gas/oil ratio, oil formation volume factor, and undersaturated oil compressibility. J. Pet. Sci. Eng. 17, 291–302.

Elsharkawy, A.M., Alikhan, A.A., 1999. Models for predicting the viscosity of Middle East crude oils. Fuel 78 (8), 891–903.

Elsharkawy, A.M., 2001. Characterization of the plus fraction and prediction of the dew point pressure for gas condensate reservoirs. In: Paper SPE 68666 Presented at the SPE Western Regional Meeting, 26-30 March, Bakersfield, California.

Farshad, F.F., Leblance, J.L., Garber, J.D., Osorio, J.G., 1996. Empirical PVT correlations for Colombian crude oils. In: Paper SPE 36105 Presented at SPE Latin America/Caribbean Petroleum Engineering Conference, 23-26 April, Port of Spain, Trinidad and Tobago.

Glaso, O., 1980. Generalized pressure-volume-temperature correlations. JPT 32, 785–795.

Godwin, O.N., 2012. A new analytical method for predicting dew-point pressure for gas condensate reservoirs. In: Paper SPE 162985 Presented at Nigeria Annual International Conference and Exhibition, 6-8 August, Lagos, Nigeria.

Hanafy, H.H., Macary, S.M., ElNady, Y.M., Bayomi, A.A., and El Batanony, M.H., 1997. Empirical PVT correlations applied to Egyptian crude oils exemplify significance of using regional correlations. In: Paper SPE 102240 Presented at 1997 SPE International Symposium on Oilfield Chemistry, 18-21 February, Houston, Texas.

Kamaria, A., Sattaria, M., Mohammad, A.H., Ramjugernatha, D., 2016. Rapid model for the estimation of dew point pressures in gas condensate systems. J. Taiwan Inst. Chem. Eng. 60, 258–266.

Kartoatmodjo, T.R.S., Schmidt, Z., 1991. New correlations for crude oil physical properties. In: Paper SPE 23556-MS. Society of Petroleum Engineers, Richardson, TX, United States, unsolicited paper.

Kartoatmodjo, T.R.S., Schmidt, Z., 1994. Large data bank improves crude physical property correlation. Oil Gas J. 4, 51–55.

Labedi, R.M., 1982. PVT Correlations of the African Crudes (Ph.D. thesis). Colorado School of Mines, Leadville, Colorado, United States.

Labedi, R.M., 1990. Use of production data to estimate the saturation pressure solution GOR and chemical composition of reservoir fluids. In: Paper SPE 21164 Presented at the SPE Latin America Petroleum, 14-19 October, Rio de Janeiro, Brazil.

Labedi, R.M., 1992. Improved correlations for predicting the viscosity of light crudes. J. Pet. Sci. Eng. 8, 221–234.

Lasater, J.A., 1958. Bubble point pressure correlations. J. Pet. Technol. 10 (5), 65–67. SPE-957-G.

Macary, S.M., El-Batanoney, M.H., 1992. Derivation of PVT correlations for the Gulf of Suez crude oils. In: Proc., 11th EGPC Petroleum Exploration and Production Conference, Cairo, Egypt, Vol 36, (1993) no. 6, p. 472–478.

Marruffo, I., Maita, J., Him, J., Rojas, G., 2001. Statistical forecast models to determine retrograde dew pressure and C7 + percentage of gas condensates on basis of production test data of eastern Venezuelan reservoirs. In: Paper SPE 69393 Presented at SPE Latin American and Caribbean Petroleum Engineering Conference. 25-28 March, Buenos Aires, Argentina.

Naseri, A., Nikazar, M., Mousavi-dehghani, S.A., 2005. A correlation approach for prediction of crude oil viscosity. J. Pet. Sci. Eng. 47, 163–174.

Nassar, I.S., El-Banbi, A.H., Sayyouh, H.M., 2013. modified black oil PVT properties correlations for volatile oil and gas condensate reservoirs. In: Paper SPE 164712 Presented at the North Africa Technical Conference and Exhibition, 15-17 April, Cairo, Egypt. doi:10.2118/164712-MS.

Nemeth, L.K., Kennedy, H.T., 1967. A correlation of dewpoint pressure with fluid composition and temperature. SPE J. 7 (2), 99–104.

Ng, J.T.H., Egbogah, E.O., 1983. An improved temperature-viscosity correlation for crude oil systems. In: Paper PETSOC-83-34-32 Presented at the Annual Technical Meeting, 10-13 May, Banff, Canada.

Omar, M.I., Todd, A.C., 1993. Development of new modified black oil correlations for Malaysian crudes. In: Paper SPE 25338 Presented at the 34th Annual SPE International Conference and Exhibition, February 8-10, Singapore.

Ovalle, A.P., Lenn, C.P., McCain Jr., W.D., 2007. Tools to manage gas/condensate reservoirs; novel fluid-property correlations on the basis of commonly available field data. SPE Res. Eval. Eng. 10 (6), 687–694.

Petrosky, G.E., 1990. PVT Correlations for Gulf of Mexico Crude Oils (MS thesis). University of Southwestern Louisiana, Lafayette, Louisiana, United States.

Piper, L.D., McCain, W.D. Jr., Corredor, J.H., 1993. Compressibility factors for naturally occurring petroleum gases. In: Paper SPE 26668 Presented at the SPE Annual Technical

Conference and Exhibition, 3-6 October, Houston, Texas. Available from: https://doi.org/10.2118/26668-MS.

Selim, A.M., 2018. Development of application maps for PVT Properties Correlations as Guidelines for Correlations Selection (MS thesis). Cairo University, Cairo, Egypt.

Standing, M.B., 1947. A Pressure-Volume-Temperature Correlation for Mixtures of California Oils and Gases. Drilling and Production Practice. American Petroleum Institute, New York, United States, 1st January.

Vasquez, M., Beggs, H.D., 1980. Correlation for fluid physical property predictions. JPT 32, 968–970.

Velarde, J., Blasingame, T.A., and McCain, W.D., Jr., 1997. Correlation of black oil properties at pressures below bubble point pressure - a new approach. In Paper PETSOC-97-93 Presented at the Petroleum Society of Canada Annual Technical Meeting, 8-11 June, Calgary, Alberta, Canada. Available from: https://doi:10.2118/97-93.

PROBLEMS

9.1 For the following basic information available for different oils, use the worksheets in Appendix F, Worksheets for Oil PVT Correlations Selection, to select the appropriate correlations to use for every PVT property.

	Sample 1	Sample 2	Sample 3	Sample 4	Sample 5	Sample 6	Sample 7
API	40.7	15	22.2	32.3	28	41	36
Gas Sp. Gr.	0.79	0.86	0.76	0.78	0.88	0.89	0.81
GOR, scf/STB	900	20	140	2300	1800	3900	3300
T, °F	190	160	190	210	211	280	240

9.2 For Samples 2, 5, and 7 above; calculate the bubble point pressure from the selected correlation(s). Also, calculate solution GOR and oil formation volume factors from the selected correlations and plot them. Comment on the results.

9.3 Use the basic input data for Samples 6 and 7 and the process of elimination to eliminate PVT correlations from Appendix C, Oil Correlations Range of Applicability. Reduce the available correlations to the ones that can be used based on the data ranges (limits) for every correlation.

Chapter 10

Artificial Neural Network Models for PVT Properties

In the last few decades, artificial neural network (ANN) models have been used in many applications in the petroleum industry (Mohaghegh et al., 2011; Cranganu et al., 2015). ANN models have been developed for PVT properties prediction; model recognition in well test analysis; identification of faults in sucker-rod pumping systems; log response estimation; reservoir rock properties prediction; multiphase flow pressure drop calculation; identification of infill drilling locations in unconventional reservoirs; selection of optimum hydraulic fracturing parameters, and many more applications.

ANN applications in prediction of PVT properties have received significant attention in recent years. In several investigations, the capabilities of ANN have been exploited to predict PVT properties using different data sets collected from PVT reports and/or the literature. As with the PVT correlations, ANN models can predict the PVT properties within the limits of the data used to develop them.

BASICS OF ARTIFICIAL NEURAL NETWORK MODELS

Neural network calculations use two calculation passes: (1) forward pass for prediction purposes and (2) backward pass for training purposes. The ANN model structure consists of several components: (1) input nodes; (2) hidden node(s); (3) activation function for the hidden node(s); (4) transformation function for the output node(s); (5) output node(s); (6) objective function; (7) optimization algorithm; and (8) training algorithm.

Fig. 10.1 illustrates the eight components of the neural network and the internal communication paths between components. X1, X2, X3, X4, ..., Xn represent the input values, while a1, a2, a3, a4, ..., an represent the output values of the first hidden layer. Weights w1,1; w1,2; ...;w1,n; w2,1; w2,2; ...; w2,n; ... are assigned for the connections between each input node and the first hidden layer nodes. The output values of the first hidden layer are usually calculated as the summation of the connected input node values with each multiplied times its weight. The values of the first hidden layer are then treated with activation function (function "F"). h1, h2, h3, h4, ..., hn values represent the output of the activation function for the first hidden layer. The

PVT Property Correlations. DOI: https://doi.org/10.1016/B978-0-12-812572-4.00010-2

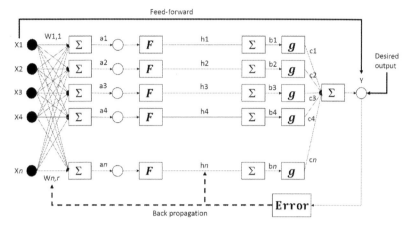

FIGURE 10.1 Basic neural network components with communication paths.

hidden nodes and activation function can be used for one or more hidden layers to successively calculate the values of the following nodes. In the illustrative figure, if more than one hidden layer is used, b1, b2, b3, b4, ..., bn represent the output values from the last hidden layer. These values are input to the transformation function (function "g"). The transformation function is used to treat the b1−bn values and output values c1, c2, c3, c4, ..., cn. The weighted sum of these values will give the output of the ANN (represented by "Y"). The calculation process to generate the output values (Y) from the input values (X1−Xn) is termed the "feed-forward" calculation.

During the training of the ANN, the calculated output (Y) is compared with actual values of output from the experimental dataset. The difference between the calculated output (Y) and actual output is evaluated to represent the ANN error (also known as "loss value"). The objective function (or "loss function") is calculated from the errors of every training record. A training algorithm (e.g., back propagation, reinforcement, and recurrent) is used to redistribute the total ANN error onto the weights connecting the nodes. Two controls on the weight adjustment process can be used: the "learning rate" and the "decay rate." The learning rate (or "learning factor") is used to control the maximum weight change occurring during the training process. The decay rate (or "decay factor") is used to control the minimum change in the weight value. An optimization routine is used to minimize the error for multiple training records until the ANN weights satisfy a predefined error tolerance. The computation procedure that includes error estimation, training technique, and optimization is termed the "feed-backward" calculation.

Fig. 10.2 shows a descriptive diagram for a hidden node that takes input from four input nodes (I1, I2, I3, and I4) and calculates output (O). In the calculation process, a transformation function is used. The common transformation function uses the weighted sum (the values of input nodes are multiplied times their weights and summed up). The calculated value can be

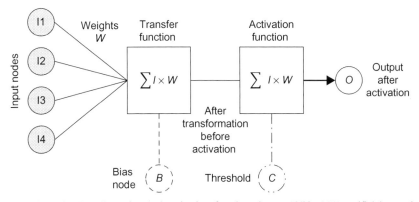

FIGURE 10.2   Transformation and activation functions in an ANN. *ANN*, artificial neural network.

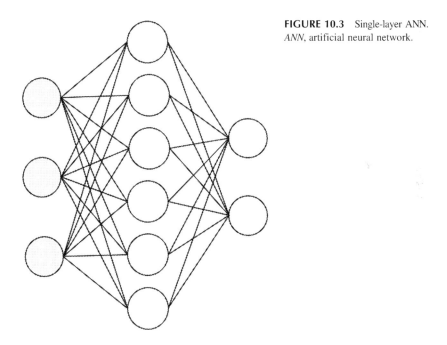

FIGURE 10.3   Single-layer ANN. *ANN*, artificial neural network.

treated with an activation function to compute the output value. An option in the transformation calculation is to use a bias node for the first hidden layer nodes. The bias node value or bias factor scales up the weighted sum to its value. This bias factor guarantees that the values in the input nodes are maintained for subsequent calculations. Another optional control is the activation threshold. The activation threshold is a value used to guarantee that the calculated output from the node is higher than or equal to the threshold value, ensuring that only significant outputs are carried to the next layer.

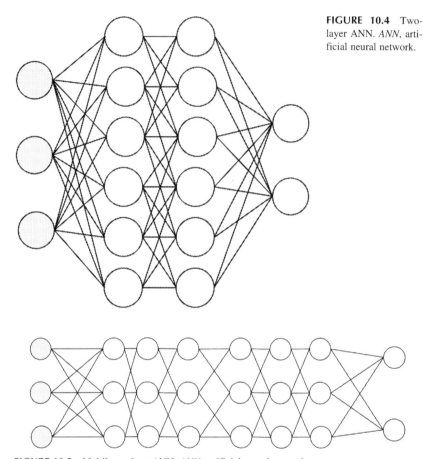

FIGURE 10.4 Two-layer ANN. *ANN*, artificial neural network.

FIGURE 10.5 Multilayer deep ANN. *ANN*, artificial neural network.

The same basic calculations are performed for any type or structure of ANN. Figs. 10.3–10.5 represent three different ANN structures. The first is an example of an ANN with one hidden layer; the second represents a network with two hidden layers, while the third represents a deep network with multiple hidden layers. The input nodes are multiple in each of the networks. The output nodes can be one or more. In general, additional hidden layers are needed to decipher complex relations between inputs and output(s). However, additional hidden layers require also that large number of training records be available.

The earlier ANN example models show differences not only in the number of hidden layers, but also in node communication. The first ANN is characterized by the concept of "full communication," whereby all nodes in the input layer communicate with all nodes in the hidden layer, and all nodes in the hidden layer communicate with all nodes in the output layer. The second and third ANN models are examples of partial communication networks.

ARTIFICIAL NEURAL NETWORK MODELS FOR PREDICTING PVT PROPERTIES

Many attempts to predict PVT properties using ANN models are documented in the petroleum literature. Investigators tried different ANN models to predict properties such as equilibrium constants (K values), oil PVT properties, dew point pressure, and gas z-factor. Tables 10.1a and 10.1b summarize many of the attempts at predicting different PVT properties using ANN models.

Habiballah et al. (1996) used neural network with back propagation and scaled conjugate gradient optimization algorithm to predict K values for fourteen multicomponent hydrocarbon mixtures. They used a database of more than 3000 points and constructed two ANN to predict two parameters that can be used to calculate the K value. The input nodes to the two ANN models included the values of pressure and temperature. The two networks had two hidden layers each.

Gharbi and Elsharkawy (1997) developed neural-network-based models for the prediction of PVT properties of crude oils from the Middle East. They used back propagation technique with momentum. Their dataset contained nearly 500 experimentally obtained data of bubble point pressures and oil formation volume factors at the bubble point.

Elsharkawy (1998) introduced ANN models to calculate formation volume factor, solution gas−oil ratio, oil viscosity, saturated oil density, undersaturated oil compressibility, and evolved gas gravity. Elsharkawy used back propagation technique with radial basis transformation function. Four input nodes were assumed for reservoir pressure, temperature, stock tank oil gravity, and separator gas gravity. The accuracy of the proposed network models to predict PVT properties for crude oils and gas systems was compared to that of numerous published PVT correlations. The ANN models were found to be superior.

Elsharkawy and Gharbi (2001) compared neural network models with models constructed according to classical regression techniques in order to estimate oil viscosity using crude oils from Kuwait. Results of their work show that viscosity models developed using ANN were more accurate than viscosity models developed using regression techniques. Their ANN model had four input nodes, one hidden layer, and one output layer Training and testing data were around 700 points.

Hajizadeh (2007) introduced new models for viscosity prediction using back propagation neural network. He used a genetic algorithm technique to optimize the weights of the neural network. The dataset consisted of 89 samples from Iranian oil PVT reports.

Dutta and Gupta (2010) expanded the work on Hajizadeh (2007) and developed ANN models for estimating bubble point pressure, solution gas−oil ratio, saturated and undersaturated oil formation volume factor, and saturated and undersaturated oil viscosity for Indian crude oils. They used back propagation technique with genetic algorithm for optimization.

TABLE 10.1a Comparison Between Different ANN Models for Predicting PVT Properties

ID	Author	Predicted Property/Properties	Input Layer Nodes	Output Layer Nodes	No. Hidden Layers
1	Adeeyo (2016)	Pb and oil FVF at Pb	4	1	1
2	Ahmadi et al. (2015)	Pb	NA	NA	NA
3	Al-Gathe et al. (2015)	Oil viscosity	4	1	2
4	Alimadadi (2011)	OFVF, oil density	24	2	1
5	Al-Marhoun and Osman (2002)	Pb and oil FVF at Pb	4	1	1
6	Al-Shammasi (2001)	Pb	4	1	2
7	Dutta and Gupta (2010)	Pb, Rs, oil FVF, oil viscosity	NA	NA	NA
8	Elsharkawy (1998)	Bo, Rs, oil viscosity, oil density, Co and gas Sp. Gr.	4	1	2
9	Elsharkawy and Gharbi (2001)	Oil viscosity	4	1	2
10	Gharbi and Elsharkawy (1997)	Pb and oil FVF at Pb	4	1	2
11	González et al. (2003)	Pd	15	2	1
12	Habiballah et al. (1996)	V/L equilibrium K values	2	2	2
13	Ikiensikimama (2012)	Oil FVF	4	1	1
14	Oloso (2009)	Oil viscosity and Rs	12	1	2
15	Osman and Abdel-Aal (2002)	Pb and oil FVF at Rs	4	1	NA
16	Osman et al. (2001)	Pb and oil FVF at Pb	4	1	1
17	Shateri et al. (2015)	z-Factor	2	1	1

TABLE 10.1b Comparison Between Different ANN Models for Predicting PVT Properties

ID	Author	Training Algorithm	Transformation Function	Training and Testing Space	MAPE	MAE
1	Adeeyo (2016)	Back propagation	NA	500	NA	Pb: 7.6%
						Oil FVF: 0.001%
2	Ahmadi et al. (2015)	Back propagation (PSO)	NA	NA	NA	Pb: 5%
3	Al-Gathe et al. (2015)	ANFIS	Fuzzy logic	2500	Oil viscosity: 0.03%	Oil viscosity: 5.54%
4	Alimadadi (2011)	Back propagation	NA	138	NA	Oil FVF: 1.75%
						Oil density: 0.49%
5	Al-Marhoun and Osman (2002)	Back propagation	Sigmoid	283	NA	Pb: 5.8%
						Oil FVF: 0.52%
6	Al-Shammasi (2001)	Back propagation	Sigmoid	1709	Pb: 19.2%	Pb: 15.08%
7	Dutta and Gupta (2010)	Back propagation (GA)	Sigmoid	1177	NA	Oil viscosity: 10.89%
8	Elsharkawy (1998)	Back propagation (RBF)	Radial	90	Rs: 4.53%	Rs: 5.22%
					Oil FVF: 0.53%	Oil FVF: 0.86%
					Oil viscosity: 8.72%	Oil viscosity: 7.58%
					Oil density: 0.4%	Oil density: 0.56%
					Co: 5.98%	Co: 8.86%

(Continued)

TABLE 10.1b (Continued)

ID	Author	Training Algorithm	Transformation Function	Training and Testing Space	MAPE	MAE
9	Elsharkawy and Gharbi (2001)	Back propagation	Sigmoid	700	NA	Oil viscosity: 2.3%
10	Gharbi and Elsharkawy (1997)	Back propagation (MBP)	Sigmoid	520	Pb: 7% Oil FVF: 2.3%	Pb: 0.33% Oil FVF: 0.09%
11	González et al. (2003)	Back propagation	Sigmoid	802	Pd: 8.74%	NA
12	Habiballah et al. (1996)	Back propagation	Sigmoid	3000	K: 6%	NA
13	Ikiensikimama (2012)	Back propagation	Sigmoid	802	Oil FVF: 1.1%	NA
14	Oloso (2009)	Hybrid (FN-SVM)	Sigmoid	99	NA	Oil Viscosity: 10.24% Rs: 12.7%
15	Osman and Abdel-Aal (2002)	NA	NA	283	NA	Pb: 5.62% Oil FVF: 0.86%
16	Osman et al. (2001)	Back propagation	Sigmoid	803	NA	Pb: 1.79% Oil FVF: NA
17	Shateri et al. (2015)	Back propagation (WGRBFN)	Wilcoxon generalized radial	978	NA	z-Factor: 2.3%

ANFIS, adaptive network-based fuzzy inference system; PSO, particle swarm optimization; WGRBFN, Wilcoxon generalized radial basis function network.
FN-SVM, Fuzzy Network with Support Vector Machine; GA, Genetic Algorithm; MAE, Mean Absolute Error; MAPE, Mean Absolute Percent Error; MBP, Multiple-layer Back Propagation; RBF, Radial Basis Function; PSO, Particle Swarm Optimization

Hemmati-Sarapardeh et al. (2014) analyzed a large database and evaluated three of the most accurate correlations for each region for comparison with ANN models. They used dead oil viscosity, viscosity below bubble point, viscosity at bubble point, and undersaturated oil viscosity properties in their comparison study. They developed four ANN models for Iranian oil reservoirs.

Ahmadi et al. (2015) developed an ANN model for prediction of bubble point pressure using back propagation technique. They used particle swarm optimization (PSO) algorithm for error minimization in calculating the network weights. Their dataset from the literature included 123 records from Farasat et al. (2013).

Osman et al. (2001) presented an ANN model for prediction of bubble point pressure and formation volume factor at the bubble point. They used 803 published data points from the Middle East, Malaysia, Colombia, and Gulf of Mexico reservoirs. They used back propagation technique and one hidden layer of five nodes. The inputs were the four parameters routinely used for each network.

Al-Marhoun and Osman (2002) presented another study that included new models to predict bubble point pressure and formation volume factor at bubble point pressure. They used 293 data sets collected from Saudi reservoirs. The two ANN models used four inputs, one hidden layer with five nodes, and one output node for each model. They compared their ANN models with published PVT correlations and concluded that the ANN models were superior.

González et al. (2003) used neural network models to predict dew point pressures in retrograde gas reservoirs. Their model predicted dew point pressure with accuracy of 8.74%. Their input layer nodes included data for temperature, hydrocarbon, and nonhydrocarbon compositions, molecular weight, and specific gravity of heptane plus fraction. The study used back propagation technique and one hidden layer.

Oloso et al. (2009) used a new approach for predicting a curve of oil viscosity and solution gas–oil ratio. In this approach, PVT properties can be predicted over the entire pressure range (rather than at specified pressure points). The models were built by use of support vector regression and functional network with ANN. Smooth curves for the PVT properties were produced.

Al-Shammasi (2001) conducted a study on neural network models for estimation of bubble point pressure and oil formation volume factor at and below bubble point using global data sets. The usual four inputs and two hidden layers were used. Performance of ANN models was compared with published correlations for bubble point pressure and formation volume factor.

Alimadadi et al. (2011) used a committee machine type ANN to build two models for oil formation volume factor and oil density. The input parameters were component mole %, solution gas–oil ratio, bubble point pressure, reservoir pressure, API, and temperature. The number of input nodes to their ANN were therefore 24. They used one hidden layer.

Ikiensikimama and Azubuike (2012) used 802 data points collected from Niger Delta region of Nigeria to build an ANN model for oil formation volume factor. They used the usual four inputs in the nodes of the input layer and one hidden layer.

Shateri et al. (2015) used a dataset of 978 data points to develop an ANN model for gas compressibility factor. Their model was compared to nine empirical correlations and five equations of state. The model predicted the compressibility factor of natural gases with an average absolute percent relative error of 2.3%. Their network had two input nodes and one hidden layer. Back propagation technique was used with the Wilcoxon generalized radial basis function network (WGRBFN).

Adeeyo (2016) developed models for prediction of bubble point pressure and formation volume factor at bubble point pressure for Nigerian crude oils. The data sets consisted of 2114 bubble point pressure and 2024 oil formation volume factor values at bubble point pressure. One hidden layer was used. The number of neurons in the hidden layer was varied until stable results were obtained. Comparison was made of the ANN models with other published correlations, and the ANN models were concluded to be superior.

A variety of ANN models were developed for prediction of PVT properties for oils and gases from basic inputs. The models used data sets of different sizes, with the majority using the back propagation technique. In addition, the majority of the models developed have only one hidden layer, with several ANN models having two hidden layers.

ARTIFICIAL NEURAL NETWORK MODELS DESIGN

The basic ANN model structure includes an input layer with a number of input nodes (determined by the number of input variables). The ANN also contains an output layer with defined output nodes in addition to a user-defined arbitrary number of hidden layers (with each hidden layer consisting of an arbitrary number of nodes) determined by the problem size. The following problem illustrates typical calculations of an ANN model.

Problem 1—Simple Artificial Neural Network Calculations

Assume the basic ANN structure in Fig. 10.6 to be an example of a single-layer neural network with two input nodes, one hidden layer with two nodes, and two output nodes. The hidden layer is connected to a bias node (B1), and the output layer is connected to another bias node (B2). The initialization and output values for the ANN training are given in

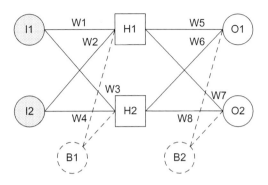

FIGURE 10.6 ANN structure for Problem 1. *ANN*, artificial neural network.

TABLE 10.2 Training Record for the ANN in Problem 1

Input Nodes	Output Nodes	Bias Nodes
I1: 0.03	O1: 0.01	B1: 0.3
I2: 0.20	O2: 0.99	B2: 0.7

TABLE 10.3 Weight Initialization Values for the ANN in Problem 1

Hidden Layer Weights		Output Layer Weights	
W1: 0.12	W3: 0.20	W5: 0.40	W7: 0.55
W2: 0.15	W4: 0.40	W6: 0.20	W8: 0.50

Table 10.2. Use learning rate of 0.6. Perform the calculations to train the ANN model.

Solution of Problem 1

The calculation process of the ANN consists of four steps: (1) to initialize the network weights (assign values for the weights connecting the input layer nodes to the hidden layer nodes and the weights connecting the hidden layer nodes to the output nodes); (2) to perform the feed-forward calculations; (3) to calculate the network error; and (4) to perform the feed-backward calculations. Steps 2−4 are repeated until the calculated error satisfies the network convergence criteria. Table 10.3 provides the initial weights for both the hidden layer and output layer nodes.

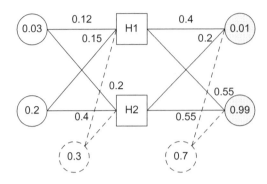

FIGURE 10.7 The example ANN structure with weight values after initialization. *ANN*, artificial neural network.

Artificial Neural Network Initialization Calculations

Fig. 10.7 shows the initial values for the eight weights in the ANN. The selection of the weights can be arbitrary or follow a certain logic. The initial weights assigned to this network were randomly selected.

Feed-Forward Calculations

The hidden layer node values are calculated using the total summation of the input node values multiplied by their assigned weights. This process is termed "transformation." The bias node with a weight of 1.0 is also added to the summation. The use of bias nodes is optional. Note that other techniques can be used to perform the transformation calculations; however, the weight sum technique is the most common. Eq. (10.1) shows the basic formula of the hidden node value determination through the total summation.

$$H1 = W1 \times I1 + W2 \times I2 + B1 \times 1 \tag{10.1}$$

$$H1 = 0.12 \times 0.03 + 0.15 \times 0.2 + 0.3 \times 1 = 0.334$$

Each node in the hidden layer will undergo the activation function calculations. In this example, a sigmoid S-shape function is used. Eq. (10.2) shows the sigmoid function form and the example calculation for node H1.

$$H1out = \frac{1}{1 + e^{-H1}} = \frac{1}{1 + e^{-0.334}} = 0.583 \tag{10.2}$$

With completion of the same calculations for node H2, the following values are obtained:

$$H2 = 0.386$$
$$H2out = 0.595$$

Similar calculations are performed for the output layers (using the hidden layer node values as inputs). Table 10.4 summarizes the results of the first iteration step for the output nodes (i.e., hidden and output).

TABLE 10.4 First Iteration Results

Hidden Node	Hidden After Activation	Calculated Output	Output After Activation	Output Nodes (Required)
H1: 0.334	H1out: 0.583	O1: 1.052	O1out: 0.741	O1: 0.01
H2: 0.386	H2out: 0.595	O2: 1.348	O2out: 0.794	O2: 0.99

Total Error Calculations

The objective function can have different forms. For this example, Eq. (10.3) is used to calculate the error for the first iteration.

$$E_{total} = \sum 0.5 \times (required - output)^2 \qquad (10.3)$$

For the first output node, the target value is 0.01 while the calculated value is 0.745. Therefore the error in output node (O1) is

$$Error\ O1 = \sum 0.5 \times (0.01 - 0.741)^2 = 0.267$$

The second output node (O2) error is

$$Error\ O2 = 0.019$$

The total error for the neural network is the sum of the output node errors. For the first iteration, the total error is

$$E_{total} = Error\ O1 + Error\ O2$$
$$E_{total} = 0.267 + 0.019 = 0.287 \qquad (10.4)$$

Feed-Backward Calculations

The objective of the feed-backward calculations is to update each weight in the network so that by minimization of the total error of the network as a whole, the network output approaches the required output. This commonly used technique of distributing the error over the network weights is known as the back propagation algorithm.

The back propagation algorithm calculates the variation of each weight value based on the effect of this weight on the total network error value. In other words, the new, adjusted weight is equal to the previous weight value minus the variation of the total error with respect to the weight value $\partial E_{total}/\partial w$.

Applying the chain rule, the change in total error with respect to the network weights is given by

$$\frac{\partial E_{\text{total}}}{\partial w} = \frac{\partial E_{\text{total}}}{\partial O} \times \frac{\partial O}{\partial N} \times \frac{\partial N}{\partial w} \qquad (10.5)$$

By calculation of the delta weight variation for the W5 (as an example), the three terms in Eq. (10.5) are evaluated as follows:

$$\frac{\partial E_{\text{total}}}{\partial O1} = \frac{\partial}{\partial O1}\left[0.5(\text{RequiredO1} - \text{OutputO1})^2 + 0.5(\text{TargetO2} - \text{OutputO2})^2\right] \qquad (10.6)$$

$$\frac{\partial E_{\text{total}}}{\partial O1} = 2 \times 0.5(\text{RequiredO1} - \text{OutputO1})^{2-1} \times (-1) + 0 \qquad (10.7)$$

It follows that

$$\frac{\partial E_{\text{total}}}{\partial O1} = -(\text{RequiredO1} - \text{Output O1}) = -(0.01 - 0.741) = 0.731 \qquad (10.8)$$

The second term represents the partial derivative for the output node with respect to its activation function. It is given by the following equation:

$$\frac{\partial O}{\partial N} = \frac{\partial}{\partial N}\left(\frac{1}{1 + e^{-O1}}\right) = O1 \times (1 - O1) \qquad (10.9)$$

$$\frac{\partial O}{\partial N} = 0.745 \times (1 - 0.741) = 0.192$$

The last term is to calculate the partial derivative of the output node with respect to the weight change. It is calculated as follows:

$$N = \text{W5} \times \text{H1out} + \text{W6} \times \text{H2out} + \text{B2} \times 1 \qquad (10.10)$$

$$\frac{\partial N}{\partial \text{W5}} = \text{H1out} + 0 + 0 = \text{H1out} \qquad (10.11)$$

Therefore

$$\frac{\partial N}{\partial \text{W5}} = \text{H1out} = 0.583$$

The error influenced by W5 is given by the multiplication of the three partial derivatives. With the use of Eq. (10.5), it follows that

$$\frac{\partial E_{\text{total}}}{\partial \text{W5}} = 0.731 \times 0.192 \times 0.583 = 0.082$$

To decrease the error between the required value and the calculated value from first iteration, the new weight (W5) should be decreased by this error variation (optionally multiplied by a learning rate). The learning rate for this example is assumed 0.6.

$$W_5^{new} = W5 - learn_{rate} \times \frac{\partial E_{total}}{\partial W5} \tag{10.12}$$

$$W_5^{new} = 0.4 - 0.6 \times 0.082 = 0.351$$

The process is continued in the same way to calculate the other weights for the same layer (i.e., W6, W7, and W8).

The same process of weight adjustment is repeated for the preceding layer (i.e., W1, W2, W3, and W4) by use of the same total error value and the values of the hidden layer nodes (i.e., H1out and H2out). Notice that the adjustments of weights between the hidden layer nodes and input layer nodes are affected by all output nodes errors. Therefore, the first term of Eq. (10.5) is calculated using Eq. (10.13) to consider the effect of all output nodes on these weights.

$$\frac{\partial E_{total}}{\partial O} = \frac{\partial E_{total}}{\partial O1} + \frac{\partial E_{total}}{\partial O2} \tag{10.13}$$

The eight weights after the first iteration are given in Table 10.5.

The new weights are used in another cycle of feed-forward calculations. The total error is calculated for the second iteration, and again redistributed by use of the feed-backward calculations. The process of feed-forward and feed-backward is repeated until the total error is minimized to a very small value predefined by the user.

After updating all the weights, the 0.03 and 0.2 original input node values with the original (initialized) weights gave a total network error of 0.286564. After adjustment of the weights in the first iteration, the 0.03 and 0.2 inputs gave a total error of 0.277983 in the feed-forward cycle. The difference between the calculated error after adjustment of the weights and the calculated error by use of the initial weights does not seem significant. However, after the process of feed-forward and feed-backward calculation is repeated 3,000 times (for example), the error reaches 1.0E-9 (which is significantly low). At this point, when the inputs of 0.03 and 0.2 are fed forward to the

TABLE 10.5 First Iteration Weight Values

W1	W2	W3	W4
0.120	0.150	0.200	0.400
W5	**W6**	**W7**	**W8**
0.351	0.151	0.561	0.561

TABLE 10.6 Final Weight Values

W1	W2	W3	W4
0.974	0.259	0.053	−0.197
W5	W6	W7	W8
−3.868	−5.295	5.933	0.655

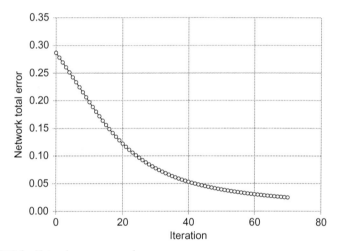

FIGURE 10.8 Network error progression.

ANN with the new weights (after 3,000 iterations), the two output nodes generate 0.01004482 (compared to 0.01 required value) and 0.99000002 (compared to 0.99 required value). The final weights (after 3000 iterations) are given in Table 10.6. The progression of the ANN error for the first 70 iterations is shown in Fig. 10.8.

ARTIFICIAL NEURAL NETWORK OPTIMIZATION

Several parameters can be used to optimize an ANN for the calculation of PVT properties. A key requirement is to have many valid data records with input and output values. The ANN parameters that can be optimized for a particular ANN include the ANN topography (structure or layout); number of hidden layers; number of nodes per hidden layer; layer connections; initialization; choice of transfer and activation functions; objective function; and training and running control. The following paragraphs summarize the use of these parameters.

Artificial Neural Network Topography

The different ANN topographies and structures are numerous. Some known network structures perform better in certain applications (e.g., prediction, optimization, clustering, noise removal, etc.). The most commonly used ANN structure for prediction problems (including PVT properties prediction) is the multilayer back propagation structure. The self-organizing map is one of the most common ANN structures for clustering applications. Clustering techniques can be used to develop expert systems to automatically select the best PVT correlations, for example.

Number of Hidden Layers

An ANN can contain any number of hidden layers. The number of hidden layers can range from zero to a very large number. When multiple hidden layers are used, the ANN is usually referred to as a deep network. The more hidden layers, the higher the capabilities of the network to discover the hidden relations between inputs and outputs. However, for multilayer networks, a sufficient number of training records should be available. Training time will also increase with the number of hidden layers and the number of nodes in each layer.

Nodes per Layer

The number of nodes in the input and output layers is usually determined by the number of inputs and outputs in the problem under investigation. The number of nodes in the hidden layer(s) can be arbitrary. However, the number of nodes in every hidden layer depends on the ANN application and on the size of the training data. Use of a different combination of hidden layers and variation of the number of nodes per layer can be investigated to optimize the network. It is usually recommended, however, to make the number of nodes per hidden layer a function of the number of nodes in the input layer. The optimum number of nodes in every hidden layer is a problem-specific. It is usually determined by trail and error, although some automatic algorithms have been recently used to find the optimum ANN structure. Availability of a large number of training records allows the choice of more nodes in the hidden layer(s) and usually produces networks that are more accurate.

Layer Connection

One of the ANN parameters that can be used to tune the ANN for a particular application is the type of layer/node connection. The three distinct types of connection are (1) full, (2) customized, and (3) recurrent. The type of connection defines which nodes are connected to one another in the calculation of connection weights. The selection among different connection layouts depends on the network application and configuration.

Weight Initialization

The first step of the ANN calculation is to initialize connection weights. Several techniques can be used for weight initialization. A straightforward way to initialize network weights is to assume them to be constant values. They can also be initialized using random values, or a booster algorithm. The booster algorithm initializes the weights through random forward calculations with random sets of the training and testing data. Weight initialization can speed up the convergence of the ANN during training.

Transformation Function

The transformation function is used to compute the value at a particular node from the values and weights of the nodes connected to it. Of the several types of transformation function, the most commonly used is the summation of the connected nodes times their weights. Another transformation function is to take the maximum (or minimum, or average, etc.) of the node values times the weights (for the nodes connected to the node to be calculated).

Activation Function

The activation function is used to convert the calculated hidden layer nodes values into different values through typical algebraic functions. Examples of activation functions include Cube, Elu, Hardsigmoid, Hardtanh, Identity, Leakyrelu, Rational-tanh, Relu, Rrelu, Sigmoid, Softmax, Softplus, Softsign, and Tanh. The objective of the activation function is to maintain the convergence of the ANN. The activation function can be applied to the output nodes as well. The most commonly used activation function in PVT prediction applications is the Sigmoid function.

Objective Function

The objective function represents the measure of difference between the actual output values of the training dataset and the output node values calculated by the ANN. Several types of error minimization functions are available. The mean squared error is the most commonly used objective function. Other objective functions include exponential log likelihood, cross entropy, multiclass cross entropy, negative log likelihood, root mean squared cross entropy, and many others. The selection of which objective function to use in a particular ANN is a function of the problem, number of output nodes, and output node internode dependency.

Optimization Algorithm

The learning process of the ANN requires thousands or sometimes millions of error minimization iterations. An optimization routine is used to find the weights of the ANN. The choice of the optimization algorithm depends on the nature of the objective function and the engineering problem or application. Examples of common optimization algorithms used in ANN include gradient directive methods, nonlinear least squared, genetic algorithms, likelihood statistical methods, and many others.

Training Control Parameters

During the training process, some parameters are used to minimize the training time as much as possible, with the same level of required error quality. The parameters are learning rate, weight decay factors, and node momentum. Learning rate is used to control the maximum change allowed for any of the node weights during the training process. The decay factor is used to control the minimum change of the weights. The momentum of the node is used to signify or designify the node value after or before activation.

Running Control Parameters

Running control parameters are parameters that can be used to terminate the training process. Running control parameters include maximum number of iteration, minimum required network error, and minimum tolerance in weight value changes. Running control parameters affect the time required for training and ANN prediction accuracy.

FUTURE DEVELOPMENT AND EXPECTATIONS

Although many ANN models have been developed to predict different PVT properties over the past several years, additional models are expected to appear in the literature. Other competing artificial intelligence (AI) and machine learning techniques (e.g., relevance vector machines, support vector machines, and random forests) are anticipated to receive growing attention in prediction of PVT properties. Like the correlation models developed by regression techniques, ANN models and others will be applicable within the ranges of the data that are used to develop them.

Improved models are expected to appear when more PVT data records are incorporated in the training of these models. The data must be quality controlled, and relations between input and output values must exist. The strength of the ANN models lies in their ability to find the relations between inputs and outputs. These relations may not be obvious or clear (i.e., not showing a trend).

ANN models consist of many components, with each component having several alternatives for performing calculations. These countless combinations and permutations offer investigators a multitude of optimization opportunities. In addition, the development of AI techniques is currently receiving considerable attention. Alternatives and options will continue to appear, and general advancements in AI are expected to yield improved models for PVT properties prediction.

The main disadvantage of ANN models remains the lack of available details for models developed. In general, investigators do not publish the ANN weights, and therefore the results cannot be reproduced. PVT correlations results, however, can be easily reproduced.

NOMENCLATURE

a   weighted summation
AI   artificial intelligence
b   input for the output node
B   bias node
c   output after the transformation function
C   threshold
F   activation function
G   transformation function
h   activated weighted summation
I   input node
K_i equilibrium constant for component (*i*)
O   output node
PSO   particle swarm optimization
SOM   self-organizing maps
W   weight value
X   input value
Y   output value

REFERENCES

Adeeyo, Y.A. 2016. Artificial neural network modelling of bubblepoint pressure and formation volume factor at bubblepoint pressure of Nigerian crude oil. In: Paper SPE-184378-MS Presented at the SPE Nigeria Annual International Conference and Exhibition, 2−4 August, Lagos, Nigeria. https://doi.org/10.2118/184378-MS.

Ahmadi, M.A., Pournik, M., Shadizadeh, S.R., 2015. Toward connectionist model for predicting bubble point pressure of crude oils: application of artificial intelligence. Petroleum 1, 307−317. Available from: https://doi.org/10.1016/j.petlm.2015.08.003.

Al-Gathe, A.A., Abd-El Fattah, K.A., El-Banbi, A.H., El-Metwally, K.A., 2015. A hybrid neuro-fuzzy approach for black oil viscosity prediction. Int. J. Innov. Appl. Stud. 13 (4), 946−957.

Alimadadi, F., Fakhri, A., Farooghi, D., Sadati, H., 2011. Using a committee machine with artificial neural networks to predict PVT properties of Iran crude oil. SPEREE 14. Available from: https://doi.org/10.2118/141165-PA.

Al-Marhoun, M.A. and Osman, E.A. 2002. Using artificial neural networks to develop new PVT correlations for Saudi crude oils. In: SPE-78592 Presented at the Abu Dhabi International Petroleum Exhibition and Conference, 13−16 October, Abu Dhabi, UAE. https://doi.org/10.2118/78592-MS.

Al-Shammasi, A.A., 2001. A review of bubblepoint pressure and oil formation volume factor correlations. Soc. Pet. Eng. . Available from: https://doi.org/10.2118/71302-PA.

Cranganu, C., Luchian, H., Breaban, M.E. (Eds.), 2015. Artificial Intelligent Approaches in Petroleum Geosciences. Springer International Publishing, Switzerland, ISBN: 978-3-319-16530-1.

Dutta, S., Gupta, J.P., 2010. PVT correlations for Indian crude using artificial neural networks. Pet. Sci. Eng. 72, 93−109.

Elsharkawy, A.M. 1998. Modelling the properties of crude oil and gas systems using RBF network. In: Paper SPE 49961 Presented at the SPE Asia Pacific Oil & Gas Conference, 12−14 October, Perth, Australia. https://doi.org/10.2118/49961-MS.

Elsharkawy, A.M., Gharbi, R.B.C., 2001. Comparing classical and neural regression techniques in modelling crude oil viscosity. Adv. Eng. Softw. 32 (3), 215−224.

Farasat, A., et al., 2013. Toward an intelligent approach for determination of saturation pressure of crude oil. Fuel Process. Technol. 115, 201−214.

Gharbi, R.B. and Elsharkawy, A.M. 1997. Neural network model for estimating the PVT properties of Middle East crude oils. In: Paper SPE 37695 Presented at the Middle East Oil Show and Conference, 15−18 March, Bahrain. https://doi.org/10.2118/37695-MS.

Gonzálcz, A., Barrufet, M.A., Startzman, R., 2003. Improved neural-network model predicts dewpoint pressure of retrograde gases. J. Pet. Sci. Technol. 37, 183−194.

Habiballah, W., Startzman, R.A., Barrufet, M.A., 1996. Use of neural networks for prediction of vapor/liquid equilibrium K values for light—hydrocarbon mixtures. SPE Res. Eng. J. 11, 121−126.

Hajizadeh, Y. 2007. Viscosity prediction of crude oils with genetic algorithms. In: Paper SPE-106763-MS Presented at the Latin American & Caribbean Petroleum Engineering Conference, 15−18 April, Buenos Aires, Argentina.

Hemmati-Sarapardeh, A., Shokrollahi, A., Tatar, A., Gharagheizi, F., Mohammadi, A.H., Naseri, A., 2014. Reservoir oil viscosity determination using a rigorous approach. Fuel 116, 39−48.

Ikiensikimama S.S. and Azubuike, I.I. 2012. Modeling approach for Niger-delta oil formation volume factor prediction using artificial neural network. In: SPE-162987 Presented at the Nigeria Annual International Conference and Exhibition, 6−8 August, Lagos, Nigeria. https://doi.org/10.2118/162987-MS.

Mohaghegh, S.D., Al-Fattah, S., Popa, A. (Eds.), 2011. Artificial Intelligence & Data Mining Applications in the E&P Industry. SPE Reprint Series, SPE, Richardson, TX, USA, ISBN: 978-1-61399-064-3.

Oloso, M., Khoukhi, A., Abdulraheem, A. and Elshafei, M. 2009. Prediction of crude oil viscosity and gas/oil ratio curves using recent advances to neural networks. In: Paper SPE-125360-MS Presented at the SPE/EAGE Reservoir Characterization and Simulation Conference, 19−21 October, Abu Dhabi, UAE. https://doi.org/10.2118/125360-MS.

Osman, E.A. and Abdel-Aal, R.E. 2002. Abductive networks: a new modeling tool for the oil and gas industry. In: SPE-77882-MS Presented at the SPE Asia Pacific Oil and Gas Conference and Exhibition, 8−10 October, Melbourne, Australia. https://doi.org/10.2118/77882-MS.

Osman, E.A., Abdel-Wahhab, O.A. and Al-Marhoun, M.A. 2001. Prediction of oil PVT properties using neural networks. In: SPE-68233-MS Presented at the SPE Middle East Oil Show, 17−20 March, Manama, Bahrain. https://doi.org/10.2118/68233-MS.

Shateri, M., Ghorbani, S., Hemmati-Sarapardeh, A., Mohammadi, A.H., 2015. Application of Wilcoxon generalized radial basis function network for prediction of natural gas compressibility factor. J. Taiwan Inst. Chem. Eng. 50, 131−141.

FURTHER READING

Esfahani, S., Baselizadeh, S., Hemmati-Sarapardeh, A., 2015. On determination of natural gas density: least square support vector machine modeling approach. J. Nat. Gas Sci. Eng. 22, 348−358.

Gholami, R., et al., 2014. Applications of artificial intelligence methods in prediction of permeability in hydrocarbon reservoirs. J. Pet. Sci. Eng. 122, 643−656.

Hemmati-Sarapardeh, A., Aminshahidy, B., Pajouhandeh, A., Yousefi, S.H., Hosseini-Kaldozakh, S.A., 2016. A soft computing approach for the determination of crude oil viscosity: light and intermediate crude oil systems. J. Taiwan Inst. Chem. Eng. 59, 1−10.

Mohagheghian, E., Zafarian-Rigaki, H., Motamedi-Ghahfarrokhi, Y., Hemmati-Sarapardeh, A., 2015. Using an artificial neural network to predict carbon dioxide compressibility factor at high pressure and temperature. Korean J. Chem. Eng. 32 (10), 2087−2096.

Osman, E.-S.A. and Al-Marhoun, M.A., 2005. Artificial neural networks models for predicting PVT properties of oil field brines. In: SPE-93765-MS Presented at the SPE Middle East Oil and Gas Show and Conference, 12−15 March, Bahrain. https://doi.org/10.2118/93765-MS.

PROBLEMS

10.1 Given the following ANN structure, compute the network output using unity for the weights as initial guess. For three iterations, calculate the new weight values associated with the ANN error at each iteration to match the given output. Assume the input nodes values to be: 0.1, 0.2, and 0.3, the bias node value is 0.5 and the required output node value is 1.0.

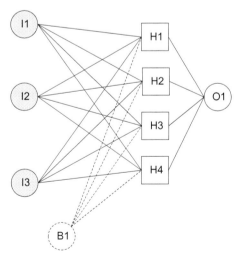

10.2 Given the following ANN structure, use the basic field input data given in the table below to compute the network weights to match the network output of the oil formation volume factor. The training data is given in the following table.

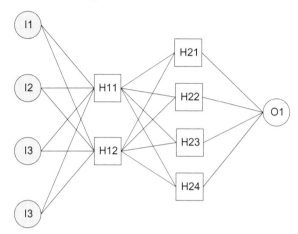

I1	I2	I3	I4	O1
API	Gas Sp. Gr.	Total GOR, scf/STB	Temperature, °F	Oil FVF, rbbl/STB
23	0.8	800	180	1.3
15	0.7	100	160	1.1
35	0.8	2900	240	2.1

Appendix A

Oil Correlations Formulae

This appendix contains a large number of oil correlations formulae. The correlations are presented in alphabetical order. The correlations use oil field system of units. Correlations for each PVT (pressure, volume, and temperature) property are grouped under the following names:

- Bubble point pressure (psia)
- Solution gas−oil ratio (scf/STB)
- Saturated oil formation volume factor (rbbl/STB)
- Saturated oil compressibility (psi^{-1})
- Saturated oil density (lbm/ft^3)
- Saturated oil viscosity (cp)
- Stock-tank gas−oil ratio (scf/STB)
- Under saturated oil formation volume factor (rbbl/STB)
- Under saturated oil compressibility (psi^{-1})
- Under saturated oil density (lbm/ft^3)
- Under saturated oil viscosity (cp)
- Dead oil viscosity (cp)
- Free gas specific gravity
- Total formation volume factor (rbbl/STB)
- Modified black oil (volatile oil)

BUBBLE POINT PRESSURE

1. Al-Marhoun (1985)

$$X = R_s^{0.722569} \frac{\gamma_o^{3.046590}}{\gamma_g^{1.879109}} (T + 459.67)^{1.302347}$$

$$p_b = -64.13891 + 7.202362 \times 10^{-3} X - 2.278475 \times 10^{-9} X^2$$

where $11309.1 \leq X \leq 1541298.4$

2. Al-Marhoun (1988)

$$p_b = \frac{5.38088 \times 10^{-3} R_s^{0.715082} \gamma_o^{3.1437} (T + 459.67)^{1.32657}}{\gamma_g^{1.87784}}$$

3. Almehaideb (1997)

$$p_b = -620.592 + 6.23087 R_s \frac{\gamma_{API}}{\gamma_g B_o^{1.38559}} + 2.89868 T$$

where

$$B_o = 1.1222018 + \frac{1.410 \times 10^{-6} R_s T}{\gamma_{API}^2}$$

4. Al-Najjar et al. (1988)

$$p_b = a_1 \left(\frac{R_s}{\gamma_g} \right)^{a_2} e^{\left[a_3 \left(\gamma_{API}/(T+459.67) \right) \right]}$$

Coefficient	$\gamma_{API} \leq 30$	$\gamma_{API} \geq 30$
A	7.920	30.910
B	1.025	0.816
C	−24.244	−19.748

5. Al-Shammasi (1999)

$$p_b = \frac{\gamma_o^{5.527215} \left[\gamma_g R_s (T + 459.67) \right]^{0.783716}}{e^{(1.841408 \gamma_g \gamma_o)}}$$

6. Asgarpour et al. (1989)

$$p_b = a_1 \gamma_g^{a_2} \gamma_{API}^{a_3} T^{a_4} \left(\frac{R_s}{5.614583} \right)^{a_5}$$

where T has units of °C

Coefficient	Cardium/Viing	D-2/Nisku	D-3/Leduc
a_1	70.9815	83.7883	193.770
a_2	−0.0101	−0.4114	0.0928
a_3	−0.2514	−0.2697	−1.1369
a_4	0.4593	0.2810	0.7899
a_5	0.5093	0.6259	0.6519

$$P_b = 27.16\left[\left(\frac{A}{B}\right) \times \left(\frac{C}{D}\right) - 30.28\right]$$

$$A = 3.4394 \times R_s^{0.57102}$$

$$B = 0.56807 \times \gamma_g^{0.922092}$$

$$C = 3.73877 \times T^{0.2304}$$

$$D = 6.27605 \times API^{0.42469}$$

7. Bolondarzadeh et al. (2006)

Extra heavy oil (°API ≤ 10)

$$p_b = 10.7025\left(\frac{R_s}{\gamma_g}\right)^{0.8986} 10^{(0.001567 - 0.00169\gamma_{API})}$$

Heavy oil (10 ≤ °API ≤ 22.3)

$$p_b = \left[\frac{56.434 R_s}{\gamma_{gc}\, 10^{10.9267(\gamma_{API}/(T+459.67))}}\right]^{0.8294}$$

Medium oil (22.3 ≤ °API ≤ 31.1)

$$p_b = \left[\frac{R_s}{0.10084\, \gamma_{gc}^{0.2556} 10^{[7.4576(\gamma_{API}/(T+459.67))]}}\right]^{1.0134}$$

8. De Ghetto et al. (1995)

Light oil (°API ≥ 31.1)

$$p_b = \left[\frac{R_s}{0.01347 \gamma_{gc}^{0.3873} 10^{[12.753(\gamma_{API}/(T+459.67))]}} \right]^{0.8536}$$

Agip model

$$p_b = \left[\frac{37.966 R_s}{\gamma_{gc} 10^{[9.441(\gamma_{API}/(T+459.67))]}} \right]^{0.8669}$$

Heavy oils

$$P_b = 15.7286 \left[\left(\frac{R_s}{\gamma_g} \right)^{0.7885} \times \frac{10^{0.0020 \times T}}{10^{0.0142 \times °API}} \right]$$

Medium oils

$$P_b = \left[\frac{R_{sf}}{0.09902 \gamma_{gcorr}^{0.2181} \times 10^{7.2153 \times °API/(T+460)}} \right]^{0.9997}$$

$$\gamma_{gcorr} = \gamma_g P_{sep} \left[1 + 0.1595 \times °API^{0.4078} \times T_{sep}^{-0.2466} \log \left(\frac{P_{sep}}{114.7} \right) \right]$$

Light oils

$$P_b = 31.7648 \left[\left(\frac{R_s}{\gamma_g} \right)^{0.7857} \times \frac{10^A}{10^B} \right]$$

$$A = 0.0009T, B = 0.0148 \times °API$$

Agip's sample

$$P_b = 21.4729 \left[\left(\frac{R_s}{\gamma_g} \right)^{0.7646} \times \frac{10^A}{10^B} \right]$$

$$A = 0.00119T, B = 0.0101 \times °API$$

9. Dindoruk and Christman (2001)

$$A = \frac{(1.42828e^{-10} \times T^{2.844591797} + (-6.74896 \times e^{-4}) \times API^{1.225226436})}{\left(0.03383304 + \left((2 \times R_s^{0.272945957})/\gamma_g^{-0.084226069}\right)\right)}$$

$$P_b = 1.869979257 \times \left(\frac{R_s^{1.221486524}}{\gamma_g^{1.370508349}} \times 10^A + 0.011688308\right)$$

10. Doklah and Osman (1992)

$$p_b = \frac{8363.876 R_s^{0.724047} \gamma_o^{0.107991}}{\gamma_g^{1.01049}(T + 459.67)^{0.952584}}$$

11. Elam (1957)

$$p_b = \frac{R_s^{0.702}}{\gamma_g^{0.514}} e^{(0.00348T - 0.0282\gamma_{API} + 3.58)}$$

12. El-Banbi et al. (2006)

$$P_{sat} = [A_0 \times R_{si}^{A1}] \times [(X+Y)^{A2}] \times [STO^{A3}] \times [Tr^{A4}]$$

$$X = \frac{SG1}{Psep1}$$

$$Y = \frac{SG2}{Psep2}$$

13. Elmabrouk et al. (2010)

$$p_b = R_{SP}^{0.683} P_{SP}^{0.18} \gamma_{oST}^{-4.98} T_R^{0.658}$$

14. Elsharkawy and Alikhan (1997)

For $\gamma_{API} > 30$

$$p_b = \left[\frac{R_s}{\gamma_g^{0.04439} \gamma_{API}^{1.1394} 10^{(8.392 \times 10^{-4}T - 2.188)}}\right]^{1.0551194}$$

For $\gamma_{API} \le 30$

$$p_b = \left[\frac{R_s}{\gamma_g 10^{(0.4636\gamma_{API}/T - 1.2179)}}\right]^{0.847271}$$

15. Farshad et al. (1996)

Model #1

$$p_b = 46.14 \left[\frac{R_s^{0.6343}}{\gamma_g^{1.15036} 10^{(7.97 \times 10^{-3} \gamma_{API} - 3.35 \times 10^{-4} T)}} - 7.2818 \right]$$

$a_1 = 33.22$

$a_2 = 0.8283$

$a_3 = 0.000037$

$a_4 = 0.0142$

Model #2

$$P_b = a_1 \left[\frac{R_s}{\gamma_g} \right]^{a2} 10^{(a3 \times T - a4 \times API)}$$

$a_1 = 0.3058$

$a_2 = 1.9013$

$a_3 = 0.26$

$a_4 = -1.378$

$a_5 = 1.053$

$a_6 = 0.00069$

$a_7 = 0.0208$

$$G = \gamma_g^{a_4} R_s^{a5} 10^{(a6 \times T - a7 \times API)}$$

$$P_b = 10^{(a_1 + a_2 \log G - a_3 (\log G)^2)}$$

16. Glaso (1980)

Nonvolatile oil

$$X = \left(\frac{R_s}{\gamma_g}\right)^{0.816} \left(\frac{T^{0.172}}{\gamma_{API}^{0.989}}\right)$$

Volatile oil

$$X = \left(\frac{R_s}{\gamma_g}\right)^{0.816} \left(\frac{T^{0.130}}{\gamma_{API}^{0.989}}\right)$$

$$p_b = 1.7669 + 1.7447\log X - 0.30218(\log X)^2$$

17. Hanafy et al. (1997)

$$K = Exp(0.00077T - 0.0097API - 0.4003\gamma_g)$$

$$P_b = 204.257K\left[R_s^{0.51} - 4.7927\right]$$

18. Hassan (2011)

$$p_b = 18.2\left[\left(\frac{R_s}{\gamma_g}\right)^{0.83} 10^{(0.00091T - 0.0125\gamma_{API})} + 2.2\right]$$

19. Hemmati and Kharrat (2007)

$$P_b = 10.4566 \times \left[\left(\frac{R_s}{\gamma_g}\right)^X \times \text{antilog}(0.0008 \times T - 0.0098 \times API) - 8.6817\right]$$

$$X = 1.5897 - (0.2735\beta_o) - (0.4429\gamma_g) + (0.04692\beta_o^2) + A$$

$$A = \left(0.1440 \times \gamma_g^2\right) - 0.1596 \times \left(\frac{1}{\gamma_g/\beta_o}\right)$$

20. Ikiensikimama and Ogboja (2009)

$$P_b = \frac{P^*(T + 635.4152349)}{\gamma_g}$$

$$P^* = 0.243181338 - (2.316548789Q) + 10.60657909 \times Q^{1.51803045}$$

$$S = (47.57094772 - 0.677706662API)^{1.53093561}$$

$$Q = \frac{N}{M}$$

$$N = \frac{R_s}{336.0064009}, M = \left(\frac{R_s}{336.0064009}\right) + \left(6.7063984 \times \frac{\gamma_o}{S}\right)$$

21. Kartoatmodjo and Schmidt (1991)

$$p_b = \left[\frac{R_s}{A \gamma_{gc}^B \, 10^{\left[C \left(\gamma_{API}/(T+459.67) \right) \right]}} \right]^D$$

$$\gamma_{gc} = \gamma_g \left[1 + 0.1595 \frac{\gamma_{API}^{0.4078}}{T_{sp}^{0.2466}} \log\left(\frac{p_{sp}}{114.7} \right) \right]$$

Coefficient	$\gamma_{API} \leq 30$	$\gamma_{API} \geq 30$
A	0.05959	0.03150
B	0.7972	0.7587
C	13.1405	11.289
D	0.998602	0.914328

22. Kartoatmodjo and Schmidt (1994)

$API \leq 30$

$$P_b = \left(\frac{R_s}{\left(a1 \gamma_{g100}^{a2} 10^{\left(a3^\circ API/(T+460) \right)} \right)} \right)^{a4}$$

$a_1 = 0.05958 \qquad a_2 = 0.7972$
$a_3 = 13.1405 \qquad a_4 = 0.9986$

$API > 30$

$$P_b = \left(\frac{R_s}{\left(a1 \gamma_{g100}^{a2} 10^{\left(a3^\circ API/(T+460) \right)} \right)} \right)^{a4}$$

$a_1 = 0.0315 \qquad a_2 = 0.7587$
$a_3 = 11.2895 \qquad a_4 = 0.9143$

$$\gamma_{g100} = \gamma_{gsep.} \left(1 + 0.1595^\circ API^{0.4078} * T_{sep.}^{-0.2466} \log\left(\frac{P_{sep.}}{114.7} \right) \right)$$

23. Khairy et al. (1998)

$$p_b = 49.3647 \frac{R_s^{0.5774} T^{0.6641}}{\gamma_g^{1.4676} \gamma_{API}^{1.0305}}$$

24. Khamehchi and Ebrahimian (2009)

$$P_b = 107.93 R_s^{0.9129} \gamma_g^{-0.666} T^{0.2122} API^{-1.08}$$

25. Labedi (1982)

$$p_b = \frac{6.0001}{\gamma_{gsp}} \left[\frac{R_s^{0.6714} (T/\gamma_{API})^{0.7097} T_{sp}^{0.08929}}{10^{(7.995 \times 10^{-5} R_s)}} \right]$$

26. Labedi (1990)

$$p_b = 21.38 \left(\left(\frac{R_s}{\gamma_g} \right)^{0.83} \times 10^{(0.00091 T - 0.0125^\circ API)} \right)^{0.9653}$$

27. Lasater (1958)

$$p_b = \frac{p_f(T + 459.67)}{\gamma_g}$$

$$x_g = \left[1 + \frac{\gamma_o}{7.521 \times 10^{-6} R_s M_o} \right]^{-1}$$

$$p_f = e^{((x_g - 0.15649)/0.33705)} - 0.59162$$

$$p_b = \frac{p_f(T + 459.67)}{y_g}$$

$$P_f = 0.3841 - 1.2008 \gamma_g + 9.648 \gamma_g^2$$

$$M_o = 725.321 - 16.0333 \times {}^\circ API + 0.09524 \times {}^\circ API^2$$

$$y_g = \frac{\left(\frac{R_s}{379.5} \right)}{\left(\frac{R_s}{350} + \frac{350}{M_o} \right)}$$

28. Levitan and Murtha (1999)

$$p_b = 14.7 \left(\frac{R_s}{\gamma_g} \right)^{0.85} \gamma_o^5 \left(\frac{T + 459.67}{519.67} \right)^{1.5}$$

29. Macary and El-Batanoney (1992)

$$p_b = 204.257 e^{(7.7 \times 10^{-4}T - 9.7 \times 10^{-3}\gamma_{API} - 0.4003\gamma_g)} \left(R_s^{0.51} - 4.7927\right)$$

30. Mazandarani and Asghari (2007)

$$P_b = 1.09373 \times 10^{-4} \times R_s^{0.5502} \times \gamma_g^{-1.71956} \times \gamma_o^{2.5486} \times (T+460)^{2.0967}$$

31. McCain (1991)

$$\ln P_b = 7.475 + 0.713Z + 0.0075Z^2$$

$$Z = \sum_{n=1}^{4} Z_n$$

$$Z_1 = -5.48 - 0.0378\ln R_{sb} + 0.281(\ln R_{sb})^2 - 0.0206(\ln R_{sb})^3$$

$$Z_2 = 1.27 - 0.0449\gamma_{API} + 4.36 \times 10^{-4}(\gamma_{API})^2 - 4.76 \times 10^{-6}(\gamma_{API})^3$$

$$Z_3 = 4.51 - 10.84\gamma_{gsp} + 8.39\left(\gamma_{gsp}\right)^2 - 2.34\left(\gamma_{gsp}\right)^3$$

$$Z_4 = -0.7835 + 6.23 \times 10^{-3}T_F - 1.22 \times 10^{-5}(T_F)^2 + 1.03 \times 10^{-8}(T_F)^3$$

32. Mehran et al. (2006)

$$P_b = 3.146 \times R_s^{0.8035} \times \gamma_g^{-1.3114} \times \gamma_o^{3.3925} \times T^{0.3466}$$

33. Moradi et al. (2010)

$$P_b = -65.853149M + 0.00040668902M^2 - 0.00000015472455M^3$$

$$M = 1.1038\log(API) * \left(\frac{141.5}{\gamma_o} - 131.5\right)^{1.1038} * B$$

$$B = exp\left(-1.8406 \times \gamma_g \times \gamma_o\right) * \left(R_s^* (T+460) * \gamma_{lg}\right)^{0.68875}$$

34. Movagharnejad and Fasih (1999)

$$P_b = 86.233589 + 0.576808 \times 10^{-5} \times C - 3.1535 \times 10^{-14} \times X^2$$

$$X = R_s^{1.084269} \times \gamma_g^{0.688056} \times \gamma_o^{-0.765396} \times T^{1.632212}$$

35. Obomanu and Okpobiri (1987)

$$p_b = \left(\frac{R_s T^{0.497} 10^{0.811}}{1.01136371\gamma_g^{2.15}\gamma_{API}^{1.27}}\right)^{1.0787}$$

36. Okoduwa and Ikiensikimama (2010)

API $\leq$ 21 (heavy oil)

$$P_b = 4.58925593 R_s^{0.92334377} \gamma_g^{-2.5394869} \gamma_o^{5.68277772} T^{0.01059824}$$

21 $\leq$ API $\leq$ 26 (medium oil)

$$P_b = 10.6356181 R_s^{1.01965369} \gamma_g^{-0.32111265} \gamma_o^{6.86132951} T^{0.00182483}$$

26 $\leq$ API $\leq$ 35 (blend oil)

$$P_b = 0.00007735 R_s^{0.95071520} \gamma_g^{-1.61775550} \gamma_o^{1.81805246} T^{1.72440768}$$

35 $\leq$ API $\leq$ 45 (light oil)

$$P_b = \frac{P^*(T + 635.415989)}{\gamma_g}$$

$$P^* = 0.32747598 - (2.26538201 Q) + 10.6528063 \times Q^{1.54502947}$$

$$S = (47.5717698 - 0.68631461 API)^{1.52728352}$$

$$Q = \frac{N}{M}$$

$$N = \frac{R_s}{336.006423}, M = \left(\frac{R_s}{336.0064009}\right) + \left(306.706423 \times \frac{\gamma_o}{S}\right)$$

API $\geq$ 45 (very light oil)

$$P_b = 0.12357843 R_s^{0.69387390} \gamma_g^{-1.56602923} \gamma_o^{9.49138112} T^{1.07875303}$$

37. Omar and Todd (1993)

$$X = 1.4256 - 0.2608 B_o - 0.4596 \gamma_g + 0.04481 B_o^2 + 0.2360 \gamma_g^2 - \frac{01.077}{\gamma_g B_o}$$

$$p_b = 18.2 \left[\left(\frac{R_s}{\gamma_g}\right)^X 10^{(0.000917 T - 0.0125 API)} + 1.4\right]$$

38. Owolabi (1984)

$$p_b = 55.0 + 0.8643 \left[\left(\frac{R_s}{\gamma_g}\right)^{1.255} \frac{T^{0.172}}{\gamma_{API}^{0.178}}\right]$$

$$p_b = -987.56359 + 179.58816 \left[\left(\frac{R_s}{\gamma_g}\right)^{0.48088266} \frac{T^{0.093538150}}{\gamma_{API}^{0.01648326}}\right]$$

39. Petrosky (1990)

$$p_b = 112.727 \left[\frac{R_s^{0.57421}}{\gamma_g^{0.8439} 10^{\left(7.916 \times 10^{-4} \gamma_{API}^{1.5410} - 4.561 \times 10^{-5} T^{1.3911}\right)}} - 12.340 \right]$$

40. Petrosky and Farshad (1993)

$$P_b = 112.727 \times \left[\left(\left(\frac{R_s^{0.5774}}{\gamma_g^{0.8439}} \right) \times 10^A \right) - 12.340 \right]$$

$$A = 4.561 \times 10^{-5} \times T^{1.3911} - 7.916 \times 10^{-4} \times {}^\circ API^{1.5410}$$

41. Petrosky and Farshad (1998)

$$P_b = 112.727 \times \left[\frac{R_s^{0.57421}}{\gamma_g^{0.8439} \times 10^{\left(7.916 \times 10^{-4} \times API^{1.5410} - 4.561 \times 10^{-5} \times (T-460)^{1.3911}\right)}} - 12.340 \right]$$

42. Standing (1947)

$$p_b = 18.2 \left[\left(\frac{R_s}{\gamma_g} \right)^{0.83} 10^{\left(0.00091T - 0.0125\gamma_{API}\right)} - 1.4 \right]$$

$$P_b = 18.2 \left[\left(\frac{R_s}{\gamma_g} \right)^{0.83} (10)^a - 1.4 \right]$$

$$a = 0.00091(T_R - 460) - 0.0125(API)$$

43. Vasquez and Beggs (1980)

$$p_b = \left[A \left(\frac{R_s}{\gamma_{gc}} \right) 10^{\left(\frac{B \times API}{T + 459.67} \right)} \right]^C$$

$$\gamma_{gc} = \gamma_g \left[1 + (0.5912 \times 10^{-4}) \gamma_{API} T_{sp} \log \left(\frac{p_{sp}}{114.7} \right) \right]$$

Coefficient	$\gamma_{API} \leq 30$	$\gamma_{API} > 30$
A	27.64	56.06
B	−11.172	−10.393
C	0.9143	0.8425

44. Velarde et al. (1997)

$$p_b = 1091.47\left[R_s^{0.081465}10^{\left(\frac{0.013098T^{0.282372}-8.2\times10^{-6}\cdot^{2.176124}_{API}}{\gamma_g^{0.161488}}\right)}-0.740152\right]^{5.354891}$$

$$P_b = 1091.47\left[R_{sb}^{0.081465}\gamma_{gs}^{-0.161488}10^X-0.740152\right]^{5.354891}$$

$$X = (0.013098T^{0.282372}) - (8.2\times10^{-6}\times API^{2.176124})$$

$$A = 9.73\times10^{-7}\gamma_g^{1.1672608}\gamma_{API}^{0.929870}T^{0.247235}p_b^{1.056052}$$

$$B = 0.022339\gamma_g^{-1.004750}\gamma_{API}^{0.337711}T^{0.132795}p_b^{0.302065}$$

$$C = 0.725167\gamma_g^{-1.485480}\gamma_{API}^{-0.164741}T^{-0.091330}p_b^{0.047094}$$

$$p_r = \frac{p}{p_b}$$

$$R_s = R_{sb}\left[Ap_r^B+(1-A)p_r^C\right]$$

45. Boukadi et al. (2004)

$$\log(6.894757\times P_b) = -172.29 + A + B + C + D - E + F - G - H - I - J$$

$$A = 148.41\times\log(R_s\times0.1801175)$$

$$B = 404.22\times\log(\gamma_o)$$

$$C = 968.94\times\log(\gamma_g)$$

$$D = 30.24\times\log\left(\frac{T-32}{1.8}\right)$$

$$E = 1.66\times\left[\log(R_s\times0.1801175)\times\log(\gamma_o)\right]$$

$$F = 3.06\times\left[\log(R_s\times0.1801175)\times\log(\gamma_g)\right]$$

$$G = 25.28\times\left[\log(R_s\times0.1801175)\times\log(\gamma_g)\times\log\left(\frac{T-32}{1.8}\right)\right]$$

$$H = 17.14 \times \log(\gamma_o) \times \log(\gamma_g)$$

$$I = 69.21 \times \log(\gamma_o) \times \log\left(\frac{T-32}{1.8}\right)$$

$$J = 168.71 \times \log(\gamma_g) \times \log\left(\frac{T-32}{1.8}\right)$$

SOLUTION GAS–OIL RATIO

1. Al-Marhoun (1988)

2. Al-Shammasi (1999)

$$R_s = 1.4903 \times 10^3 \times \gamma_g^{2.6260} \times P_b^{1.3984} \times API^{-4.3963} \times (T+460)^{-1.86}$$

$$R_s = \left[\left(\frac{\beta_o - 1 - 0.000650\left(\frac{T-60}{\gamma_o}\right)}{0.000412}\right)\right] \times \gamma_o$$

3. Casey and Cronquist (1992)

$$R_s = (1 - R_{pD})R_{sb}$$

$$P_D = \frac{P - 14.7}{P_b - 14.7}$$

$$R_{pD} = a_1 + a_2 \times P_D + a_3 \times P_D^2 + a_4 \times P_D^3 + a_5 \times P_D^4 + a_6 \times P_D^5$$

$a_1 = 0.99632$ $a_3 = 1.7964$ $a_5 = 4.3031$

$a_2 = -1.3078$ $a_4 = -4.1124$ $a_6 = -1.6743$

4. De Ghetto et al. (1995)

Extra heavy oil

$$R_s = \gamma_g \left[\left(\frac{P_b}{10.7025}\right) \times 10^A\right]^{1.1128}$$
$$A = 0.01694 \times API - 0.00156 \times T$$

Heavy oil

$$R_s = \left(\frac{P_b^{1.2057} \gamma_{gcorr}}{56.434}\right) \times 10^{\left[\frac{10.9267 \times °API}{T+460}\right]}$$

$$\gamma_{gcorr} = \gamma_g \gamma_{sep} \times \left[1 + 0.5912 \times API \times T_{sep} \times \log\left(\frac{P_{sep}}{114.7}\right) \times 10^{-4}\right]$$

Medium oils

$$R_s = 0.1008 \gamma_{gcorr}^{0.2556} \times P_b^{0.9868} \times 10^{[7.4576 \times \,^{\circ}API)/(T+460)]}$$

$$\gamma_{gcorr} = \gamma_g P_{sep} \left[1 + 0.1595 \times \,^{\circ}API^{0.4078} \times T_{sep}^{-0.2466} \log\left(\frac{P_{sep}}{114.7}\right)\right]$$

Light oils

$$R_s = 0.01347 \gamma_{gcorr}^{0.3873} \times P_b^{1.1715} \times 10^{[12.753 \times \,^{\circ}API)/(T+460)]}$$

$$\gamma_{gcorr} = \gamma_g P_{sep} \left[1 + 0.1595 \times \,^{\circ}API^{0.4078} \times T_{sep}^{-0.2466} \log\left(\frac{P_{sep}}{114.7}\right)\right]$$

Agip's model

$$R_s = \left(\frac{P_b^{1.1535} \gamma_{gcorr}}{56.434}\right) \times 10^{[9.441 \times API)/(T+460)]}$$

$$\gamma_{gcorr} = \gamma_g \gamma_{sep} \times \left[1 + 0.5912 \times API \times T_{sep} \times \log\left(\frac{P_{sep}}{114.7}\right) \times 10^{-4}\right]$$

$$A = \frac{\left(a_1 \times \gamma_g^{a_2} + a_3 \times T^{a_4}\right)}{\left(a_5 + \left((2 \times API^{a_6})/P_b^{a_7}\right)\right)^2}$$

$$Rs = \left[\left(\frac{P_b}{a_8} + a_9\right) \times \gamma_g^{a_{10}} \times 10^A\right]^{a_{11}}$$

a_1	4.86996×10^{-6}
a_2	5.730982539
a_3	9.92510×10^{-3}
a_4	1.776179364

5. Dindoruk and Christman (2001)

a_5	44.2500268
a_6	2.702889206
a_7	0.744335673
a_8	3.359754970
a_9	28.10133245
a_{10}	1.579050160
a_{10}	0.928131344

6. Elsharkawy and Alikhan (1997)

API ≤ 30

$$R_s = \gamma_g P_b^{1.18026} \times 10^{-1.2179+0.4636(API/T)}$$

API ≥ 30

$$R_s = \gamma_g^{0.04439} P_b^{0.94776} \times °API^{1.1394} \times 10^{-2.188+0.0008392T}$$

7. Farshad et al. (1996)

$$R_s = \frac{a \times \gamma_{gc} \times P_b^{a_2} \times 10^{(a_3 \times API + a_4 \times T)}}{1 + a_5(\gamma_o/T_{sep})}$$

$a_1 = 0.01456$	$a_3 = 0.1714$	$a_5 = -24.663$
$a_2 = 1.2073$	$a_4 = -0.0000446$	

8. Glaso (1980)

$$\gamma_c = \gamma_{gsep} + 15.5727\left(\frac{\gamma_o}{T_{sep}}\right) \times \log\left(\frac{P_{sep}}{114.7}\right)$$

$$R_s = \gamma_g \times \left[\frac{API^{0.989}}{(T-460)^{0.172}} P_b^*\right]^{1.225}$$

$$P_b^* = 10^x$$

$$x = 2.8869 - [14.1811 - 3.3093 \times \log p]^{0.5}$$

	P_b is unknown	P_b is known

9. Hanafy et al. (1997)

P_b is unknown

$R_{sfi} = 69 + 1.071 \times R_{sf1}$

$R_{si} = 23.94 + 1.101 \times R_{sfi}$

P_b is known

$R_{si} = -49.069 + 0.312 \times P_b$

10. Hassan (2011)

$R_s = 0.0006 P_b^{0.856} \gamma_g^{0.531} T^{1.829\circ} API^{1.462} P^A R_s^{-2.116}$

11. Hemmati and Kharrat (2007)

$R_s = 3.867 P_b^{0.306} \gamma_g^{-0.0831} T^{-0.306\circ} API^{-0.288} R_s^{0.525}$

12. Kartoatmodjo and Schmidt (1994)

$R_s = \left[0.1769 \gamma_g^{1.0674} \gamma_o^{-5.0956} T^{-0.1294} P\right]^{1.0857}$

$R_s = a_1 \times \gamma_{g100}^{a_2} \times P_b^{a_3} \times 10^{a_4 \times \left(API/(T+460)\right)}$

$\gamma_{g100} = \gamma_{gsep} \left(1 + 0.1595 \times API^{0.4078} \times T_{sep}^{-0.2466} \times \log\left(\frac{P_{sep}}{114.7}\right)\right)$

API > 30	API ≤ 30
$a_1 = 0.0315$	$a_1 = 0.05958$
$a_2 = 0.7587$	$a_2 = 0.7972$
$a_3 = 1.0937$	$a_3 = 1.0014$
$a_4 = 11.2895$	$a_4 = 13.1405$

13. Khamehchi and Ebrahimian (2009)

$R_s = 0.0103 P_b^{1.014} \gamma_g^{0.719} T^{-0.223} API^{1.182}$

14. Lasater (1958)

$R_s = \dfrac{\gamma_o \times 132,775 \times \gamma_o}{M_o(1 - \gamma_g)}$

$M_o = a_1 - a_2 \times API + a_3 \times API^2$

a_1	725.32143
a_2	16.03333
a_3	0.09524

15. Macary and El-Batanoney (1992)

$$R_s = \left[0.0049\left(\frac{P}{K}\right) + 4.7927\right]^{1.9606}$$

$$K = \exp(a_1 \times T - a_2 \times {}^\circ API - a_3 \times \gamma_g)$$

$a_1 = 0.00077 \quad a_2 = 0.0097 \quad a_3 = 0.4003$

16. Mazandarani and Asghari (2007)

$$R_s = 994.3718\gamma_g^{2.113367} P^{1.4558}\gamma_o^{-5.48944}(T+460)^{-1.90488}$$

17. Petrosky and Farshad (1993)

$$R_s = \left[\left(\frac{P}{112.727}+12.340\right) \times \gamma_g^{0.8439} \times 10^x\right]^{1.73184}$$

$$x = 7.916 \times 10^{-4} \times API^{1.5410} - 4.561 \times 10^{-5} \times (T-460)^{1.3911}$$

18. Standing (1947)

$$R_s = 30.7343 \times 10^{-3} \times \gamma_g \times P_b^{1.2048} e^{(-2.5245 \times 10^{-3} \times T + 34.677 \times 10^{-3} \times API)}$$

19. Standing (1981)

$$R_s = \gamma_g \times \left[\left(\frac{\rho}{18.2}+1.4\right) \times 10^A\right]^{1.2048}$$

$$A = 0.0125\,{}^\circ API - 0.00091\,T$$

20. Vasquez and Beggs (1980)

$$R_s = a_1 \gamma_g P_b^{a_2} e^{((a_3 \times API)/(T+460))}$$

API $\leq$ 30	API > 30
$a_1 = 0.0362$	$a_1 = 0.0178$
$a_2 = 1.0937$	$a_2 = 1.1870$
$a_3 = 25.7240$	$a_3 = 23.9310$

21. Velarde et al. (1997)

$$R_s = \left[\left(\left(\frac{P_b}{1091.47}\right)^{0.18675} + 0.740152\right)/A\right]$$

$$A = \gamma_{gs}^{-0.161488} \times 10^X$$

$$X = \left[0.013098 \times T^{0.282372}\right] - B$$

$$B = \left[8.2 \times 10^{-6} \times {}^\circ API^{2.176124}\right]$$

$$R_s = R_{sb} \times R_{sr}$$

$$R_{sr} = a_1 \times P_r^{a_2} + (1 - a_1) \times P_r^{a_3}$$

$$P_r = \frac{P - 14.7}{(P_b - 14.7)}$$

$$a_1 = A_0 \times \gamma_g^{A_1} \times {}^\circ API^{A_2} \times T^{A_3} \times (P_b - 14.7)^{A_4}$$

$$a_2 = B_0 \times \gamma_g^{B_1} \times {}^\circ API^{B_2} \times T^{B_3} \times (P_b - 14.7)^{B_4}$$
$$a_3 = C_0 \times \gamma_g^{C_1} \times {}^\circ API^{C_2} \times T^{C_3} \times (P_b - 14.7)^{C_4}$$

$A_0 = 9.73 \times 10^{-7}$	$B_0 = 0.022339$	$C_0 = 0.725167$
$A_1 = 1.672608$	$B_1 = -1.004750$	$C_1 = -1.485480$
$A_2 = 0.929870$	$B_2 = 0.337711$	$C_2 = -0.164741$
$A_3 = 0.247235$	$B_3 = 0.132795$	$C_3 = -0.091330$
$A_4 = 1.056052$	$B_4 = 0.302065$	$C_4 = 0.047094$

SATURATED OIL FORMATION VOLUME FACTOR

1. Abdul-Majeed and Salman (1988)

$$X = \frac{R_s^{1.20}}{\gamma_g^{0.1470} \gamma_o^{5.2220}}$$

$$B_{ob} = 0.9657876 + 4.8141 \times 10^{-5} X$$
$$- 6.89870 \times 10^{-10} X^2 + 7.73 \times 10^{-4} T$$

2. Ahmed (1989)

$$B_{ob} = -0.12869353 + R_s^{0.023484894} \frac{\gamma_{API}^{0.015966573}}{\gamma_g^{0.021946351}} - 4.5243973$$

$$\times 10^{-4} T + 3.9063637 \times 10^{-6} T^2 - \frac{5.5542509}{T}$$

$$- 5.7603220 \times 10^{-6} p - 3.9528992 \times 10^{-9} p^2$$

$$+ \frac{16.289473}{p} + 3.8718887 \times 10^{-4} R_s + 7.0703685$$

$$\times 10^{-8} R_s^2 - \frac{1.4358395}{R_s}$$

3. Al-Marhoun (1985)

$$X = R_s^{0.501538} \gamma_{g}^{-0.145526} \gamma_{o}^{-5.220726}$$

$$B_{ob} = 0.574095 + 7.723532$$

$$\times 10^{-4} (T + 459.67) + 2.454005 \times 10^{-3} X + 3.727676 \times 10^{-5} X^2$$

4. Al-Marhoun (1988)

$$X = R_S^{0.742390} \gamma_g^{0.323294} + \gamma_o^{-1.202040}$$

$$B_{ob} = 0.497069 + 8.62963$$

$$\times 10^{-4} (T + 459.67) + 1.82594 \times 10^{-3} X + 3.18099 \times 10^{-6} X^2$$

5. Al-Marhoun (1992)

$$B_{ob} = 1.0 + 1.7334210 \times 10^{-4} R_s + 2.20163 \times 10^{-4} R_s \frac{\gamma_g}{\gamma_o} + 4.29258$$

$$\times 10^{-6} R_s (T - 60)(1 - \gamma_o) + 5.28707 \times 10^{-4} (T - 60)$$

6. Al-Marhoun (2003)

$$a = 0.742390$$

$$b = 0.323294$$

$$c = -1.202040$$

$$f = R_s^a \times \gamma_g^b \times \rho_o^c$$

$$B_o = 0.497069 + 0.000862963 \times T + 0.00182594 \times f$$

$$+ 0.0000318099 \times f^2$$

7. Almehaideb (1997)

$$B_{ob} = 1.22018 + \frac{1.41 \times 10^{-6} R_s T}{\gamma_o^2}$$

8. Al-Najjar et al. (1988)

$$B_{ob} = 0.96325 + 4.9 \times 10^{-4} \left[R_s \left(\frac{\gamma_g}{\gamma_o}\right)^{0.5} + 1.25T \right]$$

9. Al-Shammasi (1999)

$$B_{ob} = 1 + 5.53 \times 10^{-7} R_s(T - 60) + \frac{1.81 \times 10^{-4} R_s}{\gamma_o}$$
$$+ \frac{4.49 \times 10^{-4}(T - 60)}{\gamma_o} + \frac{2.06 \times 10^{-4} R_s \gamma_g}{\gamma_o}$$

10. Arps (1962)

$$B_o = 1.05 + 0.0005 \times R_s$$

11. Asgarpour et al. (1989)

$$X = a_1 \gamma_g^{a2} \gamma_{API}^{a3} T^{a4} \left(\frac{R_s}{5.614583}\right)^{a5}$$

where T has units of °C

$$B_{ob} = a_6 + a_7 X + a_8 X^2 + a_9 X^3$$

Coefficient	Cardium/Viking	D-2/Nisku	D-3/Leduc
a_1	0.1203	0.251	0.1941
a_2	0.0645	0.0724	0.0136
a_3	02452	2.75×10^{-3}	0.0912
a_4	0.1118	0.1538	0.159
a_5	0.2321	0.2235	0.211
a_6	1.3420	−2.3211	0.8603
a_7	−0.5811	7.039	0.7341
a_8	0.1830	−5.0060	−0.9378
a_9	0.1656	1.330	0.4686

12. Bolondarzadeh et al. (2006)

$$B_o = a_1 + a_2\left((a_3 R_s^{a_4}) \times a_5 \times \gamma_g \times a_6 \times a_7 \times \text{API} \times a_8\right)$$
$$+ a_9 \times T \times a_{10} \times P_b \times a_{11}$$

$a_1 = 0.930471$

$a_2 = -0.973481$

$a_3 = 0.080264$

$a_4 = 1.140597$

$a_5 = -0.01037$

$a_6 = 0.146902$

$a_7 = 5.59574$

$a_8 = -0.05807$

$a_9 = -0.02428$

$a_{10} = 8.291315$

$a_{11} = 0.140489$

13. Casey and Cronquist (1992)

$$B_o = B_{ob} - (B_{ob} - B_{oa}) \times B_{oD}$$

$$B_{oD} = a_1 + a_2 \times P_D + a_3 \times P_D^2 + a_4 P_D^3 + a_5 P_D^4 + a_6 P_D^5$$

$$B_{oa} = 1 + \frac{(b_1 + b_2 \times T + b_3 \times T^2)}{100}$$

$$B_{ob} = B_{oa} \times (a_7 + a_8 \times R_{sb} + a_9 \times R_{sb}^2 + a_{10} \times R_{sb}^3)$$

$$b_1 = a_{11} + a_{12} \times \text{API} + a_{13} \times \text{API}^2$$

$$b_2 = a_{14} + a_{15} \times {}^\circ\text{API} + a_{16} \times {}^\circ\text{API}^2$$

$$b_3 = a_{17} + a_{18} \times {}^\circ\text{API} + a_{19} \times \text{API}^2$$

P_D = The diemsntionless pressure, $P_D = \dfrac{P - 14.7}{P_D - 14.7}$

At bubble point pressure $P_D = 1$

$a_1 = 0.98949$	$a_8 = 4.340923 \times 10^{-4}$	$a_{15} = 6.2049 \times 10^{-5}$
$a_2 = -1.8061$	$a_9 = 6.960178 \times 10^{-8}$	$a_{16} = -7.3771 \times 10^{-6}$
$a_3 = 4.4637$	$a_{10} = -1.088361 \times 10^{-11}$	$a_{17} = 4.9566 \times 10^{-5}$
$a_4 = -9.6368$	$a_{11} = -1.6066$	$a_{18} = -5.4219 \times 10^{-7}$
$a_5 = 9.3994$	$a_{12} = 3.3943 \times 10^{-3}$	$a_{19} = 4.2836 \times 10^{-9}$
$a_6 = -3.4122$	$a_{13} = -5.0443 \times 10^{-4}$	
$a_7 = 1.006933$	$a_{14} = 2.2789 \times 10^{-2}$	

14. Dindoruk and Christman (2001)

$$X = \frac{\left[\left(R_s^{2.510755}/\gamma_g^{4.852538}\gamma_o^{11.835}\right) + 1.365428 \times 10^5 (T-60)^{2.25288} + 10.0719 R_s\right]}{\left[5.352624 + 2R_s^{0.639052}\gamma_g^{-0.900074}(T-60)\right]^{0.450849}}$$

$$B_{ob} = 0.9871766 + 7.865146 \times 10^{-4} X + 2.689173 \times 10^{-6} X^2$$
$$+ 1.100001 \times 10^{-5}(T-60)\frac{\gamma_{API}}{\gamma_g}$$

15. Doklah and Osman (1992)

$$X = R_s^{0.773572}\frac{\gamma_g^{0.404020}}{\gamma_o^{0.882605}}$$

$$B_{ob} = 4.31935 \times 10^{-2} + 1.56667 \times 10^{-3}(T + 459.67)$$
$$+ 1.39775 \times 10^{-3} X + 3.80525 \times 10^{-6} X^2$$

16. Elam (1957)

$$B_{ob} = e^{\left[-0.0355 + 3.55 \times 10^{-4} R_s(\gamma_g/\gamma_o)^{0.5} + 7.1 \times 10^{-4} T\right]}$$

17. Elmabrouk et al. (2010)

$$\beta_o = 1.6624 + 0.00051 R_{sp} + 0.00015 P_{sp} - 0.802 \gamma_{osl} + 0.00057 T_R$$

18. Elsharkawy and Alikhan (1997)

$$B_{ob} = 1.0 + 4.0428 \times 10^{-4} R_s + 6.3802 \times 10^{-4}(T-60) + 7.8$$
$$\times 10^{-7} R_s(T-60)\frac{\gamma_g}{\gamma_o}$$

19. Farshad et al. (1996)

$$X = R_s^{0.5956} \frac{\gamma_g^{0.2369}}{\gamma_o^{1.3282}} + 0.0976T$$

20. Glaso (1980)

$$B_{ob} = 1.0 + 10^{[-2.6541 + 0.5576\log(X) + 0.3331\log(X)^2]}$$

$$X = R_s \left(\frac{\gamma_g}{\gamma_o}\right)^{0.526} + 0.968T$$

$$\log(B_{ob} - 1) = -6.58511 + 2.91329\log X - 0.27683(\log X)^2$$

21. Hanafy et al. (1997)

$$\beta_o = 0.0006 \times R_s + 1.079$$

22. Hemmati and Kharrat (2007)

$$\beta_o = 1 + 10^A$$

$$A = -4.6862 + 1.5959\log(M) - 0.0566(\log M)^2$$

$$M = R_s \times \left[\frac{\gamma_g}{\gamma_o}\right]^{0.0.5946} + 1.7439T$$

23. Karimnezhad et al. (2014)

$$\beta_o = 9.710^{-7} \left[\frac{R_s \gamma_g}{\gamma_o} + (R_s(T + 460))\right] + 1.0367$$

$$\beta_o = 1.66 \left(\frac{R_s \gamma_g}{\gamma_o}\right)^{-0.03} + [0.0000044(R_s(T + 460))^{0.894}]$$

$$\beta_o = 1.66 \left(\frac{R_s \gamma_g}{\gamma_o}\right)^{-0.03} + [0.0000044(R_s(T + 460))^{0.894}]^H$$

$$H = (T + 460)^{-0.044} \gamma_g^{-1.52} \gamma_o^{0.161}$$

24. Kartoatmodjo and Schmidt (1991)

$$B_{ob} = 0.98496 + 1.0 \times 10^{-4} \left(R_s^{0.755} \gamma_{gc}^{0.25} \frac{}{\gamma_o^{1.5}} + 0.45T\right)^{1.5}$$

$$\gamma_{gc} = \gamma_g \left[1 + 0.1595 \frac{\gamma_{API}^{0.4078}}{T_{sp}^{0.2466}} \log\left(\frac{p_{sp}}{114.7}\right)\right]$$

25. Kartoatmodjo and Schmidt (1994)

$$\beta_o = a_1 + a_2 \left(R_s^{a_3} \left(\frac{\gamma_{g100}^{a_4}}{\gamma_o^{a_5}} \right) + a_6 T \right)^{a_7}$$

$a_1 = 0.98496, a_2 = 10^{-4}, a_3 = 0.755, a_4 = 0.25,$
$a_5 = 1.5, a_6 = 0.45, a_7 = 1.5$

26. Khairy et al. (1998)

$$B_{ob} = 0.773413 + 7.05341 \times 10^{-4} R_s + 0.18669 \gamma_g - 9.2589$$
$$\times 10^{-4} \gamma_{API} + 4.41 \times 10^{-4} T$$

27. Labedi (1982)

$$B_{ob} = 0.9976 + 5.273 \times 10^{-4} R_{sb} + 2.6636$$
$$\times 10^{-8} (T - 60) (\gamma_{API} p_{sp}) + 1.6982 \times 10^{-5} \gamma_{API} (T - 60)$$

Black oil formulation where $B_o \leq 1.758$

$$X_1 = \frac{3.61 R_s^{0.4625} B_{ob}^{13.398} p_{sp}^{0.0775}}{10^{(3.231 B_{ob})}}$$

$$B_o = B_{ob} - X_1 \left(1 - \frac{p}{p_b} \right)$$

Volatile oil formulation where $B_o \geq 1.758$

$$X_2 = 1.6339 - 9.152 \times 10^{-4} R_{sb} + 1.584 \times 10^{-7} R_{sb}^2$$

$$B_o = B_{ob} - \left(1 - \frac{p}{p_b} \right)^{X_2}$$

28. Levitan and Murtha (1999)

$$B_{ob} = 1.0 + 0.0005 R_s \left(\frac{\gamma_g}{\gamma_o} \right)^{0.25} + \frac{0.2(T - 60)}{519.67 \gamma_g \gamma_o}$$

29. Macary and El-Batanoney (1992)

$$B_{ob} = (1.0031 + 0.0008T) e^{(0.0004 R_s + 0.0006(\gamma_o / \gamma_g))}$$

30. Mazandarani and Asghari (2007)

$$\beta_o = 0.99117 + 0.00021 R_s - 2.32 \times 10^{-6} \frac{R_s - \gamma_g}{\gamma_o} + 0.0071(T - 60) - N$$

$$N = 4.30 \times 10^{-7} (T - 60) (1 - \gamma_g)$$

31. McCain (1991)

$$B_{ob} = \frac{\rho_{STO} + 0.01357 \times R_{sb} \times \gamma_g}{\rho_{oRb}}$$

32. Mehran et al. (2006)

$$\beta_o = 1 + 10^A$$

$$A = -4.7486 + 1.587\log(M) - 0.0495(\log M)^2$$

$$M = R_s \times \left[\frac{\gamma_g}{\gamma_o}\right]^{0.4211} + 2.035T$$

33. Moradi et al. (2013)

$$\beta_o = 0.965278 + \left[0.0001\,API^{0.0672605}\,\gamma_g^{-0.465317}\,N\right]$$

$$N = \left[R_s\left(\frac{\gamma_g}{\gamma_o}\right)^{0.643141} + 2.27448T\right]^{1.15416}$$

34. Obomanu and Okpobiri (1987)

If $\gamma_o < 0.876$

$$B_{ob} = 0.3321 + 1.404154 \times 10^{-4} R_s + 4.1588128 \times 10^{-4} R_s\frac{\gamma_g}{\gamma_o} + 1.15861 \times 10^{-5}(T + 459.67)$$

If $\gamma_o \geq 0.876$

$$B_{ob} = 1.0232 + 2.725 \times 10^{-5}\left[R_s\left(\frac{\gamma_g + T}{\gamma_o}\right)^{0.79}\right]$$

35. Omar and Todd (1993)

$$X = 1.1663 + 7.62 \times 10^{-4}\frac{\gamma_{API}}{\gamma_g} - 0.0339\gamma_g$$

$$B_{ob} = 0.972 + 1.47 \times 10^{-4}\left[R_s\left(\frac{\gamma_g}{\gamma_o}\right)^{0.5} + 1.25T\right]^X$$

36. Owolabi (1984)

$$B_{ob} = 0.9871 + 4.0689 \times 10^{-4}\left[R_s\left(\frac{\gamma_g}{\gamma_o}\right)^{0.526} + 1.25T\right]$$

37. Petrosky (1990)

$$B_{ob} = 1.0113 + 7.2046 \times 10^{-5}\left(\frac{R_s^{0.3738}\gamma_g^{0.2914}}{\gamma_o^{0.6265}} + 0.24626T^{0.5371}\right)^{3.0936}$$

38. Standing (1947)

$$B_{ob} = 0.972 + 1.47 \times 10^{-4} \left[R_s \left(\frac{\gamma_g}{\gamma_o} \right)^{0.5} + 1.25T \right]^{1.175}$$

39. Standing (1981)

$$\beta_o = 0.9759 + 0.00012 \left[R_s \left(\frac{\gamma_g}{\gamma_o} \right)^{0.5} + 1.25T \right]^{1.2}$$

40. Sulaimon et al. (2014)

$$B_o = A_1 - A_2 \times \gamma_o + A_3 \times \gamma_g + A_4 \times R_s - A_5 \times \gamma_o \times \gamma_g + A_6 \\ \times \gamma_o \times R_s + A_7 \times \gamma_g \times R_s + A_8 \gamma_o^2 - A_9 \times \gamma_g^2 - A_{10} \times R_s^3$$

$A_1 = 1.08199630980282$	$A_5 = 0.00402938539792537$ $\quad$ $A_9 = 0.11166839287066$
$A_2 = 0.00805775753122104$	$A_6 = 9.11296341546098E - 06$ $\quad$ $A_{10} = 5.24239118889319E - 08$
$A_3 = 0.29401277167818$	$A_7 = 0.0002074878721653173$
$A_4 = 9.94971241197314E - 5$	$A_8 = 0.0001336099203028163$

41. Vasquez and Beggs (1980)

$$B_{ob} = 1 + a_2 R_s + (T - 60) \left(\frac{\gamma_{API}}{\gamma_{gc}} \right) (a_2 + a_3 R_s)$$

Coefficient	$\gamma_{API} \leq 30$	$\gamma_{API} > 30$
a_1	4.677×10^{-4}	4.670×10^{-4}
a_2	1.751×10^{-5}	1.100×10^{-5}
a_3	-1.8106×10^{-8}	1.337×10^{-9}

42. Velarde et al. (1997)

Trial estimate of ρ_{po}

$$\rho_{po} = -49.8930 + 85.0149\gamma_{gs} - 3.70373\gamma_{gs}\rho_{po} \\ + 0.047982\gamma_{gs}\rho_{po}^2 + 2.98914\rho_{po} - 0.035689\rho_{po}^2$$

Then calculate ρ_{po}, iterate until convergence is obtained

$$\rho_{po} = \frac{R_s \gamma_{gs} + 4600\gamma_o}{73.71 + R_s \left(\gamma_{gs}/\rho_a \right)}$$

$$\Delta\rho_p = (0.167 + 16.181 \times 10^{-0.0425\rho_{po}}) \left(\frac{p}{1000}\right)$$
$$- 0.01(0.299 + 263 \times 10^{-0.0603\rho_{po}}) \left(\frac{p}{1000}\right)^2$$

$$\Delta\rho_T = \left[\left(0.00302 + 1.505\left(\rho_{po} + \Delta\rho_p\right)^{-0.951}\right)\right](T-60)^{0.938}$$
$$+ \left[0.0216 - 0.0233 \times 10^{-0.0161(\rho_{po}+\Delta\rho_p)}\right](T-60)^{0.475}$$

$$\rho_o = \rho_{po} + \Delta\rho_p - \Delta\rho_T$$
$$B_{ob} = \frac{62.42797\gamma_o + 0.01357R_s\gamma_g}{\rho_o}$$

SATURATED OIL COMPRESSIBILITY

1. Ahmed (1989)

$$C_o = \frac{e^{(-1.8473 \times 10^{-4}p)}}{24841.0822 + 14.07428745R_s}$$

2. Al-Marhoun (1992)

$$X = -1.3668 \times 10^{-5}R_s - 1.95682 \times 10^{-8}R_s^2 + 2.408026$$
$$\times 10^{-2}\gamma_g - 9.26019 \times 10^{-8}(T+459.67)^2$$

$$C_o = \frac{Xln(p/p_b)}{(p_b - p)}$$

$$C_o = e^{(a_1 + (a_2/\gamma_{ob}) + (a_3(p-p_b)/\gamma_{ob}^3) + (a_4/(T+460)))}$$

$a_1 = -14.1042$

$a_2 = 2.7314$

$a_3 = -56.0605 \times 10^{-6}$
$a_4 = -580.8778$

3. Almehaideb (1997)

$$C_o = \frac{-70603.2 + 98.404R_s + 378.266T - 6102.03\gamma_g + 755.345\gamma_{API}}{10^6(p + 3755.53)}$$

4. Calhoun (1947)

$$C_o = \frac{(68.399 - 99.967\gamma_{ob} + 28.852\gamma_{ob}^2)}{1 \times 10^6}$$

5. De Ghetto et al. (1995)

Extra heavy oil (°API ≤ 10)

$$C_o = \frac{-889.6 + 3.1374R_s + 20T - 627.3\gamma_{gc} - 81.4476\gamma_{API}}{10^5 p}$$

Heavy oil (10 ≤ °API ≤ 22.3)

$$C_o = \frac{2.9646R_s + 25.5439T - 1230.5\gamma_{gc} + 41.91\gamma_{API} - 2841.8}{10^5 p}$$

Medium oil (22.3 ≤ °API ≤ 31.1)

$$C_o = \frac{2.2246R_{sb} + 26.0644T - 2080.823\gamma_{gc} - 9.6807\gamma_{API} - 705.288}{10^5 p}$$

Light oil (°API ≥ 31.1)

$$C_o = \frac{B_{ob}^{1.8789}\gamma_{API}^{0.3646}T^{0.1966}}{10^{6.1646}} - \left(1 - \frac{p_b}{p}\right)\frac{B_{ob}^{3.9392}T^{1.3649}}{10^{8.98}}$$

Agip model

$$C_o = \frac{4.133R_{sb} + 22.12T - 1323.8\gamma_{gc} + 10.5\gamma_{API} - 1682.8}{10^5 p}$$

6. Dindoruk and Christman (2004)

$$X_1 = \frac{R_s^{0.980922372}\gamma_g^{0.021003077}}{\gamma_o^{0.38486128}} + 20.00006358(T - 60)^{0.300001059} - 0.876813622R_s$$

$$X_2 = \frac{X_1^{1.759732076}}{\left(2.749114986 + \left(2R_s^{-1.713572145}/\gamma_g^{9.99932841}\right)(T - 60)\right)^2}$$

$$C_o = 10^{-6}(4.487462368 + 0.005197040X_2 + 1.258 \times 10^{-5}X_2^2)$$

7. Elsharkawy and Alikhan (1997)

$$C_o = \frac{-27321 + 33.784 R_s + 238.81 T}{10^6 P}$$

8. Farshad et al. (1996)

$$X = R_s^{0.1982} T^{0.6685} \gamma_g^{-0.21435} \gamma_{API}^{1.0116} p^{-0.1616}$$

9. Kartoatmodjo and Schmidt (1991)

$$C_o = 10^{(-5.4531 + 5.03 \times 10^{-4} X - 3.5 \times 10^{-8} X^2)}$$

$$C_o = \frac{10^{[0.83415 + 0.5002 \log(R_s) + 0.3613 \log(\gamma_{API}) + 0.7606 \log(T) - 0.35505 \log(\gamma_{gc})]}}{10^6 p}$$

10. Kartoatmodjo and Schmidt (1994)

$$\gamma_{gc} = \gamma_g \left(1 + \left(0.1595 \left(\frac{{}^\circ API^{0.4078}}{T_{sep}^{0.2466} \log(P_{sep}/114.7)} \right) \right) \right)$$

$$C_o = a_1 \times 10^{-6} R_s^{a_2} \times P \times T^{a_3} \gamma_{gc}^{a_4}$$

$a_1 = 0.83415$
$a_2 = 0.5002$
$a_3 = 0.7606$
$a_4 = -0.35505$

11. Labedi (1982)

$$C_{ob} = 10^{-7.902} B_{ob}^{1.3221} \gamma_{API}^{1.0503} T^{0.5369}$$

For pressures above bubble point

12. McCain (1991)

$$C_o = C_{ob} - \left(10^{-8.202} B_{ob}^{2.9650} T^{1.2194} \right) \left(1 - \frac{p_b}{p} \right)$$

$$c_o = \exp^{(-7.573 - 1.450 \times \log(P) + 0.383 \times \log(P_b) + 1.402 \times \log(T) + 0.256 \times \log(API) + 0.449 \times \log(R_{sb}))}$$

13. Petrosky (1990)

$$C_o = \frac{1.705 \times 10^{-7} R_s^{0.69357} \gamma_g^{0.1885} \gamma_{API}^{0.3272} T^{0.6729}}{p^{0.5906}}$$

14. Petrosky and Farshad (1993)

$$C_o = 1.705 \times 10^{-7} R_s^{0.69357} \gamma_g^{0.1885} {}^\circ API^{0.3272} T^{0.6729} p^{-0.5906}$$

15. Standing (1947)

$$C_o = 10^{-6} e \left[\frac{\rho_{ob} + 4.347 \times 10^{-3} (p - p_b) - 79.1}{7.141 \times 10^{-4} (p - p_b) - 12.938} \right]$$

16. Vasquez and Beggs (1980)

$$C_o = \frac{5R_s + 17.2T - 1180\gamma_{gc} + 12.61\gamma_{API} - 1.433}{10^5 p}$$

$$\gamma_{gc} = \gamma_g \left(1 + \left(5.912 \times 10^{-5} \times °\mathrm{API} \times T_{sep} \log\left(\frac{P_{sep}}{114.7}\right)\right)\right)$$

$$C_o = \frac{\left(-1433 + 5R_{sp} + 17.2T - 1180\gamma_{gc} + 12.61°\mathrm{API}\right)}{10^5 P}$$

SATURATED OIL DENSITY

1. Ahmed (1989)

$$\rho_o = \left(\frac{0.0763R_s\gamma_g + 350.4\gamma_o}{0.0027R_s + 2.4893\gamma_o + 3.491}\right)$$

2. Hanafy et al. (1997)

$$\rho_{ob} = \left(2.366 - \left(\frac{1.358}{\beta_{ob}}\right)\right)^{-1}$$

3. Standing (1947)

$$\rho_o = \left(\frac{(62.4\gamma_o + 0.0136R_s\gamma_g)}{0.972 + 0.000147\left(R_s\left(\gamma_g/\gamma_o\right)^{0.5} + 1.25T\right)^{1.175}}\right)$$

SATURATED OIL VISCOSITY

1. Abu-Khamsim and Al-Marhoun (1991)

$$\ln\mu_{ob} = -2.652294 + 8.484462\rho_{ob}^4$$

2. Al-Khafaji et al. (1987)

$$X_1 = \log(R_s)$$

$$\mu_{ob} = A\mu_{od}^{B^{-}}$$

$$A = 0.247 + 0.2824X_1 + 0.5657X_1^2 - 0.4065X_1^3 + 0.0631X_1^4$$

$$B = 0.894 + 0.0546X_1 + 0.07667X_1^2 - 0.0736X_1^3 + 0.01008X_1^4$$

3. Almehaideb (1997)

$$\mu_{ob} = \frac{6.59927 \times 10^5}{R_s^{0.597627} \, T^{0.941624} \, \gamma_g^{0.555208} \, \gamma_{API}^{1.487449}}$$

4. Beggs and Robinson (1975)

$$A = \frac{10.715}{(R_s + 100)^{0.515}}$$

$$B = \frac{5.44}{(R_s + 150)^{0.338}}$$

$$\mu_{ob} = A\mu_{od}^B$$

5. Bergman (2004)

$$A = e^{[4.768 - 0.8359 \ln(R_s + 300)]}$$

$$B = 0.555 + \frac{133.5}{R_s + 300}$$

6. Chew and Connally (1959)

$$\mu_{ab} = A\mu_{od}^B$$

$$A = 0.20 + \frac{0.80}{10^{(0.00081 R_s)}}$$

$$B = 0.43 + \frac{0.57}{10^{(0.00072 R_s)}}$$

$$\mu_{ob} = A\mu_{od}^B$$

7. De Ghetto et al. (1995)

Extra heavy oil ($^\circ$API $\leq$ 10)

$$X_1 = 0.5798 + 0.3432 \times 10^{(-8.1 \times 10^{-4} R_s)}$$

$$X_2 = \left[-0.0335 + 1.0785 \times 10^{(-8.45 \times 10^{-4} R_s)} \right] \mu_{od}^{X_1}$$

$$\mu_{ob} = 2.3945 + 0.8927 X_2 + 1.567 \times 10^{-3} X_2^2$$

Heavy oil ($10 \leq\,^\circ$API $\leq$ 22.3)

$$X_1 = 0.4731 + 0.5158 \times 10^{(-8.1 \times 10^{-4} R_s)}$$

$$X_2 = \left[0.2478 + 0.6114 \times 10^{(-8.45 \times 10^{-4} R_s)} \right] \mu_{od}^{X_1}$$

$$\mu_{ob} = 0.6311 + 1.078X_2 - 3.653 \times 10^{-3}X_2^2$$

Medium oil ($22.3 \leq {}^\circ API \leq 31.1$)

$$X_1 = 0.3855 + 0.5664 \times 10^{(-8.1 \times 10^{-4}R_s)}$$

$$X_2 = \left[0.2038 + 0.8591 \times 10^{(-8.45 \times 10^{-4}R_s)}\right]\mu_{od}^{X_1}$$

$$\mu_{ob} = 0.0132 + 0.9821X_2 - 5.215 \times 10^{-3}X_2^2$$

Light oil (${}^\circ API \geq 31.1$)

$$A = \frac{25.1921}{(R_s + 100)^{0.6487}}$$

$$\mu_{ab} = A\mu_{od}^B$$

$$B = \frac{2.7516}{(R_s + 150)^{0.2135}}$$

Agip model

$$X_1 = 0.172 + 0.7881 \times 10^{(-3.96 \times 10^{-4}R_s)}$$

$$X_2 = \left[0.1615 + 0.7024 \times 10^{(-5.83 \times 10^{-4}R_s)}\right]\mu_{od}^{X_1}$$

$$\mu_{ob} = -0.032124 + 0.9289X_2 - 2.865 \times 10^{-2}X_2^2$$

$$A = \frac{a_1}{e^{(a_2R_s)}} + \frac{a_3R_s^{a_4}}{e^{(a_5R_s)}}$$

$$B = \frac{a_6}{e^{(a_7R_s)}} + \frac{a_8R_s^{a_9}}{e^{(a_{10}R_s)}}$$

$$\mu_{ab} = A\mu_{od}^B$$

where

$a_1 = 1.0$

$a_2 = 4.740729 \times 10^{-4}$

$a_3 = -1.023451 \times 10^{-2}$

8. Dindoruk and Christman (2001)

$a_4 = 6.600358 \times 10^{-1}$

$a_5 = 1.075080 \times 10^{-5}$

$a_6 = 1.0$

$a_7 = -2.191172 \times 10^{-5}$

$a_8 = -1.660981 \times 10^{-2}$

$a_9 = 4.2331779 \times 10^{-1}$

$a_{10} = -2.273945 \times 10^{-4}$

9. Elsharkawy and Alikhan (1997)

$$A = \frac{1241.932}{(R_s + 641.026)^{1.12410}}$$

$$\mu_{ab} = A\mu_{od}^B$$

$$B = \frac{1768.841}{(R_s + 1180.335)^{1.06622}}$$

10. Hanafy et al. (1997)

$$\mu_{ob} = e^{(7.296\rho_{ob}^3 - 3.095)}$$

11. Kartoatmodjo and Schmidt (1991)

$$X_1 = 0.43 + 0.5165 \times 10^{(-8.1 \times 10^{-4} R_s)}$$

$$X_2 = \left[0.2001 + 0.8428 \times 10^{(-8.45 \times 10^{-4} R_s)}\right]\mu_{od}^{X_1}$$

$$\mu_{ob} = -0.06821 + 0.9824 X_2 + 4.034 \times 10^{-4} X_2^2$$

12. Khamehchi and Ebrahimian (2009)

$$\mu_{ob} = 6.28 \times 10^6 \times R_s^{-0.0266} \times T^{-1.032} \times API^{-2.786}$$

13. Khan et al. (1987)

$$\mu_{ob} = \frac{0.09\gamma_g^{0.5}}{R_{sb}^{1/3}\left[(T+459.67)/459.67\right]^{4.5}(1-\gamma_o)^3}$$

If $p \leq p_b$

$$\mu_o = \frac{\mu_{ob}e^{[-2.5 \times 10^{-4}(p-p_b)]}}{(p/p_b)}$$

14. Labedi (1982)

$$\mu_{ob} = \frac{10^{(2.344-0.03542\gamma_{API})}\mu_{od}^{0.6447}}{p_b^{0.426}}$$

At $p \leq p_b$

15. Labedi (1992)

$$\mu_o = \frac{\mu_{ob}}{1 - \left(10^{-3.876}p_b^{0.5423}\gamma_{API}^{1.1302}\right)\left(1 - p/p_b\right)}$$

$$\mu_{ob} = \frac{1.303\mu_{od}^{0.6447}}{10^{(2.133\times10^{-4}p_b)}}$$

At $p \geq p_b$

16. Naseri et al. (2005)

$$\mu_o = \frac{\mu_{ob}}{1 - \left(2.861\times10^{(1.076\times10^{-4}p_b)}\right)\left(1 - p/p_b\right)}$$

$$\mu_{ob} = 10^{1.1145} \times P_b^{-0.4956} \times \mu_{od}^{0.9961}$$

17. Petrosky (1990)

$$A = 0.1651 + \frac{0.6165}{10^{(6.0866\times10^{-4}R_s)}}$$

$$B = 0.5131 + \frac{0.5109}{10^{(1.1831\times10^{-3}R_s)}}$$

$$\mu_{ob} = A\mu_{od}^B$$

18. Petrosky and Farshad (1995)

$a_1 = 0.1651$

$a_2 = 0.6165$

$a_3 = -6.0866 \times 10^{-4}$

$a_4 = 0.5131 \quad a_5 = 0.5109$

$a_6 = -0.0011831$

$A = a_1 + a_2 \times 10^{a_3 \times R_s}$

$B = a_4 + a_5 \times 10^{a_6 \times R_s}$

$\mu_{ob} = A \times \mu_{od}^B$

19. Standing (1981)

$$\mu_{ob} = 10^a \times \mu_{od}^b$$

$$a = R_s \times (2.2 \times 10^{-7} \times R_s - 7.4 \times 10^{-4})$$

$$b = 8.62 \times 10^{-5} \times R_s + 0.25 \times 10^{1.1 \times 10^{-3} \times R_s} + 0.062 \times 10^{-3.74 \times 10^{-3} \times R_s}$$

STOCK-TANK GAS–OIL RATIO

1. Elmabrouk et al. (2010)

$$R_{st} = P_{sp}^{1.02} \gamma_{ost}^{-9.47} T_{sp}^{-0.524}$$

2. Rollins et al. (1990)

$$\log R_s = 0.4896 - 4.9161 \log \gamma_{osep} + 3.469 \log \gamma_{sep} + C$$

$$C = 1.50 \log P_{sep} - 0.9213 \log T_{sep}$$

$$\log(R_{ST}) = a_1 + a_2 \log(\gamma_{oST}) + a_3 \log(\gamma_{gSP}) + a_4 \log(P_{SP}) + a_5 \log(T_{sp})$$

where

$a_1 = 0.3818$
$a_2 = -5.506$
$a_3 = 2.902$
$a_4 = 1.327$
$a_5 = -0.7355$

3. Valkó and McCain (2003)

$$\ln R_{ST} = 3.955 + 0.83Z - 0.024Z^2 + 0.075Z^3$$

$$Z = \sum_{n=1}^{3} Z_n$$

$$Z_n = C0_n + C1_n \text{VAR}_n + C2_n \text{VAR}_n^2$$

UNDER-SATURATED OIL FORMATION VOLUME FACTOR

1. Ahmed (1992)

$$\beta_o = \beta_{ob} \times e^{\left[D\left[e^{(a \times p)} - e^{(a \times p_b)}\right]\right]}$$

where

$$D = (4.588893 + 0.0025999 R_s)^{-1}$$

$$a = -0.00018473$$

2. Al-Marhoun (1992)

$$\beta_o = \beta_{ob} \left(\frac{P}{P_b}\right)^C$$

$$C = a_1 R_s + a_2 R_s^2 + a_3 \gamma_g + a_4 \times (T + 460)^2$$

where

$$a_1 = -1.3668 \times 10^{-5}$$
$$a_2 = -1.95682 \times 10^{-8}$$
$$a_3 = 0.02408026$$
$$a_4 = -9.26019 \times 10^{-8}$$

3. General equation

$$B_o = B_{ob} e^{-C_o(P - P_b)}$$

UNDER SATURATED OIL COMPRESSIBILITY

1. Ahmed (1989)

$$c_o = \frac{\exp(-1.8473 \times 10^{-4 \times P})}{24841.0822 + 14.07428745 \times R_s}$$

2. Al-Marhoun (1992)

$$x = -1.3668 \times 10^{-5} \times R_s - 1.95682 \times 10^{-8} \times R_s^2 + 2.408026$$
$$\times 10^{-2} \times \gamma_g - 9.26019 \times 10^{-8}(T + 459.67)^2$$

$$c_o = \frac{x \times \log(P/P_b)}{P_b - P}$$

3. Almehaideb (1997)

$$c_o = \frac{-70{,}603.2 + 98.404 \times R_s + 378.266 \times T - 6102.03 \times \gamma_g + 755.345 \times API}{10^6 \times P}$$

4. De Ghetto et al. (1995)

Heavy oil °API <= 10

$$c_o = \frac{\left(-889.6 + 3.1374 \times R_s + 20 \times T - 627.3 \times \gamma_{gc} - 81.4476 \times API\right)}{10^5 \times P}$$

Heavy oil 10 < °API < 22.3

$$c_o = \frac{\left(2.9646 \times R_s + 25.5439 \times T - 1230.5 \times \gamma_{gc} + 41.91 \times API - 2841.8\right)}{10^5 \times P}$$

Medium oil 22.3 < °API < 31.1

$$c_o = \frac{\left(2.2246 \times R_s + 26.0644 \times T - 2080.823 \times \gamma_{gc} - 9.6807 \times API - 705.288\right)}{10^5 \times P}$$

Light oil °API > 31.1

$$c_o = \left(\frac{B_{ob}^{1.8789} \times API^{0.3646} \times T^{0.1966}}{10^{6.1646}}\right) - \left[1 - \frac{P}{P_b}\right] \times \left[\frac{B_{ob}^{3.9392} \times T^{1.349}}{10^{8.98}}\right]$$

5. Dindoruk and Christman (2004)

$$A = \frac{\left(\left(R_s^{a_1} \gamma_g^{a_2}/\gamma_o^{a_3}\right) + (a_4(T-60)^{a_5} + a_6 R_s)\right)^{a_7}}{\left(a_8 + (2R_s^{a_9}/(\gamma_g^{a_{10}} \times (T-60)))\right)^2}$$

$$c_{obp} = (a_{11} + a_{12} \times A + a_{13}A^2) \times 10^{-6}$$

a_1	0.980922372
a_2	0.021003077
a_3	0.338486128
a_4	20.00006358
a_5	0.300001059
a_6	-0.876813622
a_7	1.759732076
a_8	2.749114986
a_9	-1.713572145
a_{10}	9.999932841
a_{11}	4.487462368
a_{12}	0.005197040
a_{13}	0.000012580

6. Elsharkawy and Alikhan (1997)

$$c_o = \frac{-27,321 + 33.784 \times R_s + 239.81 \times T}{10^6 \times P}$$

7. Farshad et al. (1996)

$$c_o = \left(\frac{B_{op}^{1.8789} \times API^{0.3646} \times T^{0.1966}}{10^{6.1646}}\right) - \left(1 - \frac{P}{P_b}\right) \times \left(\frac{B_{op}^{3.9392} \times T^{1.349}}{10^{8.98}}\right)$$

8. Kartoatmodjo and Schmidt (1994)

$$\gamma_{gc} = \gamma_g \times \left(1 + 0.1595 \times \frac{API^{0.4078}}{T_{sp}^{0.2466}} \times \log\left(\frac{P_{sp}}{114.7}\right)\right)$$

$$c_o = \frac{10^{(0.83415 + 0.5002 \times \log(R_s) + 0.3613 \times \log(API) + 0.7606 \times \log T) - 0.35505 \times \log(\gamma_{gc}))}}{10^6 \times P}$$

9. Labedi (1982)

$$c_{ob} = 10^{-7.902} \times B_{ob}^{1.3221} \times API^{1.0503} \times T^{-0.5369}$$

10. McCain (1991)

$$c_o = c_{ob} \times (10^{-8.202} \times B_{ob}^{2.9650} \times T^{1.2194})\left(1 - \frac{P_b}{P}\right)$$

$$c_o = \exp^{(-7.633 - 1.497 \times \log(P) + 1.115 \times \log(T) + 0.533 \times \log(API) + 0.184 \times \log(R_{sb}))}$$

11. Petrosky and Farshad (1993)

$$c_o = 1.705 \times R_{sb}^{0.69357} \times \gamma_g^{0.1885} \times API^{00.3272} \times (T - 460)^{0.6729} \times P^{-0.5906}$$

12. Spivey et al. (2007)

$$C_{ofb} = e^{(2.434+0.475Z+0.048Z^2-ln(10^6))}$$

$$Z = \sum_{n=1}^{6} Z_n$$

$$Z_n = C0_n + C1_n VAR_n + C2_n VAR_n^2$$

N	VAR$_n$	C0$_n$	C1$_n$	C2$_n$
1	ln°API	3.011	−2.6254	0.497
2	lnγ_{gsp}	−0.0835	−0.259	0.382
3	lnP_b	3.51	−0.0289	−0.0584
4	lnp/p_b	0.327	−0.608	0.0911
5	lnR_s	−1.918	−0.642	0.154
6	lnT_R	2.52	−2.73	0.429

13. Standing (1947)

$$c_o = 10^{-6} \exp\left(\frac{\rho_{ob} + 4.347 \times 10^{-3} \times (P - P_b) - 79.1}{7.141 \times 10^{-4} \times (P - P_b) - 12.938} \right)$$

14. Vasquez and Beggs (1980)

$$\gamma_{gc} = \gamma_g \times \left(1 + \left(5.912 \times 10^{-5} \times API \times T_{sep} \times \log\left(\frac{P_{sep}}{114.7}\right)\right)\right)$$

$$c_o = \left(\frac{-1.433 + 5 \times R_{sb} + 17.2 \times (T - 460) - 1180 \times \gamma_{gc} + 12.61 \times API}{10^5 \times P} \right)$$

UNDER SATURATED OIL DENSITY

1. General equation

$$\rho_o = \rho_{ob} \times \exp(c_o(P - P_b))$$

UNDER SATURATED OIL VISCOSITY

1. Abdul-Majeed et al. (1990)

$$X = 1.9311 - 0.8941(\ln R_s) - 0.001194\gamma_{API}^2 + 0.0092545\gamma_{API}(\ln R_s)$$

$$\mu_o = \mu_{ob} + 1000 \times 10^{[X - 5.2106 + 1.11\log[6.8945757(p - p_b)]]}$$

$$\mu_o = \mu_{ob} + 10^{[A - 5.2106 + 1.11\log(p - p_b)]}$$

2. Al-Khafaji et al. (1987)

$$A = 1.9311 - 0.89941(\ln R_s) - 0.001194\gamma_{API}^2 + 0.0092545\gamma_{API}(\ln R_s)$$

$$X = -0.3806 - 0.1845\gamma_{API} + 0.004034\gamma_{API}^2 - 3.716 \times 10^{-5}\gamma_{API}^3$$

3. Almehaideb (1997)

$$\mu_o = \mu_{ob} + 10^{[X + 1.11\log[0.07031(p - p_b)]]}$$

$$\mu_o = \mu_{ob}\left(\frac{p}{p_b}\right)^{\left(0.134819 + 1.94345 \times 10^{-4}R_s - 1.93106 \times 10^{-9}R_s^2\right)}$$

4. Beal (1946)

$$\mu_o = \mu_{ob} + [0.001(p - p_b)](0.024\mu_{ob}^{1.6} + 0.038\mu_{ob}^{0.56})$$

5. Beggs and Robinson (1975)

$$m = 2.6 \times p^{1.187} \times \exp^{(-11.513 + (-8.98 \times 10^{-5} \times P))}$$

$$\mu_o = \mu_{ob} \times \left(\frac{P}{P_b}\right)^m$$

6. Bergman (2004)

$$\mu_o = \mu_{ob}e^{\alpha(p - p_b)^\beta}$$

$$\alpha = 6.5698 \times 10^{-7}\ln(\mu_{ob})^2 - 1.48211 \times 10^{-5}\ln(\mu_{ob}) + 2.27877 \times 10^{-4}$$

$$\beta = 2.24623 \times 10^{-2}\ln(\mu_{ob}) + 0.873204$$

7. De Ghetto et al. (1995)

Extra heavy oil ($^\circ$API $\le$ 10)

$$\mu_o = \mu_{ob} - \left(1 - \frac{p}{p_b}\right)\frac{10^{-2.19}\mu_{od}^{1.055}p_b^{0.3132}}{10^{(0.0099\gamma_{API})}}$$

Heavy oil ($10 \le {}^\circ$API ≤ 22.3)

$$\mu_o = 0.9886\mu_{ob} + 0.002763(p - p_b) \times (-0.01153\mu_{ob}^{1.7933} + 0.0316\mu_{ob}^{1.5939})$$

Medium and light oil (°API ≥ 22.3)

$$\mu_o = \mu_{ob} - \left(1 - \frac{p}{p_b}\right) \frac{10^{-3.8055} \mu_{od}^{1.4131} p_b^{0.6957}}{10^{(-0.00288\gamma_{API})}}$$

Agip model

$$\mu_o = \mu_{ob} - \left(1 - \frac{p}{p_b}\right) \frac{10^{-1.9} \mu_{od}^{0.7423} p_b^{0.5026}}{10^{(0.0243\gamma_{API})}}$$

8. Dindoruk and Christman (2001)

$$X = a_1 + a_2 \log(\mu_{ob}) + a_3 \log(R_s) + a_4 \mu_{ob} \log(R_s) + a_5 (p - p_b)$$

$$\mu_o = \mu_{ob} + a_6 (p - p_b) 10^X$$

where

$a_1 = 0.776644115$

$a_2 = 0.987658646$

$a_3 = -0.190564677$

$a_4 = 0.009147711$

$a_5 = -0.000019111$

$a_6 = 0.000063340$

9. Elsharkawy and Alikhan (1999)

$$\mu_o = \mu_{ob} + \frac{10^{-2.0771}(p - p_b)\mu_{od}^{1.129279}}{\mu_{ob}^{0.40712} p_b^{0.7941}}$$

10. Kartoatmodjo and Schmidt (1991)

$$\mu_o = 1.00081\mu_{ob} + 1.127 \times 10^{-3}(p - p_b)$$
$$\left(-6.517 \times 10^{-3}\mu_{ob}^{1.8148} + 0.0038\mu_{ob}^{1.59}\right)$$

11. Khan et al. (1987)

$$\mu_o = \mu_{ob} e^{9.6 \times 10^{-5}(p - p_b)}$$

12. Kouzel (1965)

$$\mu_o = \mu_{ob} 10^{[(p-p_b)(0.0239+0.01638\mu_{ob}^{0.278})/1000]}$$

13. Kouzel (1997)

$$\mu_o = \mu_{ob} 10^{[(p-p_b)(-0.0102+0.04042\mu_o^{0.181})/1000]}$$

14. Labedi (1982)

$$\mu_o = \mu_{ob} + \frac{\mu_{od}^{0.9036} p_b^{0.6151}}{10^{(2.488+0.01976\gamma_{API})}} \left(\frac{p}{p_b} - 1\right)$$

15. Labedi (1992)

$$\mu_o = \mu_{ob} + 0.0483 \mu_{od}^{0.7374} \left(\frac{p}{p_b} - 1 \right)$$

16. Naseri et al. (2005)

$$a = 1.5029 \times 10^{-5} + 1.602 \times 10^{-5} \times \mu_{od} + 1.73695 \times \mu_{od}^2 - 4.2347 \times 10^{-6} \times \mu_{od}^{-3}$$

$$\mu_o = \mu_{od} + a \times (P - P_b)$$

17. Petrosky (1990)

$$X_1 = \log(\mu_{ob})$$

$$X_2 = -1.0146 + 1.3322 X_1 - 0.4876 X_1^2 - 1.15036 X_1^3$$

$$\mu_o = \mu_{ob} + 1.3449 \times 10^{-3} (p - p_b) 10^{X_2}$$

18. Petrosky and Farshad (1995)

$$a = -1.0146 + 1.3322 \times \log(\mu_{ob}) - 0.4876 \times (\log(\mu_{ob}))^2 - 1.15036 \times (\log \mu_{ob})^3$$

$$\mu_o = \mu_{ob} + 1.3449 \times 10^{-3} \times (PP_b) \times 10^a$$

19. Twu (1985)

$$\mu_{od} = \gamma_{oT} v_T$$

$$\gamma_{oT} = 0.999012 \gamma_{o60} VCF_T$$

$$VCF_T = e^{[-\alpha_{o60}\Delta T(1 + 0.8\alpha_{o60}\Delta T)]}$$

$$\alpha_{60} = \frac{K_0 + K_1 \gamma_{o60}}{\gamma_{o60}^2}$$

$$K_0 = 3.4175 \times 10^{-4}$$

$$K_1 = -4.542 \times 10^{-5}$$

For generalized crude oils

$$K_0 = 2.5042 \times 10^{-4}$$

$$K_1 = 8.3020 \times 10^{-5}$$

$$\Delta T = T - 60$$

20. Vasquez and Beggs (1980)

$$\mu_o = \mu_{ob} \left(\frac{p}{p_b} \right)^{\left[2.6 p^{1.187} 10^{(-3.9 \times 10^{-5} p - 5)} \right]}$$

DEAD OIL VISCOSITY

1. Al-Khafaji et al. (1987)

$$\mu_{od} = 10^{\left[(4.9563-0.00488T)/\left(\gamma_{API}+\frac{T}{14.29}\right)^{2.709}\right]}$$

2. Andrade (1934)

$$\mu_{od} = A10^{[B/(T+459.67)]}$$

3. Beal (1946)

$$X = e^{[2.302585(0.43+(8.33/\gamma_{API}))]}$$

$$\mu_{od} = \left(0.32 + \frac{1.8 \times 10^7}{\gamma_{API}^{4.53}}\right)\left(\frac{360}{T+200}\right)^X$$

$$\mu_{do} = 10^X$$

$$X = A + B + C$$

$$A = 10.5439 - 0.44521^\circ API + 0.66470$$
$$\times 10^{-2}\,^\circ API^2 - 0.335972 \times 10^{-4}\,^\circ API^3$$

$$B = -0.470516 \times 10^{-1} \times T - 0.81557 \times 10^{-4} \times T^2 - 0.789049 \times 10^{-4} \times T^7$$

$$C = 0.1352 \times 10^{-2} \times ^\circ API - 0.1145 \times 10^{-4}$$
$$\times T - 0.955 \times 10^{-6} \times T^2 \times ^\circ API$$

4. Beggs and Robinson (1975)

$$X = 10^{(3.0324-0.020023\gamma_{API})}T^{-1.163}$$

$$\mu_{od} = 10^X - 1.0$$

5. Bennison (1998)

$$\mu_{od} = 10^{(-0.8021\gamma_{API}+23.8765)}T^{(0.31458\gamma_{API}-0.921592)}$$

$$\mu_{od} = 10^{(0.10231 \times API^2 - 3.9464 \times ^\circ API)+46.5037}$$
$$T^{(-0.04542 \times API^2 + 1.70405 \times API + 46.1918)}$$

6. Bergman (2004)

$$X = e^{[22.33-0.194\gamma_{API}+0.00033\gamma_{API}^2 - (3.2-0.0185\gamma_{API})\log(T+310)]}$$

$$\mu_{od} = 10^X - 1$$

7. De Ghetto et al. (1995)

Extra heavy oil (°API ≤ 10)

$$X = 10^{[1.90296 - 1.2619 \times 10^{-2} \gamma_{API} - 0.61748 \log(T)]}$$

Heavy oil (10 ≤ °API ≤ 22.3)

$$X = 10^{[2.06492 - 1.79 \times 10^{-2} \gamma_{API} - 0.70226 \log(T)]}$$

Light oil (°API ≥ 22.3)

$$X = 10^{[1.67083 - 1.6728 \times 10^{-2} \gamma_{API} - 0.61304 \log(T)]}$$

For each evaluation of X

$$\mu_{od} = 10^X - 1.0$$

Medium oil (22.3 ≤ °API ≤ 31.1)

$$\mu_{od} = \frac{220.15 \times 10^9}{T^{3.5560}} \log(\gamma_{API})^{[12.5428 \log(T) - 45.7874]}$$

Agip model

$$X = 10^{[1.8513 - 2.5548 \times 10^{-2} \gamma_{API} - 0.56238 \log(T)]}$$

$$\mu_{od} = 10^X - 1.0$$

8. Dindoruk and Christman (2001)

$$\mu_{od} = \frac{a_3 T^{a_4} \log(\gamma_{API})^{[a_1 \log(T) + a_2]}}{a_5 p_b^{a_6} + a_7 R_s^{a_8}}$$

where

$a_1 = 14.505357625$

$a_2 = -44.868655416$

$a_3 = 9.36579 \times 10^9$

$a_4 = -4.194017808$

$a_5 = -3.1461171 \times 10^{-9}$

$a_6 = 1.517652716$

$a_7 = 0.010433654$

$a_8 = -0.000776880$

9. Elsharkawy and Alikhan (1999)

$$X = 10^{[2.16924 - 0.02525\gamma_{API} - 0.68875\log T]}$$

$$\mu_{od} = 10^X - 1$$

10. Elsharkawy and Gharbi (2001)

$$\mu_{do} = 10.7580 - 3.9145\log^\circ API - 1.9364\log T$$

11. Fitzgerlad (1994)

$$A_1 = 34.931 - 8.84387 \times 10^{-2} T_b + 6.73513 \times 10^{-5} T_b^2 - 1.01394 \times 10^{-8} T_b^3$$

$$A_2 = -2.92649 + 6.98405 \times 10^{-3} T_b - 5.09947 \times 10^{-6} T_b^2 + 7.49378 \times 10^{-10} T_b^3$$

$$\log X_1 = A_1 + A_2 K_w$$

$$\log X_2 = -1.35579 + 8.16059 \times 10^{-4} T_b + 8.38505 \times 10^{-7} T_b^2$$

$$v_{100} = X_1 + X_2$$

$$\log v_{210} = -1.92353 + 2.41071 \times 10^{-4} T_b + 0.51130\log(T_b v_{100})$$

$$X_3 = v_{100} + 0.7 + e^{\left(-1.47 - 1.84v_{100} - 0.51v_{100}^2\right)}$$

$$X_4 = v_{210} + 0.7 + e^{\left(-1.47 - 1.84v_{210} - 0.51v_{210}^2\right)}$$

$$X_5 = \frac{[\log\log(X_3) - \log\log(X_4)]}{[\log(559.67) - \log(669.67)]}$$

$$\log(X_6) = 10^{[\log\log(X_3) + X_5(\log(T+459.67) - \log(559.67))]}$$

$$\mu_{od} = \left[\gamma_o - 4.6 \times 10^{-4}(T - 60)\right]$$

$$\times \left\{X_6 - 0.7 - e^{\left[-0.7487 - 3.295(X_6 - 0.7) + 0.6119(X_6 - 0.7)^2 - 0.3193(X_6 - 0.7)^3\right]}\right\}$$

12. Glaso (1980)

$$\mu_{od} = \left(\frac{3.141 \times 10^{10}}{T^{3.444}}\right)\log(\gamma_{API})^{[10.313\log(T) - 36.447]}$$

13. Hossain et al. (2005)

$$\mu_{do} = 10^A \times T^B A$$

$$A = -0.71523 \times °API + 22.13766$$

$$B = 0.269024 \times °API + 8.268047$$

14. Ikiensikimama (2008)

$$\mu_{do} = 10^{10^D} - 1$$

$$D = 2.0930 - 0.0350 \times \log(API) - 0.6063\log(T)$$

15. Kartoatmodjo and Schmidt (1991)

$$\mu_{od} = \frac{1.6 \times 10^9}{T^{2.8177}} \log(\gamma_{API})^{[5.7526\log(T)-26.9718]}$$

16. Kartoatmodjo and Schmidt (1994)

$$\mu_{od} = 16 \times 10^8 \times T^{-2.8177} \times \log API^{(5.756\log(T)-26.9718)}$$

Medium oils

$$\mu_{od} = 220.15 \times 10^9 \times T^{-3.556} \times \log API^{(12.5428\log(T)-45.7874)}$$

Dead oil

$$\mu_{ob} = -0.06821 + 0.9824F + 0.0004034F^2$$

$$F = (0.2001 + 0.8428 \times 10^{-0.000845R_s}) \times \mu_{ob}^{(0.43+0.5165y)}$$

$$y = 10^{-0.0008R_s}$$

Under-saturated

$$\mu_o = 1.0008\mu_{ob} + 0.001127(P - P_b) \times (-0.006517\mu_{ob}^{1.8148} + 0.038\mu_{ob}^{1.59})$$

17. Kaye (1985)

API ≤ 12

$$\mu_{od} = 10^D - 1$$

$$D = T^{-0.65} \times 10^{2.203-0.025 \times API}$$

API ≥ 12

$$\mu_{od} = 10^D - 1$$

$$D = T^{-0.65} \times 10^{2.305-0.03354 \times API}$$

18. Khan et al. (1987)

$$\mu_{od} = 10^{9.224} \gamma_{API}^{-4.7013} T^{-0.6739}$$

19. Labedi (1982)

20. Labedi (1992)

$$\mu_{od} = 10^{9.37} \gamma_{API}^{-2.92} T^{-2.0356}$$

21. McCain (1991)

$$\mu_{od} = 10^{10^{[1.8653 - 0.025086 \times API - 0.5644 \times Log(T)]}} - 1$$

22. Naseri et al. (2005)

$$\mu_{do} = 10^B$$

$$B = 11.2699 - 4.2699 \log(^\circ API) - 2.052 \log(T)$$

23. Naseri et al. (2012)

$$\mu_{do} = 10^{10^B} - 1.12$$

$$B = 7.9684 - 2.7942 \times \log(^\circ API) - 1.6044 \times \log(T) + A$$

$$A = -\frac{47.3757}{T} - \frac{165.1894}{API^2}$$

24. Ng and Egbogah (1983)

$$X = 10^{[1.8653 - 2.5086 \times 10^{-2} \gamma_{API} - 0.56411 \log(T)]}$$

$$\mu_{od} = 10^X - 1.0$$

25. Oyedeko and Ulaeto (2011)

$$\mu_{do} = 10^{7.173} API^{-2.9986} T^{-1.1226}$$

26. Petrosky (1990)

$$\mu_{od} = \frac{2.3511 \times 10^7}{T^{2.8177}} \log(\gamma_{API})^{[4.59388 \log(T) - 22.82792]}$$

27. Petrosky and Farshad (1995)

$$a_1 = 4.59388$$

$$a_2 = 22.82792$$

$$a_3 = 2.3511 \times 10^7$$

$$a_4 = -2.10255$$

$$A = a_1 \times \log(T) - a_2$$

28. Standing (1947)

$$\mu_{od} = a_3 \times T^{a_4} \times (\log API)^A$$

$$\mu_{od} = 10^{10^{[1.8653 - 0.025086 \times API - 0.5644 \times Log(T)]}} - 1$$

29. Ulaeto and Oyedeko (2014)

$$\mu_{do} = X_1 \times (1 + X_2 \ln(\text{API})) + X_3 \ln(T))$$

For general crude oil

$$X_1 = \frac{47209999.96}{\text{API}^{3.185} T^{0.176}}$$

$$X_2 = 0.04545482(\ln(\text{API}) - 6.644518272)$$

$$X_2 = 0.014721569(\ln(T) - 11.67542323)$$

For typical Niger Delta crude oil

$$X_1 = \frac{52270000 \, T^{0.7327}}{\text{API}^{4.655}}$$

$$X_2 = 0.04928568(\ln(\text{API}) - 6.369426752)$$

$$X_2 = 0.015379637(\ln(T) - 11.40250855)$$

30. Whitson and Brule (2000)

$$\log \frac{\mu_{do}}{\rho_p} = \frac{1}{A} - 2.17$$

$$A = X_3 \left[K_w - \left(\frac{8.24}{\gamma_o} + 1.639 X_2 - 1.059 \right) \right]$$

$$X_1 = 1 + 8.69 \log \left[\frac{T + 459.67}{559.67} \right]$$

$$X_2 = 1 + 0.544 \log \left[\frac{T + 459.67}{559.67} \right]$$

$$X_3 = -0.1285 \left[\frac{(2.87 \times X_1 - 1) \times \gamma_o}{2.87 \times X_1 - \gamma_o} \right]$$

$$\rho_o = \frac{\gamma_o}{1 + 0.000321(T - 60) \times 10^{0.00462 \times {}^\circ \text{API}}}$$

FREE GAS SPECIFIC GRAVITY

1. McCain and Hill (1995)

$$\frac{1}{\gamma_{gR}} = \frac{a_1}{P} + \frac{a_2}{P^2} + a_3 P + \frac{a_4}{\sqrt{T}} + a_5 T + a_6 R_{sb} + a_7 {}^{\circ}API + \frac{a_8}{\gamma_{gsb}} + a_9 \gamma_{gsb}^2$$

where

$a_1 = -208.0797$

$a_2 = 22{,}885$

$a_3 = -0.000063641$

$a_4 = 3.38346$

$a_5 = -0.000992$

$a_6 = -0.000081147$

$a_7 = -0.001956$

$a_8 = 1.081956$

$a_9 = 0.394035$

TOTAL FORMATION VOLUME FACTOR

1. Al-Marhoun (1988)

$$B_t = 0.314693 + 0.106253 \times 10^{-4}$$
$$\times \left(R_s^{0.644516} \times \gamma_g^{-1.079340} \times API^{0.724874} \times T^{2.006210} \times P^{-0.761910} \right)$$
$$+ 0.18883 \times 10^{-10} \left(R_s^{0.644516} \times \gamma_g^{-1.079340} \times API^{0.724874} \times T^{2.006210} \times P^{-0.761910} \right)^2$$

2. Glaso (1980)

$$B_t = 10^{\left[(0.080135 + 0.47257 \times \log\left(\left(R_s \times (T - 460)^{0.5} \times API^{(2.9 \times 10^{(-0.00027 \times R_s)})}/API^{0.3} \right) \times P^{-1.1089} \right)) + \right.}$$
$$\left. 0.17351 \times \left[\log\left(\left(R_s \times (T - 460)^{0.5} \times API^{(2.9 \times 10^{(-0.00027 \times R_s)})}/API^{0.3} \right) \times P^{-1.1089} \right) \right]^2 \right]}$$

3. Standing (1947)

$$B_t = 10^{(-5.223 - (47.4/x))}$$

$$x = -12.22 + \log(z)$$

$$z = \left(10^{(y - (10.1 - (96.8/(6.604 + \log(p)))))} \right)$$

$$y = \left[R_s \times \frac{(T - 460)^{0.5} \times API^{2.9 \times 10^{(-0.000278 \times R_s)}}}{\gamma_g^{0.3}} \right.$$

MODIFIED BLACK OIL (VOLATILE OIL)

1. El-Banbi et al. (2006)

$$R_s = \gamma g \times \left[\left(\tfrac{p}{A1} + A2\right) \times 10^X\right]^{A3}$$

$$X = A4 \times API - A5 \times (T - 160)$$

A1	A2	A3	A4	A5
47.23306	−8.83351	1.325153	0.009176	−0.00039

$$R_s = A1 \times \gamma gs \times P^{A2} \times e^{[(A3 \times API)/T]}$$

$$\gamma gs = \gamma g \times \left[1 + 5.912 \times 10^{-5} \times API \times (T - 460) \times \log(Psep/114.7)\right]$$

A1	A2	A3
0.000547	1.607759	20.35994

$$B_o = A1 + A2 \times \left(R_s \times \sqrt{\tfrac{\gamma gs}{\gamma osc} + A3 \times (T - 460)}\right)^{A4}$$

A1	A2	A3	A4
0.839615	0.000461	2.013137	1.015821

$$R_V = \frac{A1 \times \rho_{gsc} \times (A2 \times p^2 + A3 \times p + A4)}{p_s} \times \exp\left[\frac{A5 \times CGR \times T_{sc}}{\rho_{osc} \times T \times P_{sc}}\right]$$

A1	A2	A3	A4	A5
1.225537	0.000107257	−0.19423	240.5499	8.321374

2. Nassar et al. (2013)

$$P_{pc} = A1 + A2 \times \gamma_{gs} + A3 + \gamma_{gs}^2$$

$$T_{pc} = B1 + B2 \times \gamma_{gs} + B3 + \gamma_{gs}^2$$

A1	A2	A3	B1	B2	B3
608.52	−45.8495	27.71169	29.70495	871.572	−502.604

$$B_g = \frac{5.04 \times z \times T}{p}$$

Bubble point pressure

$$P_{sat} = [A_0 \times R_{si}^{A1}] \times [(X+Y)^{A2}] \times [STO^{A3}] \times [Tr^{A4}]$$

$$X = \frac{SG1}{Psep1}$$

$$Y = \frac{SG2}{Psep2}$$

P_{sat} Correlation Parameters

Separator Stages	A0	A1	A2	A3	A4
2 Stages	0.064175	0.999745	−0.00524	−5.76605	1.726174
3 Stages	0.032743	1.056228	−0.00354	−6.21952	1.830545

Saturated gas–oil ratio

$$Rs = \frac{[(A0 \times P^2) + (A1 \times P) + A2] \times [e^{A3 \times X + A4 \times Y}]}{Psat \times e^{A5 \times V}}$$

$$X = \frac{SG1}{Psep1}$$

$$Y = \frac{SG2}{Psep2}$$

$$V = \frac{Tr}{P_{sat}} \times STO$$

R_S Correlation Parameters (With P_{sat})

Separator Stages	A0	A1	A2	A3	A4	A5
2 Stages	0.000488	0.634602	407.5338	−8.32096		3.483213
3 Stages	0.000426	0.545593	356.3234	1200.328	−524.589	4.001274

R_S Correlation Parameters (Without P_{sat})

Separator Stages	A0	A1	A2	A3	A4	A5
2 Stages	1E−07	0.000142	0.081392	−7.79741		0.000553
3 Stages	9.16E−08	0.000127	0.074463	988.2156	−433.27	0.000741

Saturated oil formation volume factor

$$B_o = \left[(A0 \times P^2) + (A1 \times P) + A2 \right] \times \left[10^{(A3 \times X + A4 \times Y)} \right] \times e^{A5 \times V}$$

B_o Correlation Parameters (With P_{sat})

Separator Stages	A0	A1	A2	A3	A4	A5
2 Stages	0.000181	0.293802	4381.696	1.63E−05		0.000654
3 Stages	0.000182	0.268328	4443.58	−1.1E−05	4.7E−05	0.000612

B_o Correlation Parameters (Without P_{sat})

Separator Stages	A0	A1	A2	A3	A4	A5
2 Stages	3.78E−08	7.4E−05	0.968007	1.38E−05		0.00063
3 Stages	3.82E−08	6.83E−05	0.983406	−1.2E−05	4.28E−05	0.000584

Under-saturated oil formation volume factor

$$B_o = (A0 \times P) + (A1 \times V) + (A2 \times P_{sat}) + (A3 \times B_{osat}) + (A4 \times X)$$

Undersat_B_o Correlation Parameters

Separator Stages	A0	A1	A2	A3	A4	A5
2 Stages	−7.1E−05	−8.6E−05	0.000102	0.951694	1.25E−06	
3 Stages	−7E−05	−8.8E−05	0.0001	0.950837	−5.4E−06	1.39E−05

$$X = \frac{Psep1}{SG1} \times STO$$

$$Y = \frac{Psep2}{SG2} \times STO$$

$$V = Tr \times STO$$

REFERENCES

Abdul-Majeed, G.H., Salman, N.H., 1988. Statistical evaluation of PVT correlations solution gas−oil ratio. Pet. Soc. Can. Available from: https://doi.org/10.2118/88-04-08.

Abdul-Majeed, G.H., Kattan, R.R., Salman, N.H., 1990. New correlation for estimating the viscosity of undersaturated crude oils. Pet. Soc. Can. Available from: https://doi.org/10.2118/90-03-10.

Abu-Khamsin, A., Al-Marhoun, M., 1991. Development of a new correlation for bubblepoint viscosity. Arab. J. Sci. Eng. 16, 99−106.

Ahmed, T.H., 1989. Hydrocarbon Phase Behavior, Vol. 7. Gulf Publishing Company, Houston, USA.

Ahmed, T., 1991. A practical equation of state. SPERE 6 (1), 137−146.

Ahmed, T., 2006. Reservoir Engineering Handbook, 3rd. ed. Gulf Publishing Company, Boston, MA, United States.

Ahmed, T., 2010. Reservoir Engineering Handbook, 4th. ed. Gulf Publishing Company, Boston, MA, United States.

Al-Khafaji, A.H., Abdul-Majeed, G.H., Hassoon, S.F., 1987. Viscosity correlation for dead, live and undersaturated crude oil. J. Petrol. Res. 6, 1−16.

Al-Marhoun, M.A., 1985. Pressure-volume-temperature correlations for Saudi crude oils. In: SPE 13718.

Al-Marhoun, M.A., 1988. PVT correlations for middle east crude oils. Soc. Pet. Eng. Available from: https://doi.org/10.2118/13718-PA.

Al-Marhoun, M.A., 1992. New correlations for formation volume factors of oil and gas mixtures. Pet. Soc. Can. Available from: https://doi.org/10.2118/92-03-02.

Al-Marhoun, M.A., 2003. The coefficient of isothermal compressibility of black oils. Soc. Pet. Eng. Available from: https://doi.org/10.2118/81432-MS.

Al-Marhoun, M.A., 2006. A New correlation for undersaturated isothermal oil compressibility, In: Paper SPE-81432-SUM, Society of Petroleum Engineers.

Almehaideb, R.A., 1997. Improved PVT correlations For UAE crude oils. Soc. Pet. Eng. Available from: https://doi.org/10.2118/37691-MS.

Al-Najjar, H.S., Al-Soof, N.B.A., Al-Khalisy, K.M., 1988. Correlations for bubble-point pressures, gas oil ratios and formation volume factors for Iraqi crude oils. J. Petrol. Res 7 (June), 13−35.

Al-Shammasi, A.A., 1999. Bubble point pressure and oil formation volume factor correlations. Soc. Pet. Eng . Available from: https://doi.org/10.2118/53185-MS.

Andrade, E.N., 1934. A theory of viscosity of liquids. Philos. Mag. 17, 497−511.

Arps, J.J., 1962. Estimation of Primary Oil and Gas Reserve. Frick, T, C., Petroleum Production Handbook, Volume II, Dallas, SPE, 1962, Chapter 37, P.I.

Asgarpour, S., McLauchlin, L., Wong, D., Cheung, V., 1989. Pressure-volume-temperature correlations for western Canadian gases and oils. JCPT 28 (4), 103−111. SPE-89-04-08.

Beal, C., 1946. Viscosity of air, water, natural gas, crude oil and its associated gases at oil field temperature and pressures. Trans AIME 165, 94−115.

Beggs, H.D., Robinson, J.R., 1975. Estimating the viscosity of crude oil systems. JPT 27, 1140−1141.

Bennison, T., 1998. Prediction of heavy oil viscosity. In: Presented at the IBC Heavy Oil Field Development Conference, London.

Bergman, D.F., 2004. Don't forget viscosity. In: Presented at the Petroleum Technology Transfer Council 2nd Annual Reservoir Engineering Symposium, Lafayette, Louisiana.

Bolondarzadeh, A., et al., 2006. The new PVT generated correlations of Iranian oil properties. In: Fourth Iranian Petroleum Engineering Student Conference.

Boukadi, F.H., Bemani, A.S., Hashemi, A., 2004. Pressure-volume-temperature empirical formulations for understaturated Omani oils. Petrol. Sci. Technol. 22, 1251−1259. Available from: https://doi.org/10.1081/LFT-200034078.

Burnett, E.S., 1936. Application of the Burnett method of compressibility determinations to multiphase fluid mixtures. J. Appl. Mech. Trans. ASME 58, A136.

Calhoun Jr., J.C., 1947. Fundamentals of Reservoir Engineering. University of Oklahoma Press, Norman, Oklahoma, p. 35.

Casey, J.M., Cronquist, C., 1992. Estimate GOR and FVF using dimensionless PVT analysis. World Oil 213, 83−87.

Chew, J., Connally Jr., C.A., 1959. A viscosity correlation for gas-saturated crude oils, Transactions of the American Institute of Mining, Metallurgical, and Petroleum Engineers, Vol. 216. Society of Petroleum Engineers of AIME, Dallas, Texas, p. 23.

De Ghetto, G., Paone, F., and Villa, M., 1995. Pressure-volume-temperature correlations for heavy and extra heavy oils. In: Presented at the SPE International Heavy Oil Symposium, Calgary, 19−21 June. SPE-30316-MS.

Dindoruk, B., Christman, P.G., 2001. PVT properties and viscosity correlations for Gulf of Mexico oils. Soc. Pet. Eng . Available from: https://doi.org/10.2118/71633-MS.

Dindoruk, B., Christman, P.G., 2004. PVT properties and viscosity correlations for Gulf of Mexico oils. Soc. Pet. Eng . Available from: https://doi.org/10.2118/89030-PA.

Doklah, M. and Osman, M., 1992. Correlation of PVT properties for UAE crudes. In: SPE Formation Evaluation. pp. 41−46.

Egbogah, E.O., Ng, J.T., 1990. An improved temperature-viscosity correlation for crude oil systems. J. Pet. Sci. Eng. 4 (3), 197−200.

Elam, F.M., (1957). Prediction of Bubble Point Pressures and Formation Volume Factors from Field Data. MS Thesis, University of Texas at Austin, Austin, Texas.

El-Banbi, A.H., Abdel Fattah, K.A., Sayyouh, M.H., 2006. New modified black oil correlations for gas condensate and volatile oil fluids. In: Paper SPE 102240 Presented at the SPE Annual Technical Conference and Exhibition, 24−27 September, San Antonio, TX.

Elmabrouk, S. Zekri, A., and Shirif, E., 2010. Prediction of bubblepoint pressure and bubblepoint oil formation volume factor in the absence of PVT analysis. In: Paper SPE 137368 Presented at the SPE Latin American & Caribbean Petroleum Engineering Conference, Lima, Peru, 1-3 Dec. doi:10.2118/137368-MS.

Elsharkawy, A.M., Alikhan, A.A., 1997. Correlations for predicting solution gas/oil ratio, oil formation volume factor, and undersaturated oil compressibility. J. Pet. Sci. Eng. 17, 291−302.

Elsharkawy, A.M., Alikhan, A.A., 1999. Models for predicting the viscosity of middle east crude oils. Fuel 78 (8), 891−903.

Elsharkawy, A.M., Gharbi, R.B.C., 2001. Comparing classical and neural regression techniques in modelling crude oil viscosity. Adv. Eng. Softw. 32 (3), 215−224.

Farshad, F.F., Leblance, J.L., Garber J.D. Osorio J.G., 1996. A new correlation for bubble point pressure according to the separator conditions. In: SPE Latin America/Caribbean Petroleum Engineering Conference, 23−26 April, Port-of-Spain, Trinidad.

Fitzgerald, D.J., 1994. A predictive method for estimating the viscosity of undefined hydrocarbon liquid mixtures (MS thesis). Pennsylvania State University, State College, Pennsylvania, United States.

Glaso, O., 1980. Generalized pressure-volume-temperature correlations. JPT 32, 785−795.

Hanafy, H.H., Macary, S.A., Elnady, Y.M., Bayomi, A.A. and El-Batanoney, M.H., 1997. Empirical PVT correlations applied to Egyptian crude oils exemplify significance of using regional correlations. In: Paper SPE 37295 Presented at the SPE International Symposium on Oilfield Chemistry, Houston.

Hassan, O.F., 2011. Correlation for solution gas-oil ratio of Iraqi oils at pressures below the bubble point pressure. Iraq. J. Chem. Petrol. Eng. 12 (2), 1−8. June 2011.

Hemmati M.N., Kharrat R., 2007. A correlation approach for prediction of crude oil PVT properties. In: 15th SPE Middle East Oil and Gas Show, Bahrain.

Hossain, M.S., Sarica, C., Zhang, H.-Q., Rhyne, L., Greenhill, K.L., 2005. Assessment and development of heavy oil viscosity correlations. Soc. Pet. Eng . Available from: https://doi.org/10.2118/97907-MS.

Ikiensikimama, S., 2008. Oil and gas PVT characterization (Ph.D. thesis). University of Lagos, Akoka, Nigeria.

Ikiensikimama, S. and Ogboja, O., 2009. New bubble point pressure empirical PVT correlation. In: Paper SPE 128893 Presented at the SPE Technical Conference and Exhibition in Abuja, Nigeria.

Karimnezhad, M., Heidarian, M., Kamari, M., Jalalifar, H., 2014. A new empirical correlation for estimating bubble point oil formation volume factor. J. Nat. Gas Sci. Eng. 18, 329−335.

Kartoatmodjo, T., Schmidt, Z., 1994. Large data bank improves crude physical property correlation. Oil Gas J 4, 51−55.

Kartoatmodjo, T.R.S. and Schmidt, Z., 1991. New Correlations for crude oil physical properties. In: Paper SPE 23556-MS. Society of Petroleum Engineers, Richardson, TX, unsolicited paper.

Kaye, S.E., 1985. Offshore California Viscosity Correlations. COFRC, TS85000940.

Khairy, M., El-Tayeb, S., Hamdallah, M., 1998. PVT correlations developed for Egyptian crudes. Oil Gas J. 96, 114−116.

Khamehchi, R., Ebrahimian, R., 2009. Novel empirical correlations for estimation of bubble point pressure, saturated viscosity and gas solubility of crude oils. Pet. Sci. 6, 86−91. Available from: https://doi.org/10.1007/s12182-009-0016-x.

Khan, S.A., Al-Marhoun, M.A., Duffuaa, S.O., Abu-Khamsin, S.A., 1987. Viscosity correlations for Saudi Arabian crude oils. Soc. Pet. Eng . Available from: https://doi.org/10.2118/15720-MS.

Kouzel, B., 1965. How pressure affects liquid viscosity. Hydrocarb. Process. Petrol. Refiner 44 (3), 120.

Kouzel, B., 1997. Standard: API - Technical Data Book Petroleum Refining. American Petroleum Institute, Washington DC, United States.

Labedi, R., 1992. Improved correlations for predicting the viscosity of light crudes. J. Pet. Sci. Eng. 8, 221−234.

Labedi, R.M., 1982. PVT Correlations of the African Crudes. PhD Thesis. 1982. PhD Thesis, Colorado School of Mines, Leadville, Colorado.

Labedi R.M., 1990. Use of production data to estimate the saturation pressure solution GOR and chemical composition of reservoir fluids. In: Presented at the SPE Latin America Petroleum. Engineering Conference Rio de Janeiro Brazil, pp: 14−19.

Lasater, J.A., 1958. Bubble point pressure correlations. J. Pet. Technol. 10 (5), 65−67. SPE-957-G.

Levitan, L.L., Murtha, M., 1999. New correlations estimate Pb. FVF. Oil Gas J. 97, 70−76.

Macary, S.M. and El-Batanoney, M.H., 1992. Derivation of PVT correlations for the Gulf of Suez crude oils. In: Proc., 11th EGPC Petroleum Exploration and Production Conference, Cairo, Egypt, Vol. 1, 374.

Mazandarani, M.T., and Asghari, S.M., 2007. Correlations for predicting solution gas–oil ratio, bubble point pressure and oil formation volume factor at bubble-point of Iran crude oils. In: European Congress of Chemical Engineering (ECCE-6) Copenhagen.

McCain, W.D., 1991. Reservoir-fluid property correlations-state of the art (includes associated papers 23583 and 23594). Soc. Pet. Eng . Available from: https://doi.org/10.2118/18571-PA.

McCain, W.D., & Hill, N.C., 1995. Correlations for liquid densities and evolved gas specific gravities for black oils during pressure depletion. In: Paper SPE 30773. Presented at the SPE Annual Technical Conference and Exhibition, Dallas. Doi: 10.2118/30773-MS.

Mehran, F. Movagharnejad, K. and Didanloo, A., 2006. New correlation for estimation of formation volume factor and bubble point pressure for Iranian oil fields. In: First Iranian Petroleum Engineering Conference, Tehran.

Moradi, B., Malekzadeh, E., Amani, M., Boukadi, F.H., and Kharrat, R., 2010. Bubble point pressure empirical correlation. In: Paper SPE 132756 Prepared for Presentation at the Trinidad and Tobago Energy Resources Conference held in Port of Spain, Trinidad.

Moradi B., Malekzadeh,E., Shoushtari, M.A., Mariyamni, A., Moradi, P., 2013. New Oil Formation Volume Factor Empirical Correlation for Middle East Crude Oils.

Movagharnejad, Fasih, 1999. The new correlation for prediction bubble point pressure and oil formation volume factor for Iranian reservoirs. Research Institute of Petroleum Industry, National Iranian Oil Company, Iran.

Naseri, A., Nikazar, M., Mousavi-dehghani, S.A., 2005. A correlation approach for prediction of crude oil viscosity. J. Pet. Sci. Eng. 47, 163–174.

Naseri, A., Yousefi, S.H., Sanaei, A., Gharesheikhlou, A.A., 2012. A neural network model and an updated correlation for estimation of dead crude oil viscosity. Braz. J. Pet. Gas ISSN: 1982-0593.

Nassar, I.S., El-Banbi, A.H., Sayyouh, M.H.M., 2013. Modified black oil PVT properties correlations for volatile oil and gas condensate reservoirs. Soc. Pet. Eng . Available from: https://doi.org/10.2118/164712-MS.

Ng, J.T.H., and Egbogah, E.O., 1983. An improved temperature-viscosity correlation for crude oil systems. In: Presented at the Annual Technical Meeting, Banff, Canada. PETSOC-83-34-32.

Obomanu, D.A., Okpobiri, G.A., 1987. Correlating the PVT properties of Nigerian crudes. J. Energy Resour. Technol. 109 (4), 214–217.

Okoduwa, I.G., and Ikiensikimama, S.S., 2010. Bubble point pressure correlations for Niger Delta crude oils. In: Paper SPE 136968 Prepared for Presentation at the 34th Annual SPE International Conference and Exhibition Held in Tinapa, Calabar, Nigeria.

Omar, M.I., and Todd, A.C., 1993. Development of new modified black oil correlations for Malaysian crudes. In: Presented at the SPE Asia Pacific Oil and Gas Conference, Singapore, 8–10 February. SPE-25338-MS.

Owolabi, O.O., 1984. Reservoir Fluid Properties of Alaskan Crudes. MS Thesis, University of Alaska, Fairbanks, Alaska.

Oyedeko, K.F., Ulaeto, U.W., 2011. Predicting the dead oil viscosity of reservoir fluids: a case study of the Niger Delta. J. Energy Technol. Policy 3 (13), 2013.

Petrosky, G.E., Farshad, F.F., 1993. Pressure-volume-temperature correlations for Gulf of Mexico crude oils. Soc. Pet. Eng . Available from: https://doi.org/10.2118/26644-MS.

Petrosky, G.E., Farshad, F.F., 1995. Viscosity correlations for Gulf of Mexico crude oils. Soc. Pet. Eng . Available from: https://doi.org/10.2118/29468-MS.

Petrosky, G.E., Farshad, F., 1998. Pressure-volume-temperature correlations for Gulf of Mexico crude oils. Soc. Pet. Eng . Available from: https://doi.org/10.2118/51395-PA.

Petrosky, G.E.J., 1990. PVT Correlations for Gulf of Mexico Crude Oils. MSC Thesis University of Southwestern Louisiana, Lafayette Louisiana USA.

Rollins, J.B., McCain Jr., W.D., Creeger, J.T., 1990. Estimation of solution GOR of black oils. J. Pet. Tech. 42, 92–94. January. Transactions, AIME. Vol. 289.

Spivey, J.P., Valko, P.P., McCain Jr., W.D., 2007. Applications of the coefficient of isothermal compressibility to various reservoir situations with new correlations for each situation. Soc. Pet. Eng. Reservoir Eval. Eng 10 (February), 43–49.

Standing, M.B., 1947. A pressure-volume-temperature correlation for mixtures of California oils and gases, drilling and production practice, American Petroleum Institute, 1st January, New York, United States.

Standing, M.B., 1962. Oil-system correlation, Petroleum production handbook. McGraw-Hill Book Co, New York City, USA.

Standing, M.B., 1979. A set of equations for computing equilibrium ratios of a crude oil/natural gas system at pressures below 1,000 psia. JPT 31 (9), 1193–1195.

Standing, M.B., 1981. Volumetric and Phase Behavior of Oil Field Hydrocarbon Systems, ninth ed Society of Petroleum Engineers of AIME, Richardson, Texas.

Sulaimon, A.A., Ramli, N., Adeyemi, B.J. and Saaid, I.M., 2014. New correlation for oil formation volume factor. In: Paper SPE 172396 Prepared for Presentation at the SPE Nigeria Annual International Conference and Exhibition Held in Lagos, Nigeria.

Twu, C.H., 1985. Internally consistent correlation for predicting liquid viscosities of petroleum fractions. Ind. Eng. Chem. Process Des. Dev. 34 (4), 1287–1293.

Ulaeto, U., Oyedeko, K.F.K., 2014. Improved dead oil viscosity model. J. Energy Technol. Policy 4 (7).

Valko, P.P., McCain Jr., W.D., 2003. Reservoir oil bubblepoint pressures revisited; solution gas-oil ratios and surface gas specific gravities. J. Pet. Sci. Eng. 37, 153–169. Available from: https://doi.org/10.1016/S0920-4105(02)00319-4.

Vasquez, M., Beggs, H.D., 1980. Correlation for fluid physical property predictions. JPT 32, 968–970.

Velarde, J., Blasingame, T.A., McCain, W.D., 1997. Correlation of black oil properties at pressures below bubble point pressure—a new approach. Pet. Soc. Can . Available from: https://doi.org/10.2118/97-93.

Whitson, C.H., Brule, M.R., 2000. Phase Behavior, 20. SPE Monograph, TX, United States, SPE Richardson.

Appendix B

Gas Correlations Formulae

Appendix B contains a large number of gas correlations formulae. The correlations are presented in alphabetical order and use oil field system of units. Correlations for each PVT (pressure, volume, and temperature) property are grouped under the following names:

- Gas Formation Volume Factor (rft^3/scf)
- Gas Compressibility (psi^{-1})
- Gas Density (lbm/ft^3)
- Gas Viscosity (cp)
- Single-Phase Z-Factor
- T_c and P_c Using Specific Gravity ($^\circ$R and psia)
- T_c and P_c for Pseudo Components ($^\circ$R and psia)
- T_c and P_c Using Composition ($^\circ$R and psia)
- Dew Point Pressure Using Specific Gravity (psia)
- Dew Point Pressure Using Composition (psia)
- Gas Condensate Specific Gravity
- Two-Phase Z-Factor
- Condensate Gas Ratio Correlations (STB/MMscf)
- Modified Black Oil (Gas Condensate)

GAS FORMATION VOLUME FACTOR

1. General equation

$$B_{gdry} = 0.0283 \frac{Z\,T}{P}$$

$$M_o = \frac{6084}{API - 5.9}$$

$$B_{g\,wet} = \frac{Z \times T}{P} \times \left[\frac{0.0282\,R_s + \left(\left(3758 \times \left(141.5/(131.5 + API)\right)\right)/M_o\right)}{\left[R_s + 1330000\left(\left(141.5/(131.5 + API)\right)\right)/M_o\right]} \right]$$

GAS COMPRESSIBILITY

1. McCain (1991)

$$\rho_{pr} = 0.27 \left[\frac{p_{pr}}{Z\,T_{pr}} \right]$$

$$C_{pr} = \frac{1}{P_{pr}} - \frac{0.27}{Z^2 T_{pr}} \left[\frac{(\partial Z/\partial P_{pr})_{T_{pr}}}{1 + (\partial P_{pr}/z)(\partial Z/\partial P_{pr})_{T_{pr}}} \right]$$

$$\left(\frac{\partial Z}{\partial \rho_{pr}} \right)_{T_{pr}} = \left(A_1 + A_2/T_{pr} + A_3/T_{pr}^3 + A_4/T_{pr}^4 + A_5/T_{pr}^5 \right) + 2\rho_{pr}\left(A_6 + A_7/T_{pr} + A_8/T_{pr}^2 \right)$$

$$- 5\rho_{pr}\left(A_7 + A_8/T_{pr}^2 \right) + \frac{2A_{10}\rho_{pr}}{T_{pr}^3}\left(1 + A_{11}\rho_{pr}^2 - A_{11}^2\rho_{pr}^4 \right) e^{\left(-A_{11}\rho_{pr}^2 \right)}$$

GAS DENSITY

1. General equation

$$\rho = \frac{P \times MW}{z \times R \times T}$$

GAS VISCOSITY

1. Carr et al. (1954)

$$\mu_{gsc} = \left(1.709 \times 10^{-5} - 2.062 \times 10^{-6} \gamma_g\right)(T - 459.67) + 8.188$$
$$\times 10^{-3} - 6.15 \times 10^{-3} \log(\gamma_g)$$

$$\ln\left(\frac{\mu_g}{\mu_{gsc}} T_{pr}\right) = a_0 + a_1 p_{pr} + a_2 p_{pr}^2 + a_3 p_{pr}^3 + T_{pr}\left(a_4 + a_5 p_{pr} + a_6 p_{pr}^2 + a_7 p_{pr}^3\right)$$
$$+ T_{pr}^2\left(a_8 + a_9 p_{pr} + a_{10} p_{pr}^2 + a_{11} p_{pr}^3\right) + T_{pr}^3\left(a_{12} + a_{13} p_{pr} + a_{14} p_{pr}^2 + a_{15} p_{pr}^3\right)$$

i	a_i
0	−2.4621182
1	2.97054714
2	−0.286264054
3	0.00805420522
4	2.80860949
5	−3.49803305
6	0.360373020
7	−0.0104432413
8	−0.793385684
9	1.39643306
10	−0.149144925
11	0.00441015512
12	0.0839387178
13	−0.186408848
14	0.0203367881
15	−0.000609579263

2. Chen and Ruth (1993)

$$\mu_g = (a_1 + a_2 T_k) - (a_3 + a_4 T_k)\sqrt{MW}$$

$$\frac{\mu}{\mu_g} = (b_1 + b_2 P_r + b_3 P_r^2)\frac{1}{T_r} + (b_4 + b_5 P_r + b_6 P_r^2)\frac{1}{T_r^4} + (b_7 + b_8 P_r + b_9 P_r^2)$$

a_1	0.38539×10^{-2}
a_2	0.00356×10^{-2}
a_3	0.04131×10^{-2}
a_4	0.00016×10^{-2}
b_1	-0.4888439
b_2	-0.0943952
b_3	0.0199591
b_4	0.8266923
b_5	1.7124100
b_6	-0.0700968
b_7	1.2076900
b_8	0.0301188
b_9	-0.0048318

3. Dean and Stiel (1965)

$$\rho_r = \frac{0.27 \times P_{pr}}{z T_{pr}}$$

$T_{pr} \le 1.5$	$T_{pr} > 1.5$
m_1	m_1
$= \left(34 \times 10^{-5} \times T_{pr}^{(8/9)}\right)/e_m$	$= \left(166.8 \times 10^{-5} \times (0.1338 T_{pr} - 0.0932)^{(5/9)}\right)/e_m$

$$e_m = \frac{5.4402 \times T_{pc}^{1/6}}{M_a^{0.5} P_c^{2/3}}$$

$$\mu_g = m_1 + \frac{10.8 \times 10^{-5} \exp(1.439 \times \rho_r) - \exp\left(-1.111 \times \rho_r^{1.888}\right)}{e_m}$$

4. Dempsey (1965)

$$\ln\left[T_{pr}\frac{\mu_g}{\mu_1}\right] = a_0 + a_1 P_{pr} + a_2 P_{pr}^2 + a_3 P_{pr}^3 + T_{pr}\left(a_4 + a_5 P_{pr} + a_6 P_{pr}^2 + a_7 P_{pr}^3\right)$$
$$+ T_{pr}^2\left(a_8 + a_9 P_{pr} + a_{10} P_{pr}^2 + a_{11} P_{pr}^3\right) + T_{pr}^3\left(a_{12} + a_{13} P_{pr} + a_{14} P_{pr}^2 + a_{15} P_{pr}^3\right)$$

$a_0 = -2.46211820$ $a_8 = -7.93385684 \times 10^{-1}$

$a_1 = 2.97054714$ $a_9 = 1.39643306$

$a_2 = -2.86264054 \times 10^{-1}$ $a_{10} = -1.49144925 \times 10^{-1}$

$a_3 = 8.05420522 \times 10^{-3}$ $a_{11} = 4.41015512 \times 10^{-1}$

$a_4 = 2.80860949$ $a_{12} = 8.39387178 \times 10^{-2}$

$a_5 = -3.49803305$ $a_{13} = -1.86408848 \times 10^{-1}$

$a_6 = 3.60373020 \times 10^{-1}$ $a_{14} = 2.03367881 \times 10^{-2}$

$a_7 = -1.044324 \times 10^{-2}$ $a_{15} = -6.09579263 \times 10^{-4}$

5. Elsharkawy (2006)

$$\Delta\mu_g = y_{H_2S}\left(-3.2268 \times 10^{-3}\log\gamma_g + 2.1479 \times 10^{-3}\right)$$
$$\Delta\mu_g = y_{CO_2}\left(6.4366 \times 10^{-3}\log\gamma_g + 6.7255 \times 10^{-3}\right)$$
$$\Delta\mu_g = y_{C_{7+}}\left(-3.2875 \times 10^{-1}\log\gamma_g + 1.2885 \times 10^{-1}\right)$$
$$\mu_g = D_1 \times 10^{-4}\exp\left(D_2 \times \rho_g^{D_3}\right)$$
$$D_1 = \frac{\left(9.379 + 0.01607 M_g\right) T^{1.5}}{\left(209.2 + 19.26 M_g + T\right)}$$
$$D_2 = 3.448 + \left(\frac{986.4}{T}\right) + 0.01009 M_g$$
$$D_3 = 2.447 - 0.224 \times A_2$$

6. Lee et al. (1966)

$$\mu_g = A(10^{-4})e^{B\rho_g^C}$$

where

$$A = \frac{(9.379 + 0.01607M)T^{1.5}}{209.2 + 19.26M + T}$$

$$B = 3.448 + \frac{986.4}{T} + 0.01009M$$

$$C = 2.447 - 0.2224B$$

7. Londono et al. (2002)

$$\mu_g = 10^{-4}\mathrm{Ke}^{\left(X\rho_g^Y\right)}$$

where

$$\mu_{gsc}^{10^4} = K = \frac{(16.7175 + 0.0419188M)T^{1.40256}}{212.209 + 18.1349M + T}$$

$$Y = 1.09809 - 0.0392851X$$

8. Londono et al. (2005)

$$\mu_g = \mu_{1\,\mathrm{atm}} + f(\rho)$$

$$f(\rho) = \frac{a + b\rho + c\rho^2 + d\rho^3}{e + f\rho + g\rho^2 + h\rho^3}$$

where

$a = a_0 + a_1 T + a_2 T^2$

$b = b_0 + b_1 T + b_2 T^2$

$c = c_0 + c_1 T + c_2 T^2$

$d = d_0 + d_1 T + d_2 T^2$

$e = e_0 + e_1 T + e_2 T^2$

$f = f_0 + f_1 T + f_2 T^2$

$g = g_0 + g_1 T + g_2 T^2$

$h = h_0 + h_1 T + h_2 T^2$

$$\ln(\mu_{1atm}) = \left[\frac{a_0 + a_1\ln(\gamma_g) + a_2\ln(T) + a_3\ln(\gamma_g)\ln(T)}{1 + b_1\ln(\gamma_g) + b_2\ln(T) + b_3\ln(\gamma_g)\ln(T)}\right]$$

i	a_i	b_i
0	-6.39821	—
1	-0.6045922	0.069718
2	0.749768	-0.1013889
3	0.1261051	-0.0215294

and

i	0	1	2
a_i	0.953363	-1.07384	0.00131729
b_i	-0.971028	11.2077	0.09013
c_i	1.01803	4.98986	0.302737
d_i	-0.990531	4.17585	-0.63662
e_i	1	-3.19646	3.90961
f_i	-1.00364	-0.181633	-7.79089
g_i	0.99808	-1.62108	0.000634836
h_i	-1.00103	0.676875	4.62481

$$\mu_{gsc}\varepsilon = 10^{-4}\left[0.807T_{pr}^{0.618} - 0.357\exp(-0.449T_{pr}) + 0.340\exp(-4.058T_{pr} + 0.018)\right]$$

$$\frac{\mu_g}{\mu_{gsc}} = 1 + \frac{A_1 P_{pr}^{1.3088}}{A_2 P_{pr}^{A_5} + \left(1 + A_3 P_{pr}^{A_4}\right)^{-1}}$$

$$\varepsilon = 0.9490\left[\frac{T_{pc}}{M^3 P_{pc}^4}\right]^{1/6}$$

9. Lucas (1981)

$$A_1 = \frac{(1.245 \times 10^{-3})\exp\left(5.1726 T_{pr}^{-0.3286}\right)}{T_{pr}}$$

$$A_2 = A_1(1.6553 T_{pr} - 1.2723)$$

$$A_3 = \frac{0.4489\exp\left(3.0578 T_{pr}^{-37.7332}\right)}{T_{pr}}$$

$$A_4 = \frac{1.7368\exp\left(2.2310 T_{pr}^{-7.6351}\right)}{T_{pr}}$$

$$A_5 = 0.9425\exp\left(-0.1853 T_{pr}^{0.4489}\right)$$

10. Jossi et al. (1962)

$$\mu_{gsc} = \frac{34.0 \times 10^{-5} \times T_r^{0.94}}{\varepsilon} \quad \text{for } T_r \leq 1.5$$

$$\mu_{gsc} = \frac{17.78 \times 10^{-5} \times (4.58 T_r - 1.67)^{5/8}}{\varepsilon} \quad \text{for } T_r > 1.5$$

$$\varepsilon = 5.4398\left[\frac{T_{pc}}{M^3 P_{pc}^4}\right]$$

$$\rho_r = \frac{0.27 P_{pr}}{Z T_{pr}}$$

$$\left[\left((\mu_g - \mu_{gsc})\varepsilon + 10^{-4}\right)^{1/4}\right] = 0.10230 + 0.023364\rho_r$$
$$+\, 0.058533\rho_r^2 - 0.040758\rho_r^3 + 0.0093324\rho_r^4$$

11. McCain (1991)
12. Standing (1947)

$$\mu_1 = \mu_{1\,\text{uncorrected}} + \Delta\mu_{N_2} + \Delta\mu_{CO_2} + \Delta\mu_{H_2S}$$

$$\mu_{1\,\text{uncorrected}} = \left(3.0764 \times 10^{-5} - 3.712 \times 10^{-6} \times \gamma_g\right)$$
$$(T - 256) + 8.188 \times 10^{-3} - 6.15 \times 10^{-3}\log(\gamma_g)$$

$$\Delta\mu_{N_2} = y_{N_2} \times \left[8.48 \times 10^{-3}\log(\gamma_g) + 9.59 \times 10^{-3}\right]$$
$$\Delta\mu_{CO_2} = y_{CO_2} \times \left[9.08 \times 10^{-3}\log(\gamma_g) + 6.24 \times 10^{-3}\right]$$
$$\Delta\mu_{H_2S} = y_{H_2S} \times \left[8.49 \times 10^{-3}\log(\gamma_g) + 3.73 \times 10^{-3}\right]$$

13. Sutton (2007)

$$\mu_g = \mu_{gsc} e^{[X v_g^Y]}$$

where

$$X = 3.47 + \frac{1588}{T} + 0.0009M$$

$$Y = 1.66378 - 0.04679X$$

$$\mu_{gsc} = 10^{-4} \left[0.807 T_{pr}^{0.618} - 0.357 e^{-(0.449 T_{pr})} + 0.34 e^{(-4.058 T_{pr})} + 0.018 \right]$$

$$\xi = 0.949 \left(\frac{T_{pc}}{M^3 p_{pc}^4} \right)^{1/6}$$

SINGLE-PHASE Z-FACTOR

1. Azizi et al. (2010)

$$Z = A + \frac{B + C}{D + E}$$

$$A = a \times T_{pr}^{2.16} + b \times P_{pr}^{1.028} + c \times P_{pr}^{1.58} \times T_{pr}^{-2.1} + d\ln\left(T_{pr}^{-0.5}\right)$$

$$B = e + f T_{pr}^{2.4} + g P_{pr}^{1.56} + h P_{pr}^{0.124} T_{pr}^{3.033}$$

$$C = i\ln\left(T_{pr}^{-1.28}\right) + j\ln\left(T_{pr}^{1.37}\right) + k\ln(P_{pr}) + l\ln\left(P_{pr}^2\right) + m\ln(P_{pr})\ln(T_{pr})$$

$$D = 1 + n T_{pr}^{5.55} + o P_{pr}^{0.68} T_{pr}^{0.33}$$

$$E = p\ln\left(T_{pr}^{1.18}\right) + q\ln\left(T_{pr}^{2.1}\right) + r\ln(P_{pr}) + s\ln\left(P_{pr}^2\right) + t\ln(P_{pr})\ln(T_{pr})$$

a	0.0373142485385592
b	-0.0140807151485369
c	0.0163263245387186
d	-0.0307776478819813
e	13843575480.943800
f	-16799138854.763700

g	1624178942.6497600
h	1370227028 1.086900
i	$-4164550 9.896474600$
j	2372499 67625.01300
K	$-24449 114791.1531$
l	19357955749.3274
m	$-12635 4717916.607$
n	623705678.385784
o	1799765 1104.3330
p	151211393445.064
q	139474437997.172
r	24233012984.0950
s	18938047327.5205
t	$-14140 1620722.689$

2. Beggs and Brill (1978)

$$Z = A + \frac{1-A}{e^B} + C \times P_{pr}^D$$

where

$$A = 1.39(T_{pr} - 0.92)^{0.5} - 0.36 T_{pr} - 0.101$$

$$B = (0.62 - 0.23 T_{pr})P_{pr} + \left(\frac{0.066}{T_{pr} - 0.86} - 0.037\right)P_{pr}^2 + \frac{0.32 P_{pr}^6}{10^E}$$

$$C = 0.132 - 0.32 \times \log(T_{pr})$$

$$D = 10^F$$

$$E = 9(T_{pr} - 1)$$

$$F = 0.3106 - 0.49 \times T_{pr} + 0.1824 \times T_{pr}^2$$

3. Burnett (1979)

$$N = (1.1 + 0.26 T_r + (1.04 - 1.42 T_r) \times U)\frac{\exp(U)}{T_r}$$

$$U = \frac{P_r}{P_r'}$$

$$P_r' = 21.46 Z' - 11.9 Z'^2 - 5.9$$

$$Z' = 0.3379 \log(\log(T_r)) + 1.091$$

$$Z = 1 + (Z' - 1)(\sin(90U))^N$$

4. Dranchuk and Abou-Kassem (1975)

$$Z = 1 + \left(A_1 + A_2/T_{pr} + A_3/T_{pr}^3 + A_4/T_{pr}^4 + A_5/T_{pr}^5\right)\rho_{pr} + \left(A_6 + A_7/T_{pr} + A_8/T_{pr}^2\right)\rho_{pr}^2$$

$$- A_9\left(A_7/T_{pr} + A_8/T_{pr}^2\right)\rho_{pr}^5 + A_{10}\left(1 + A_{11}\rho_{pr}^2\right)\left(\rho_{pr}^2/T_{pr}^3\right)e^{\left(-A_{11}\rho_{pr}^5\right)}$$

where

$$\rho_{pr} = 0.27\left[\frac{p_{pr}}{z T_{pr}}\right]$$

and

i	A_i
1	0.3265
2	− 1.07
3	− 0.5339
4	0.01569
5	− 0.05165
6	0.5475
7	− 0.7361
8	0.1844
9	0.1056
10	0.6134
11	0.721

5. Dranchuk et al. (1974)

$$Z = 1 + (A_1 + A_2/T_r + A_3/T_r^3)Y + (A_4 + A_5/T_r)Y^2 + (A_5 A_6/T_r)Y^5 + \left(A_7/T_r^3 Y^2(1 + A_8 Y^2)e^{(-A_8 Y^2)}\right)$$

$$Z = \frac{0.27 P_r}{(YT_r)}$$

where

$$f(Y) = AY^6 + BY^3 + CY^2 + DY + EY^3(1 + fY^3)e^{(-fY^2)} - G$$

i	A_i	Coefficients	Values
1	0.31506237	A	0.06423
2	-1.0467099	B	$0.5353 T_r - 0.6123$
3	-0.57832729	C	$0.3151 T_r^{-1.0467} - (0.5783/T_r^3)$
4	0.53530771		
5	-0.61232032	D	T_r
6	-0.10488813	E	$0.5783/T_r^2$
7	0.6857001	F	0.6845
8	0.68446549	G	$0.27 P_r$

6. Gopal (1977)

$0.2 \leq P_r \leq 5.4$

$1.05 \leq T_r \leq 3.0$

$Z = P_r(AT_r + B) + CT_r + D$

$5.4 \leq P_r \leq 15$

$1.05 \leq T_r \leq 3.0$

$Z = P_r(0.711 + 3.66T_r) - 1.637/(0.319T_r + 0.522) + 2.071$

P_r range between	T_r Range between	A	B	C	D
0.2 and 1.2	1.05 and 1.2	1.6643	−2.2114	−0.3647	1.4385
	1.2+ and 1.4	0.5222	−0.8511	−0.0364	1.0490
	1.4+ and 2.0	0.1391	−0.2988	0.0007	0.9969
	2.0+ and 3.0	0.0295	−0.0825	0.0009	0.9967
1.2+ and 2.8	1.05 and 1.2	−1.357	1.4942	4.6315	−4.7009
	1.2+ and 1.4	0.1717	−0.3232	0.5869	0.1229
	1.4+ and 2.0	0.0984	−0.2053	0.0621	0.8580
	2.0+ and 3.0	0.0211	−0.0527	0.0127	0.9549
2.8+ and 5.4	1.05 and 1.2	−0.3278	0.4752	1.8223	−1.9036
	1.2+ and 1.4	−0.2521	0.3871	1.6087	−1.6635
	1.4+ and 2.0	−0.0284	0.0625	0.4714	−0.0011
	2.0+ and 3.0	0.0041	0.0039	0.0607	0.7927

7. Hall and Yarborough (1973)

$$Z = \left[\frac{0.06125 P_{pr} t}{Y}\right] e^{(-1.2(1-t)^2)}$$

Reduced density $Y = -0.06125 P_{pr} t \, e^{-1.2(1-t)^2} + \dfrac{Y + Y^2 + Y^3 - Y^4}{(1-Y)^3} - (14.76t - 9.76t^2 + 4.58t^3)Y^2$

$$+ (90.7t - 242.2t^2 + 4.58t^3)Y^{(2.18+2.82t)} = 0$$

8. Hankinson et al. (1969)

A_1	2.827793	3.252838
A_2	-4.688191×10^{-1}	-1.306424×10^{-1}
A_3	-1.262288	6.449194×10^{-1}
A_4	-1.536524	-1.518028
A_5	-4.535045	-5.391019
A_6	6.895104×10^{-2}	-1.379588×10^{-2}
A_7	1.903869×10^{-1}	6.600633×10^{-2}
A_8	6.200089×10^{-1}	6.120783×10^{-1}
A_9	1.838479	2.317431
A_{10}	4.052367×10^{-1}	1.632223×10^{-1}
A_{11}	1.073574	5.660595×10^{-1}

9. Heidaryan et al. (2010)

$$Z = \ln\left(\frac{A_1 + A_3\ln(P_{pr}) + (A_5/T_{pr}) + A_7(\ln(P_{pr}))^2 + (A_9/T_{pr}^2) + (A_{11}/T_{pr})\ln(P_{pr})}{1 + A_2\ln(P_{pr}) + (A_4/T_{pr}) + A_6(\ln(P_{pr}))^2 + (A_8/T_{pr}^2) + (A_{10}/T_{pr})\ln(P_{pr})}\right)$$

$$Z = \left[\frac{0.06125 P_{pr} t}{Y}\right] e^{(-1.2(1-t)^2)}$$

Reduced density $Y = -0.06125 P_{pr} t_e^{-1.2(1-t)^2} + \dfrac{Y + Y^2 + Y^3 - Y^4}{(1-Y)^3} - (14.76t - 9.76t^2 + 4.58t^3)Y^2$

$$+ (90.7t - 242.2t^2 + 4.58t^3)Y^{(2.18+2.82t)} = 0$$

10. Kareem et al. (2016)

$$Z = \frac{DP_{pr}(1 + y + y^2 - y^3)}{(DP_{pr} + Ey^2 - Fy^G)(1-y)^3}$$

$$y = \frac{DP_{pr}}{(((1+A^2)/C) + (A^2 B/C^3))}$$

$$t = \frac{1}{T_{pr}}$$

$$A = a_1 t e^{a_2(1-t)^2} P_{pr}$$

$$B = a_3 \times t + a_4 t^2 + a_5 t^6 P_{pr}^6$$

$$C = a_9 + a_8 t P_{pr} + a_7 \times t^2 \times P_{pr}^2 + a_6 t^3 P_{pr}^3$$

$$D = a_{10} t e^{a_{11}(1-t)^2}$$

$$E = a_{12} t + a_{13} t^2 + + a_{14} t^3$$

$$F = a_{15} t + a_{16} t^2 + a_{17} t^3$$

$$G = a_{18} + a_{19} t$$

a_1	0.317842
a_2	0.382216
a_3	-7.76835
a_4	14.2905
a_5	2.18363×10^{-6}

a_6	-0.00469257
a_7	0.0962541
a_8	0.16672
a_9	0.96691
a_{10}	0.063069
a_{11}	-1.966847
a_{12}	21.0581
a_{13}	-27.0246
a_{14}	16.23
a_{15}	207.783
a_{16}	-488.161
a_{17}	176.29
a_{18}	1.88453
a_{19}	3.05921

11. Kumar et al. (2004)
Shell

$$ZA = -0.101 - 0.36 \times T_r + 1.3868 \times (T_r - 0.919)^{0.5}$$

$$ZB = 0.21 + \frac{0.04275}{T_r - 0.65}$$

$$ZC = 0.6222 - 0.224 \times T_r$$

$$ZD = \left[\frac{0.0657}{T_r - 0.86}\right] - 0.037$$

$$ZE = 0.32 \times \exp(-19.53 \times (T_r - 1))$$

$$ZF = 0.122 \times \exp(-11.3 \times (T_r - 1))$$

$$ZG = P_r \times (ZC + ZD \times P_r + ZE \times P_r^4)$$

$$Z = ZA + ZB \times P_r + (1 - ZA)\exp(-ZG) - ZF\left(\frac{P_r}{10}\right)^4$$

12. Leung (1983)

$$Z = B_{ji} \sum_{i=1}^{4} \sum_{j=1}^{4} \rho_r^{(i-1)} T_r^{(1-j)}$$

where

i	B_i
1	1.877
2	−4.936
3	8.987
4	−5.215
5	−0.656
6	3.692
7	−6.477
8	3.077
9	0.1015
10	−0.5242
11	0.8359
12	−0.3192
13	−0.00422
14	0.0205
15	−0.0288
16	0.00742

13. Nishiumi and Saito (1975)

$$Z = 1 + (A_1 - A_2/T_r - A_3/T_r^3 - A_4/T_r^4 - A_5/T_r^5)\rho_{pr}^2 + (A_6 - A_7/T_r - A_8/T_r^2 - A_9/T_r^5 - A_{10}/T_r^{24})\rho_r^2$$

$$+ A_{11}\left(A_7/T_r + A_8/T_r^2 + \frac{A_9}{T_r^5} + \frac{A_{10}}{T_r^{24}}\right)\rho_r^5 + (A_{12}/T_r^3 + A_{13}/T_r^9 + A_{14}/T_r^{18})\rho_r^2\left(1 + A_{15}\rho_r^2\right)e^{(-A_{15}\rho_r^2)}$$

i	A_i
1	0.266986
2	1.048341
3	−1.516869
4	4.435926
5	−2.407212
6	0.608967
7	0.517467
8	1.26739
9	−0.0289282
10	−0.0168404
11	2.120655
12	−05046405
13	0.1802678
14	0.0856387
15	0.4956134

14. Papay (1985)

$$Z = 1 - \left(\frac{3.52 P_r}{10^{0.9813 T_r}} \right) + (0.274 P_r^2 10^{0.8257 T_r})$$

15. Papp (1979)

$P_r \leq 5$

$$R_1 = -3.23716 \, 10^{-3} \, T_r^5 + 4.0734 \, 10^{-2} \, T_r^4 - 0.202996 \, T_r^3 + 0.494968 \, T_r^2 - 0.573187 \, T_r + 0.244275$$

$$R_2 = (17.5648 \, T_r^3 + 185.316 \, T_r^2 - 337.652 \, T_r + 420.016)^{-1}$$

$$R_3 = \frac{(86.1372 - 235.563 \, T_r + 256.036 \, T_r^2 - 135.986 \, T_r^3 + 34.2576 \, T_r^4 - 2.7089 \, T_r^5)}{T_r^3}$$

$$R_4 = 9.6283 \, T_r^2 - 19.7803 \, T_r + 11.9919$$

$$R_5 = \frac{6.83622 - 22.0656 \, T_r + 14.3817 \, T_r^2 + 2.081973 \, T_r^3}{T_r^3}$$

$$R_6 = R_3 \; R_4 - 2 \, R_5$$

$$R_7 = R_5^2$$

$$W = 1$$

$$Z = 1 + R_1 \times P_r + R_2 \times P_r^2 - WR_3 \frac{P_r}{(P_r^2 + R_5P_r + R_6)}$$

$$P_r > 5$$

$$W = 1 - (R_{10} \times (P_r - R_5)^2 \times (P_r - R_8) \times (P_r - R_9)^2)$$

$$R_8 = R_5 + \left[\frac{3.7 \times 10^{-10} \times (15.1 - R_5)}{(T_r - 1.1)^9}\right]$$

$$R_9 = 15.1 + \left[\frac{2 \times (15.1 - R_5) \times (15.1 - R_8)}{45.3 - 2 \times R_8 - R_5}\right]$$

$$R_{10} = ((15.1 - R_5)^2 \times (15.1 - R_8) \times (15.1 - R_9)^2)^{-1}$$

$$Z = 1 + R_1 \times P_r + R_2 \times P_r^2 - WR_3 \frac{P_r}{(P_r^2 + R_5P_r + R_6)}$$

$$Z = 1 + A_1 P_{pr} + A_2 P_{pr}^2 + \frac{A_3 P_{pr}^{A_4}}{T_{pr}^{A_5}} + \frac{A_6 P_{pr}^{(A_4+1)}}{T_{pr}^{A_2}} + \frac{A_8 P_{pr}^{(A_4+2)}}{T_{pr}^{(A_7+1)}}$$

16. Sanjari and Lay (2012)

	Constants for $P_{pr} \leq 3$	Constants for $P_{pr} \geq 3$
A_1	0.007698	0.015642
A_2	0.003839	0.000701
A_3	-0.462212	2.341511
A_4	1.018801	-0.657903
A_5	3.805723	8.902112
A_6	-0.087361	-1.136000
A_7	7.138305	3.543614
A_8	0.083440	0.134041

17. Sarem (1961)

$$Z = P_r(0.0395 - 0.167T_r) + 0.1118T_r + 0.019P_r^{T_r} + 0.8181$$

$$Z = P_r(1.0078 - 0.1003T_r)^{-18.0196} - \frac{1.9089}{(1.1448T_r - 0.2036)} + 0.000552P_r^{T_r} + 1.6886$$

18. Standing (1977)

$$Z = A + (1 - A)/e^B + CP_{pr}^D$$

$$A = 1.39(T_{pr} - 0.92)^{0.5} - 0.36T_{pr} - 0.101$$

$$B = (0.62 - 0.23T_{pr})P_{pr} + \left[\frac{0.066}{(T_{pr} - 0.86)} - 0.037\right]P_{pr}^2 + \frac{0.32}{10^{9(T_{pr}-1)}}P_{pr}^6$$

$$C = (0.132 - 0.32\log T_{pr})$$

$$D = 10^{(0.3106 - 0.49T_{pr} + 0.1824T_{pr}^2)}$$

T_C AND P_C USING SPECIFIC GRAVITY

1. Elsharkawy et al. (2000)

Model 1

$$P_c = 193.941 - 131.347\gamma_g + 217.144\frac{\gamma_1}{\gamma_g} + 1060.349\frac{\gamma_2}{\gamma_g} + 344.573\left(\frac{\gamma_1}{\gamma_g}\right)^2 - 60.591\left(\frac{\gamma_2}{\gamma_g}\right)^2$$

$$T_c = 195.958 - 206.121\gamma_g + 25.855\frac{\gamma_1}{\gamma_g} - 6.421\frac{\gamma_2}{\gamma_g} + 9.022\left(\frac{\gamma_1}{\gamma_g}\right)^2 + 163.247\left(\frac{\gamma_2}{\gamma_g}\right)^2$$

γ_1 = Hydrocarbon gas specific gravity (air = 1)
γ_2 = Nonhydrocarbon gas specific gravity (air = 1)

Model 2

$$P_{pc} = 787.06 - 147.34\gamma_g - 7.916\gamma_g^2$$
$$T_{pc} = 149.18 + 358.14\gamma_g - 66.976\gamma_g^2$$

2. Londono et al. (2000)

$$P_{pc} = 725.89 - 70.27\gamma_g - 9.05\gamma_g^2$$
$$T_{pc} = 40.39 + 549.47\gamma_g - 94.01\gamma_g^2$$
$$P_{pc} = 621.81 - 81.09\gamma_g - 56.51\gamma_g^2$$
$$T_{pc} = 46.91 + 542.86\gamma_g - 93.14\gamma_g^2$$

3. Piper et al. (1993)

$$P_{pc} = 798.8 - 314.9\gamma_g + 76.2\gamma_g^2 + 80.5\gamma_g^{(Y_{H_2S} + Y_{CO_2} + Y_{N_2})}$$

$$T_{pc} = 302.4 - 432.86\gamma_g - 108.56\gamma_g^2 - 193.6\gamma_g^{(Y_{H_2S} + Y_{CO_2} + Y_{N_2})}$$

4. Said and El-Banbi (2005)

$$P_{pc} = 798.8 - 314.9\gamma_g + 76.2\gamma_g^2 + 80.5\gamma_g^{(Y_{H_2S}+Y_{CO_2}+Y_{N_2})}$$

$$T_{pc} = 302.4 - 432.86\gamma_g - 108.56\gamma_g^2 - 193.6\gamma_g^{(Y_{H_2S}+Y_{CO_2}+Y_{N_2})}$$

5. Sancet (2007)

$$P_c = 653 \times \exp(-0.007427M) + 82.82$$

$$T_c = [383.5 \times \ln(M - 4.075)] - 778.5$$

6. Standing (1977)

For gas condensate

$$T_{pc} = 706 - 51.7\gamma_g - 11.1\gamma_g^2$$

$$T_{pc} = 187 + 330\gamma_g - 71.5\gamma_g^2$$

$$P_{pc} = 677 + 15\gamma_g - 37.5\gamma_g^2$$

For dry gas

$$T_{pc} = 168 + 325\gamma_g - 12.5\gamma_g^2$$

$$P_{pc} = 787.06 - 147.34\gamma_g - 7.916\gamma_g^2$$

$$T_{pc} = 149.18 + 358.14\gamma_g - 66.976\gamma_g^2$$

7. Sutton (1985)

Model 1

$$T_c = 169.2 + 349.5 \times \gamma_g - 74 \times \gamma_g^2$$

$$P_c = 756.8 - 131 \times \gamma_g - 3.6 \times \gamma_g^2$$

Model 2

$$P_{pc} = 756.8 - 131\gamma_g - 3.6\gamma_g^2$$

$$T_{pc} = 169.2 + 349.5\gamma_g - 74\gamma_g^2$$

8. Sutton (2005)

$$P_{pcHC} = 671.1 + 14\gamma_{gHC} - 34.3\gamma_{gHC}^2$$

$$T_{pcHC} = 120.1 + 429\gamma_{gHC} - 62.9\gamma_{gHC}^2$$

$$y_{HC} = 1 - y_{N_2} - y_{CO_2} - y_{H_2S}$$

$$\gamma_{gHC} = \frac{\gamma_g\left(y_{H_2S}M_{H_2S} + C_{CO_2}M_{CO_2} + y_{N_2}M_{N_2}\right)/M_{air}}{y_{HC}}$$

9. Sutton (2007)

$$P_{pcHC} = 744 - 125.4 \gamma_{gHC} + 5.9 \gamma_{gHC}^2$$

$$T_{pcHC} = 164.3 + 357.7 \gamma_{gHC} - 67.7 \gamma_{gHC}^2$$

$$y_{HC} = 1 - y_{N_2} - y_{CO_2} - y_{H_2S}$$

$$\gamma_{gHC} = \frac{\gamma_g \left(y_{H_2S} M_{H_2S} + C_{CO_2} M_{CO_2} + y_{N_2} M_{N_2} \right) / M_{air}}{y_{HC}}$$

10. Wichert and Aziz (1972)

$$\varepsilon = 120 \times \left((y_{CO_2} + y_{H_2S})^{0.9} - (y_{CO_2} + y_{H_2S})^{1.6} \right) + 15 \times \left(y_{H_2S}^{0.5} - y_{H_2S}^4 \right)$$

$$T_{pc}^* = T_{pc} - \varepsilon$$

$$P_{pc}^* = \frac{P_{pc} T_{pc}^*}{T_{pc} + y_{H_2S}(1 - y_{H_2S}) \varepsilon}$$

where

$T_{pc} = {}^\circ R$

$P_{pc} = \text{psia}$

$\varepsilon = \text{pseudocritical temperature adjustment factor}$

T_C AND P_C FOR PSEUDO COMPONENT

1. Cavett (1962)

$$T_c = a_0 + a_1 \times T_{bF} + a_2 \times T_{bF}^2 + a_3 \times API \times T_{bF} + a_4 \times T_{bF}^3 + a_5 \times API \times T_{bF}^2 + a_6 \times API^2 \times T_{bF}^2$$

$$\text{Log}(P_c) = b_0 + b_1 \times T_{bF} + b_2 T_{bF}^2 + b_3 \times API \times T_{bF} + b_4 T_{bF}^3 + b_5$$
$$\times API \times T_{bF}^2 + b_6 \times API^2 \times T_{bF} + b_7 \times API^2 \times T_{bF}^2$$

T_c critical temperature (°R)

P_c critical pressure (psia)

T_{bF} normal boiling point (°F)

API API gravity of the fraction

2. Kesler and Lee (1976)

$$\ln(P_c) = 8.3634 - \frac{0.0566}{\gamma} - \left[0.24244 + \frac{2.2898}{\gamma} + \frac{0.11857}{\gamma^2} \right] \times 10^{-3} \times T_b$$
$$+ \left[1.4685 + \frac{3.648}{\gamma} + \frac{0.47227}{\gamma^2} \right] 10^{-7} T_b^2 - \left[0.42019 + \frac{1.6977}{\gamma^2} \right] 10^{-10} T_b^3$$

$$T_c = 341.7 + 811.1 \times \gamma + [0.4244 + 0.1174 \times \gamma] T_b + \frac{[0.4669 - 3.26238 \times \gamma]10^5}{T_b}$$

3. Watansiri et al. (1985)

$$\ln(T_c) = -0.0650504 - 0.0005217T_b + 0.03095\ln[M] + 1.11067\ln(T_b)$$
$$+ M\left[0.078154\gamma^{1/2} - 0.061061 \times \gamma^{1/3} - 0.016943 \times \gamma\right]$$

$$\ln(V_c) = 76.313887 - 129.8038\gamma + 63.1750\gamma^2 - 13.175\gamma^3 + 1.10108\ln[M] + 42.1958\ln[\gamma]$$

$$\ln(P_c) = 6.6418853 + 0.01617283\left[\frac{T_c}{V_c}\right]^{0.8} - 8.712\left[\frac{M}{T_c}\right] - 0.08843889\left[\frac{T_b}{M}\right]$$

4. Magoulas and Tassios (1990)

$$T_c = -1247.4 + 0.792 \times M + 1971\gamma - \frac{27000}{M} + \frac{707.4}{\gamma}$$

$$\ln(P_c) = 0.01901 - 0.0048442 \times M + 0.13239 \times \gamma + \frac{227}{M} - \frac{1.1663}{\gamma} + 1.2702 \times \ln(M)$$

T_C AND P_C USING COMPOSITION

1. Elsharkawy et al. (2000)

$$J = \left[\sum y_i(T_c/P_c)_i\right]_{C_1-C_6} + \left\{a_0 + [a_1(yT_c/P_c)]\right\}_{C_{7+}}$$

$$K = \left[\sum y_i(T_c/P_c^{0.5})_i\right]_{C_1-C_6} + \left\{b_0 + [b_1(yT_c/P_c^{0.5})]\right\}_{C_{7+}}$$

where

I	A	B
0	-0.0416977	-0.8816945
1	0.8648146	1.0368581

$$J = \left[\sum y_i(T_c/P_c)\right]_{C_1-C_6} + \left\{a_0 + [a_1(yT_c/P_c)]_{C_{7+}} + [a_2(yT_c/P_c)]_{N_2}\right.$$
$$+ [a_3(yT_c/P_c)]_{CO_2} + [a_4(yT_c/P_c)]_{H_2S}\right\}$$

$$K = \left[\sum y_i \left(T_c/P_c^{0.5}\right)\right]_{C_1-C_6} + \left\{b_0 + \left[b_1\left(yT_c/P_c^{0.5}\right)\right]_{C_{7+}} + \left[b_2\left(yT_c/P_c^{0.5}\right)\right]_{N_2}\right.$$

$$\left. + \left[b_3\left(yT_c/P_c^{0.5}\right)\right]_{CO_2} + \left[b_4\left(yT_c/P_c^{0.5}\right)\right]_{H_2S}\right\}$$

where

i	a	b
0	-0.040279933	-0.776423332
1	0.881709332	1.030721752
2	0.800591625	0.734009058
3	1.037850321	0.909963446
4	1.059063178	0.88959152

2. Kay (1936)

$$P_{pc} = \sum_{i=1}^{N} y_i P_{ci}$$

$$T_{pc} = \sum_{i=1}^{N} y_i T_{ci}$$

$$P_p c = \exp[8.3634 - 0.0566/(\gamma\_(C7+) - (0.24244 + 2.2898/(\gamma\_(C7+)$$
$$+ 0.11857/(\gamma\_(C7+)^2)) \, [10]^{(-3)}T_B + (1.4685 + 3.648/(\gamma\_(C7+)$$
$$+ 0.42227/(\gamma\_(C7+)^2)) \, [10]^{(-7)}T_B{}^2 - (0.42019$$
$$+ 1.6977/(\gamma\_(C7+)^2)) \, [10]^{(-3)}T_B]$$

$$T_{pc} = 341.7 + 811\gamma_{C_{7+}} + (0.42344 + 0.1174\gamma_{C_{7+}})T_B + (0.4669 - 3.2623\gamma_{C_{7+}})10^5/T_B$$

$$T_p = \left[4.5579 M_{C7+}^{0.15178}\gamma_{C7+}^{0.15427}\right]^3$$

$$P_{pc} = \frac{K^2}{J^2}$$

and

$$T_{pc} = \frac{K^2}{J}$$

3. Piper et al. (1993)

$$J = \alpha_0 + \sum_{i=1}^{3} \alpha_i y_i \left[\frac{T_c}{P_c}\right]_i + \alpha_4 y_i + \alpha_5 \gamma_g^2$$

$$K = \beta_0 + \sum_{i=1}^{3} \beta_i y_i \left[\frac{T_c}{P_c}\right]_i + \beta_4 y_i + \beta_5 \gamma_g^2$$

where

$y_i \epsilon \{y_{N_2}, y_{CO_2}, y_{H_2S}\}$

i	α_i	β_i
0	1.1582E−01	3.8216E+00
1	−4.582E−01	−6.534E−02
2	−9.048E−01	−4.2113E−01
3	−6.6026E−01	−9.1249E−01
4	7.0729E−01	1.7438E+01
5	−9.9397E−02	−3.2191E+00

$$P_{pc} = \frac{K^2}{J^2}$$

$$T_{pc} = \frac{K^2}{J}$$

$$J = \alpha_0 + \sum_{i=1}^{3} \alpha_i y_i \left[\frac{T_c}{P_c}\right]_i + \alpha_4 \sum_j y_i \left[\frac{T_c}{P_c}\right]_j + \alpha_5 y_{C_{7+}} M_{C_{7+}} + \alpha_6 \left(y_{C_{7+}} M_{C_{7+}}\right)^2$$

and

$$K = \beta_0 + \sum_{i=1}^{3} \beta_i y_i \left[\frac{T_c}{\sqrt{P_c}}\right]_i + \beta_4 \sum_j y_i \left[\frac{T_c}{\sqrt{P_c}}\right]_j + \beta_5 y_{C_{7+}} M_{C_{7+}} + \beta_6 \left(y_{C_{7+}} M_{C_{7+}}\right)^2$$

where $y_i \epsilon \{y_{N_2}, y_{CO_2}, y_{H_2S}\}$

$y_i \epsilon \{y_{C_1}, y_{C_2}, \ldots, y_{nC_6}\}$

4. Piper et al. (1993)

i	α_i	β_i
0	5.2073E − 02	− 3.9741E − 01
1	1.1016E + 00	1.0503E + 00
2	8.6961E − 01	9.6592E − 01
3	7.2646E − 01	7.8569E − 01
4	8.5101E − 01	9.8211E − 01
5	2.0818E − 02	− 3.7634E − 03
6	− 1.506E − 04	− 3.7634E − 03

$$T_{pc} = \frac{K^2}{J}$$

$$P_{pc} = \frac{T_{pc}}{J}$$

where

$$J = \left(\left(\frac{1}{3}\right) \left[\sum y_i \left(\frac{T_c}{P_c}\right) \right]_i + \left(\frac{2}{3}\right) \left[\sum y_i \left(\frac{T_c}{P_c}\right)_i^{0.5} \right]^2 \right)$$

$$K = \sum \left[y_i (T_c / P_c^{0.5}) \right]_i$$

$$T_{pc} = \frac{K^2}{J}$$

$$P_{pc} = \frac{T_{pc}}{J}$$

5. Stewart et al. (1959)

6. Sutton (1985)

$$F_j = (1/3) \left[y (T_c / P_c) \right]_{C_7} + (2/3) \left[y_i (T_c / P_c)_i^{0.5} \right]_{C_{7+}}^2$$

$$E_j = 0.6081 F_j + 1.1325 F_j^2 - 14.004 F_j y_{C_{7+}} + 64.434 F_j y_{C_{7+}}^2$$

$$E_k = (T_c / P_c^{0.5})_{C_{7+}} \left[0.3129 y_{C_{7+}} - 4.8156 y_{C_{7+}}^2 + 27.3751 y_{C_{7+}}^3 \right]$$

$$J' = J - E_j$$

$$K' = K - E_k$$

DEW POINT PRESSURE USING SPECIFIC GRAVITY

1. Al-Dhamen and Al-Marhoun (2011)

$$\ln(P_d) = a_1 + a_2 \ln\left(\frac{GOR \times \gamma_g}{\gamma_{7+}}\right) + a_3 \ln(T_R) + a_4 \gamma_g + \frac{a_5}{\gamma_{cond.}} + a_6 e^{a_7 + a_8 \ln(GOR)}$$

where

a_1	18.6012
a_2	-0.1520
a_3	-0.1674
a_4	0.0685
a_5	-5.8982
a_6	-0.0559
a_7	8.496
a_8	-0.7466

2. Ovalle et al. (2007)

$$\ln P_d = 8.48 + 0.32239 Z + 0.00477 Z^2$$

where

$$Z = \sum_{n-1}^{3} Z_n$$

and

$$Z_n = CO_n + C1_n VAR_n + C2_n VAR_n^2$$

N	VAR_n	CO_n	$C1_n$	$C2_n$
1	INR_{SPld}	9.8895	-0.87528	-0.01691
2	API_d	11.7	-0.29709	0.00151
3	Υ_{gRd}	3.5202	-2.9145	-0.81744

DEW POINT PRESSURE USING COMPOSITION

1. Elsharkawy (2001)

$$P_d = A_0 + A_1 Tf + A_2 xH_2S + A_3 xCO_2 + A_4 xN_2 + A_5 xC_1 + A_6 xC_2 + A_7 xC_3 + A_8 xC_4 + A_9 xC_5$$
$$+ A_{10} xC_6 + A_{11} xC_7 + A_{12} MW_{C_{7+}} + A_{13}\gamma_{C_{7+}} + A_{14}\left(xC_{7+}MW_{C_{7+}}\right)$$
$$+ A_{17}\left\{\frac{xC_{7+}}{xC_1 + xC_2}\right\} + A_{18}\left\{xC_{7+}/(xC_2 + xC_3 + xC_4 + C_5 + C_6)\right\}$$

A_i	Value
A_0	4268.85
A_1	0.094056
A_2	−7157.87
A_3	−4540.58
A_4	−4663.55
A_5	−1357.56
A_6	−7776.10
A_7	−9967.9
A_8	−4257.10
A_9	−1417.10
A_{10}	691.5298
A_{11}	40660.36
A_{12}	205.26
A_{13}	−7260.32
A_{14}	−352.413
A_{15}	−114.519
A_{16}	8.133
A_{17}	94.916
A_{18}	238.252

2. Godwin (2012)

$$P_d = A_0 + A_1B_1 + A_2B_2 + A_3B_3 + A_4B_4 + A_5B_5 + A_6B_6 + A_7B_7 + A_8B_8 + A_9B_9 + A_{10}B_{10} + A_{11}B_{11} + A_{12}B_{12} + A_{13}B_{13} + A_{14}B_{14} + A_{15}B_{15}$$

B_i	Equals
B_1	T_f
B_2	$XH_2S + XCO_2 + XN_2$
B_3	XC_1
B_4	XC_2
B_5	XC_3
B_6	XC_4
B_7	XC_5
B_8	XC_6
B_9	XC_{7+}
B_{10}	$XC_{7+}MW_{C_{7+}}$
B_{11}	$MW_{C_{7+}}$
B_{12}	$\gamma_{C_{7+}}$
B_{13}	$XC_{7+}MW_{C_{7+}}/\gamma_{C_{7+}}$
B_{14}	$XC_{7+}/(XC_1 + XC_2)$
B_{15}	$XC_{7+}/(XC_3 + XC_4 + XC_5 + XC_6)$

A_i	Value
A_0	1
A_1	8.5
A_2	9.4
A_3	9.9
A_4	11
A_5	9.35
A_6	22.35
A_7	22.30
A_8	19.08
A_9	19.1

A_{10} 19.1
A_{11} 19.09
A_{12} 19.1
A_{13} 19.1
A_{14} 19.12
A_{15} 5

3. Humoud and Al-Marhoun (2001)

$$\ln(P_d) = \beta_0 + \beta_1 \ln(T_R) + \beta_2 \ln(R_m) + \beta_3 \ln(P_{sp} \cdot T_{sp}) + \frac{\beta_4}{T_{pr}} + \frac{\beta_5}{P_{pr}} + \frac{\beta_6}{\gamma_{C_{7+}}}$$

where

$$R_m = \frac{R_{sp} \cdot \gamma_{gsp}}{\gamma_{C_{7+}}}$$

Coefficient	Estimate
β0	43.777183
β1	-3.594131
β2	-0.247436
β3	-0.053527
β4	-4.291404
β5	-3.698703
β6	-4.590091
P_{sp}	Primary sep. pressure

4. Kamaria et al. (2016)

$$P_d = A + B$$

where

$$A = ((13.145 - 4.942ZC1 + 1961.7ZC2 - 6212.71ZC4 + 39335.07(ZC4)^{\wedge}2 + 2097ZC5 \\ - 3451.17ZC6 + 201.93Z(H2S) - 0.065224T_r))/((0.0031904T_r + 0.094398))$$

and

$$B = ((1367.4 + 9.98ZC1 [\![MW]\!] \_(C7 +) - 1697.6ZC3 - 5096.8ZC7 + 3528.07\ln(Z(C7 +)) \\ + 933.35Z([\![CO]\!] \_2) + 1909.7Z(N2))/((1.0214 - [\![SG]\!] \_(C7 +)))$$

5. Marruffo et al. (2001)

$$P_d = K1 \times \left[\frac{GCR^{K2}}{C_{7+}^{K3}} \times K8 \times {}^\circ API^{\left[K4 T_r^{K5} - K6 C_{7+}^{K7} \right]} \right]$$

where

K1 = 346.7764689
K2 = 0.0974139
K3 = −0.294782419
K4 = −0.047833243
K5 = 0.281255219
K6 = 0.00068358
K7 = 1.906328237
K8 = 8.417626216

6. Nemeth and Kennedy (1967)

$$\ln P_{DP} = A_1[C_2 + CO_2 + H_2S + C_5 + C_6 + 2(C_3 + C_4) + 0.4C_1 + 0.2N_2] + A_2 SGC_{7+} + A_3 \left[\frac{C_1}{C_{7+} + 0.002} \right]$$
$$+ A_4 T + A_5(C_{7+} M_w C_{7+}) + A_6(C_{7+} M_w C_{7+}) + A_7(C_{7+} M_w C_{7+})^2 + A_7(C_{7+} M_w C_{7+})^3 + A_8 \left[M_w C_{7+} / (SGC_{7+} + 0.0001) \right]$$
$$+ A_9 \left[M_w C_{7+} / (SGC_{7+} + 0.0001) \right]^2 + A_{10} \left[M_w C_{7+} / (SGC_{7+} + 0.0001) \right]^3 + A_{11}$$

A_i	Value
A_1	−2.0623054
A_2	6.6259728
A_3	−4.4670559 × 10^{-3}
A_4	1.0448346 × 10^{-4}
A_5	3.2673714 × 10^{-2}
A_6	−3.6453277 × 10^{-3}
A_7	7.4299951 × 10^{-5}
A_8	−1.1381195 × 10^{-1}
A_9	6.2476497 × 10^{-4}
A_{10}	−1.0716866 × 10^{-6}
A_{11}	1.0746622 × 10

7. Shokir (2008)

$$DPP = B_1 B_2 B_3 B_4$$

where

$$B_1 = 201.875481(ZC_{7+}(((T_r(((ZC_3 - (ZH_2S - ZCO_2) - (ZC_6 - (ZCO_2 - ZC_4)))$$
$$- ZC_2)) - ((ZC_4((ZCO_2 - ZC_4) - (M_wC_{7+} - ZN_2) - (M_wC_4^2 ZC_5)))) - ZC_{7+})$$
$$- (ZH_2S - ((ZN_2(T_r(ZC_1^2 - ZC_{7+}))) - (M_wC_{7+} - (ZC_2 - ZH_2Si))))) + 38456.87953ZC_6$$

$$B_2 = 0.000007((T_r(((ZCO_2 - M_wC_{7+}) - ZC_3)M_wC_{7+}))ZN_2) + 225500.9399ZC_5$$
$$- ((ZH_2S - T_r)(M_wC_{7+} - ZC_3)M_wC_{7+}) - (ZCO_2 - T_r)))$$

$$B_3 = 120586.9719ZC_1((((ZH_2SZC_3) - (ZC_5 - ZC_{7+}))ZH_2S)$$
$$- (((ZC_{7+} - ZC_1)(ZC_{7+} - ZC_6)) - (ZH_2SZN_2^2))) + 72.6908M_wC_{7+}$$

and

$$B_4 = -1962.40851(ZC_5(M_wC_{7+} - ZC_1^2)) - 253385.67764((ZC_{7+}((ZCO_2ZC_3) -$$
$$(ZC_4 - ZC_{7+}))(ZCO_2(ZC_3(ZC_3 - M_wC_{7+})))) - 13358.59271ZC_4 + 4676.933602ZC_2 - 6567.9$$

8. Al-Zahabi et al. (2015)

$$y = \beta_o + \beta_1 x_1 + \ldots + \beta_6 x_6 + \ldots + \beta_{12}x_{12}{}^* x_2 + \ldots + \beta_{56}x_5 x_6$$

where

$y = \log(\text{dew point pressure})$
$X1 = \text{temperature}$
$X2 = \log(CO_2 + 0.1)$
$X3 = \log(CH_4)$
$X3 = \log(CH_4)^2$
$X4 = \log(C_3H_8 - C_5H_{12})$
$X5 = \log(C_6H_{14}{}^+)$

Coefficient	Estimate
β_0	19.1109840
β_1	-0.0679165
β_2	-0.0162705
β_3	-6.6190184
β_4	0.5104139
β_5	1.1398989

β_6	0.6263451
β_{13}	0.0371260
β_{14}	-0.0048367
β_{23}	0.0573708
β_{26}	-0.0565329
β_{34}	0.0794272
β_{35}	-0.1985207
β_{56}	-0.1334765

GAS CONDENSATE SPECIFIC GRAVITY

1. Gold et al. (1989)

$$\gamma_w = \frac{R_1 \gamma_1 + 4602 \gamma_o + G_{pa}}{R_1 + V_{eq}}$$

where

$$G_{pa} = R_2 \gamma_2 + R_3 \gamma_3$$

$$V_{eq} = (133316 \gamma_o / M_o) + R_2 + R_3$$

and

$$M_o = [5954/(\rho - 8.811)] = [42.43 \gamma_o / (1.008 - \gamma_o)]$$

Three stage

$$G_{pa} = \beta_1 (p_{s1} - 14.65)^{\beta_2} p_{s1}^{\beta_3} \gamma_1^{\beta_4} \rho^{\beta_5} T_{s1}^{\beta_5} T_{s2}^{\beta_6} T_{s1}^{\beta_7}$$

$$V_{eq} = \beta_o + \beta_1 p_{s1}^{\beta_2} \gamma_1^{\beta_3} \rho^{\beta_4} T_{s1}^{\beta_5} T_{s2}^{\beta_6} T_{s3}^{\beta_7}$$

Two stage

$$G_{pa} = \beta_1 (p_{s1} - 14.65)^{\beta_2} p_{s1}^{\beta_3} \gamma_1^{\beta_4} \rho^{\beta_5} T_{s1}^{\beta_5} T_{s2}^{\beta_6}$$

$$V_{eq} = \beta_o + \beta_1 p_{s1}^{\beta_2} \gamma_1^{\beta_3} \rho^{\beta_4} T_{s1}^{\beta_5} T_{s2}^{\beta_6}$$

Coefficients	Three Stages		Two Stages	
	G_{pa}	V_{eq}	G_{pa}	V_{eq}
	Correlation	Correlation	Correlation	Correlation
30	–	536.56	–	651.55
31	1.0347	1.4309	0.15763	0.091752
32	0.97703	0.79225	1.3365	1.0928
33	6.5230	4.6352	7.0393	5.2244
34	1.3045	1.2639	1.3891	1.7350
35	–1.1860	–0.84710	–1.4370	–1.0843
36	0.42499	0.17103	0.77907	0.41384
37	0.11524	0.1898	–	–

$$\gamma_{gRr} = 0.81805 + 0.12085Z - 0.00851Z^2$$

where

$$Z = \sum_{n=1}^{2} Z_n$$

$$\gamma_{gRr} = \frac{\gamma_{gR}}{\gamma_{gRd}}$$

$$Z_n = CO_n + C1_n VAR_n + C2_n VAR_n^2 + C3_n VAR_n^3 + C4_n VAR_n^4$$

N	VAR_n	CO_n	$C1_n$	$C2_n$	$C3_n$	$C4_n$
1	P_r	–0.92913	2.1688	–2.73343	2.74921	0
2	Υ_{gRr}	–19.79538	84.86958	–120.65049	70.96942	–15.24672

$$\ln Y_{SPI} = 3.684 + 0.61967Z + 0.0153591Z^2$$

$$Z = \sum_{n=1}^{4} Z_n$$

$$Z_n = CO_n + C1_n VAR_n + C2_n VAR_n^2 + C3_n VAR_n^3 + C4_n VAR_n^4$$

2. Ovalle et al. (2007)

N	VAR_n	$C0_n$	$C1_n$	$C2_n$	$C3_n$	$C4_n$
1	$\ln P$	20.809	-6.7095	0.5136	0.000	0.000
2	API_d	11.175	-1.2965	0.042311	-0.0005438	2.4889×10^{-6}
3	γ_{gRd}	-13.365	27.652	-18.598	4.3658	0.000
4	T_R	-1.5309	0.0058453	1.4035×10^{-6}	0.000	0.000

TWO-PHASE Z-FACTOR

1. Rayes et al. (1992)

$$Z_{2p} = A_0 + A_1(p_r) + A_2\left(\frac{1}{T_r}\right) + A_3(p_r)^2 + A_4\left(\frac{1}{T_r}\right)^2 + A_5\left(\frac{p_r}{T_r}\right)$$

$$n_{C_{7+}} = -0.0885119 + 0.141013(\gamma_g)$$

A_0	2.24353	A_3	0.000829231
A_1	-0.0375281	A_4	1.53428
A_2	-3.56539	A_5	0.131987

CONDENSATE GAS RATIO CORRELATIONS

1. Dindoruk (2012)

$$CGR = 7.5537 \times 10^{-5} MW^4 - 9.5072 \times 10^{-3} MW^3 + 4.9304 \times 10^{-1} MW^2 - 2.563 MW - 60$$

$$CGR = 0.0861 MW^2 + 4.1229 MW - 90$$

Linear Equ.

$$CGR = 305.169 \gamma_w - 196.2995$$

Quadratic Equ.

$$CGR = 82.77885 \gamma_w^2 + 101.6806 \gamma_w - 85.3052$$

$$CGR = \frac{\left(\gamma_w - \gamma_g\right) \times 10^6}{4584 \gamma_o - 2998.42 \gamma_w (1.03 - \gamma_0)}$$

MODIFIED BLACK OIL (GAS CONDENSATE)

1. El-Banbi et al. (2006)

$$R_s = \gamma g \times \left[\left(\frac{P}{A1} + A2 \right) \times 10^X \right]^{A3}$$

$$X = A4 \times API - A5 \times (T - 160)$$

A1	A2	A3	A4	A5
0.194085	−3709.42	1.060521	−0.05022	−0.00377

$$R_s = A1 \times \gamma gs \times P^{A2} \times e^{[(A3 \times API)/T]}$$

$$\gamma_{gs} = \gamma_g \times \left[1 + 5.912 \times 10^{-5} \times API \times (T - 460) \times \log\left(\frac{Psep}{114.7} \right) \right]$$

A1	A2	A3
12.58498	1.343554	−105.487

$$Bo = A1 + A2 \times \left(R_s \times \sqrt{\frac{\gamma gs}{\gamma osc}} + A3 \times (T - 460) \right)^{A4}$$

A1	A2	A3	A4
0.96511	0.000343	1.303306	1.053171

$$R_V = \frac{A1 \times \rho_{gsc} \times (A2 \times p^2 + A3 \times p + A4)}{p_s} \times \exp\left[\frac{A5 \times CGR \times T_{sc}}{\rho_{osc} \times T \times P_{sc}} \right]$$

A1	A2	A3	A4	A5
3.458411	6.89E − 05	−0.0317	251.0827	4.174003

$$p_{pc} = A1 + A2 \times \gamma_{gs} + A3 + \gamma_{gs}^2$$
$$T_{pc} = B1 + B2 \times \gamma_{gs} + B3 + \gamma_{gs}^2$$

A1	A2	A3	B1	B2	B3
670.958	−188.078	−3.7882	480.1142	−94.2128	9.398696

$$B_g = \frac{5.04 \times z \times T}{p}$$

Dew point pressure

$$P_{sat} = [A_0 \times R_{vi}^{A1}] \times [(X+Y)^{A2}] \times [STO^{A3}] \times [T_r^{A4}]$$
$$X = \frac{SG1}{Psep1}$$
$$Y = \frac{SG2}{Psep2}$$

P_{sat} Correlation Parameters

Separator Stages	A0	A1	A2	A3	A4
2 Stages	728.7427	−0.05596	−0.00731	0.969049	0.379907
3 Stages	761.7611	−0.05725	−0.01701	0.9619	0.384164

Vaporized Oil-Gas Ratio

$$R_{vi} = [A0 \times e^{(A1 \times (X+Y))}] + [(A2 \times STO^2) + (A3 \times STO) + A4] + [A5 \times T_r]$$

Initial Condensate Gas Ration Correlation Parameters

Separator Stages	A0	A1	A2	A3	A4	A5
2 Stages	30.72076	−6.6E−07	−3.80138	6.554239	−33.523	0.000749
3 Stages	−2.7E−12	0.019049	−3.77151	6.460271	−2.74861	0.00076

2. Nassar et al. (2013)

$$R_v = [(A0 \times P^2) + (A1 \times P) + A2] + [e^{(A3 \times X + A4 \times Y)}] + [e^{(A5 \times V)}] \times R_{vi}$$

R_v Correlation Parameters (With P_{sat})

Separator Stages	A0	A1	A2	A3	A4	A5
2 Stages	8.59E − 08	−0.00019	0.605571	1.21E − 05		−0.04952
3 Stages	8.68E − 08	−0.0002	0.633988	−0.00012	0.000223	−0.04945

R_v Correlation Parameters (Without P_{sat})

Separator Stages	A0	A1	A2	A3	A4	A5
2 Stages	−4.6E − 09	0.00025	0.255068	8.55E − 06		−242.823
3 Stages	−2E − 09	0.000232	0.286397	−0.00138	0.00237	−233.698

$X = SG1 \times Psep1$
$Y = SG2 \times Psep2$
$V = \dfrac{STO}{T_r}$

Gas Formation Volume Factor

$$B_g = [A0 \times P^{A1}] \times [e^{A2 \times X + A3 \times Y}] \times [e^{A4 \times V}]$$

B_g Correlation Parameters

Separator Stages	A0	A1	A2	A3	A4	A5
2 Stages	349.836	−0.775	−8.9E − 06			0.002
3 Stages	404	−0.773	3.26E − 05	−7.5E − 05		0.002

$X = SG1 \times Psep1$
$Y = SG2 \times Pse2$
$V = STO \times T_r$

REFERENCES

Ahmed, T., 1989. Hydrocarbon Phase Behavior, Vol.7. Gulf Publishing Company, Houston, United States.

Ahmed, T., 2016. Equations of State and PVT Analysis, 1st ed Gulf Professional Publishing, United States.

Ahmed, T., 2006. Reservoir Engineering Handbook, 3rd. ed. Gulf Publishing Company, Boston, MA, United States.

Ahmed, T., 2010. Reservoir Engineering Handbook, 4th. ed. Gulf Publishing Company, Boston, MA, United States.

Ahmed, T., 2016. Equations of State and PVT Analysis, 2nd ed Gulf Professional Publishing, United States, ISBM: 9780128017524.

Al-Dhamen, M., Al-Marhoun, M., 2011. New correlations for dew-point pressure for gas condensate. Soc. Pet. Eng. Available from: https://doi.org/10.2118/155410-MS.

Alzahabi, A., El-Banbi, A.H., Trindade, A.A., Soliman, M., 2017. A regression model for estimation of dew point pressure from down-hole fluid analyzer data. J. Pet. Explor. Prod. Technol. 7 (4), 1173−1183. Available from: https://doi.org/10.1007/s13202-016-0308-9.

Azizi, N., Behbahani, R., Isazadeh, M.A., 2010. An efficient correlation for calculating compressibility factor of natural gases. J Nat Gas Chem 19, 642−645. Available from: https://doi.org/10.1016/S1003-9953(09)60081-5.

Beggs H., and Brill J., 1978. Two-Phase Flow in Pipes. The University of Tulsa, Tulsa, OK.

Burnett, R.R., 1979. Calculator gives compressibility factors. Oil Gas J. 70−74.

Carr, N.L., Kobayashi, R., Burrows, D.B., 1954. Viscosity of hydrocarbon gases under pressure. Trans. AIME 201, 264−272.

Cavett, R.H., 1962. Physical data for distillation calculations, vapor-liquid equilibria. American Petroleum Institute, Proc. 27th Mid-Year Meeting, API, San Francisco, pp. 351-366.

Chen, Z.A. and Ruth, D.W., 1993. On viscosity correlations of natural gas. In: PETSOC-93-02 Paper Presented at the Annual Technical Meeting, May 9-12, Calgary, Alberta, Canada.

Dean, D.E., Stiel, L.I., 1965. The viscosity of nonpolar gas mixtures at moderate and high pressures. AICHE J. May, 526−532.

Dempsey, J.R., 1965. Computer routine treats gas viscosity as a variable. Oil Gas J August 16, 141.

Dindoruk, B., 2012. Development of a correlation for the estimation of condensate to gas ratio (CGR) and other key gas properties from density data. Soc. Pet. Eng. Available from: https://doi.org/10.2118/160170-MS.

Dranchuk, P.M., Abou-Kassem, J.H., 1975. Calculation of z-factors for natural gases using equations of state. J. Can. Pet. Tech. 14, 34−36 (July−Sept.).

Dranchuk, P.M., Purvis, R.A., and Robinson, D.B., 1974. Computer calculation of natural gas compressibility factors using the Standing and Katz correlations. Institute of Petroleum Technical Series, No. IP74-008 1−13.

El-Banbi, A.H., Fattah, K.A., Sayyouh, M.H., 2006. New modified black-oil correlations for gas condensate and volatile oil fluids. In: SPE Annual Technical Conference and Exhibition, San Antonio, Texas, USA, September 24−27.

Elechi, V.U., Ikiensikimama S.S.,and Azubuike I.I., 2015. A correlation for estimating gas compressibility factor in the Niger delta. In: Festschrift for J. A. Ajienka, pp. 137−148.

Elsharkawy, A.M., 2001. Characterization of the plus fraction and prediction of the dew point pressure for gas condensate reservoirs. In: SPE Western Regional Meeting, 26e30 March, Bakersfield, California. SPE Paper 68666.

Elsharkawy, A.M., 2006. Efficient methods for calculations of compressibility, density and viscosity of natural gases. JCPT 45 (6), 55−61.

Elsharkawy, A.M., Hashem, Y.S.K.S., Alikhan, A.A., 2000. Compressibility factor for gas condensates. Soc. Pet. Eng. Available from: https://doi.org/10.2118/59702-MS.

Godwin, O.N., 2012. A new analytical method for predicting dew-point pressure for gas condensate reservoirs. In: Paper SPE 162985 Presented at the 2012 SPE Nigerian Annual International Conference and Exhibition, 1−4, October, Nigeria.

Gold, D.K., McCain, W.D., Jennings, J.W., 1989. An improved method for the determination of the reservoir-gas specific gravity for retrograde gases (includes associated papers 20006 and 20010). Soc. Pet. Eng. Available from: https://doi.org/10.2118/17310-PA.

Gopal, V.N., 1977. Gas Z-factor equations developed for computer. Oil Gas J. 75 (August), 58−60.

Hall, K.R., Yarborough, L., 1973. A new equation of state for Z-factor calculations. Oil Gas J Feb. 18, 82−92.

Hankinson, R.W., Thomas, L.K., Phillips, K.A., 1969. Predict natural gas properties. Hydrocarbon Process. 106−108.

Heidaryan, E., Salarabadi, A., Moghadasi, J., 2010. A novel correlation approach for prediction of natural gas compressibility factor. J. Nat. Gas Chem. 19, 189−192. Available from: https://doi.org/10.1016/S1003-9953(09)60050-5.

Humoud, A.A., Al-Marhoun, M.A., 2001. A new correlation for gas-condensate dewpoint pressure prediction. Soc. Pet. Eng. Available from: https://doi.org/10.2118/68230-MS.

Jossi, J.A., Stiel, L.I., Thodos, G., 1962. The viscosity of pure substances in the dense gaseous and liquid phases. Am. Inst. Chem. Eng. J. 8, 59−63.

Kamaria, A., Sattaria, M., Amir, H., Mohammad, A.H., Ramjugernatha, D., 2016. Rapid model for the estimation of dew point pressures in gas condensate systems. J. Taiwan Inst. Chem. Eng. 60, 258−266.

Kareem, L.A., Iwalewa, T.M., Marhoun, M., 2016. New explicit correlation for the compressibility factor of natural gas: linearized z-factor isotherms. J. Petrol. Explor. Prod. Technol. 6, 481−492. Available from: https://doi.org/10.1007/s13202-015-0209-3.

Kay, W., 1936. Gases and vapors at high temperature and pressure—density of hydrocarbon. Ind. Eng. Chem. 28 (9), 1014−1019.

Kesler, M.G., Lee, B.I., 1976. Improve prediction of enthalpy fractions. Hydrocarbon Process. 55, 153−158.

Kumar, P., Turner, D., Sloan, E.D., 2004. Thermal diffusivity measurements of porous methane hydrate and hydrate-sediment mixtures. J. Geophys. Res. 109, B01207. Available from: https://doi.org/10.1029/2003JB002763.

Lee, A.L., Gonzalez, M.H., Eakin, B.E., 1966. The viscosity of natural gas. Trans. AIME 237, 997−1000.

Leung, L.C., 1983. Numerical evaluation of the effect of simultaneous steam and carbon dioxide injection on the recovery of heavy oil. JPT 35 (9), 1591−1599. SPE-10776-PA.

Londono, F.E., Archer, R.A., Blasingame, T.A., 2002. Simplified correlations for hydrocarbon gas viscosity and gas density validation and correlation behavior using a large scale database. In: Paper SPE 75721 Presented at the SPE GasTechnology Symposium, Calgary, Canada, April 30−May 2. DOI: 10.2118/75721-MS.

Londono, F.E., Archer, R.A., Blasingame, T.A., 2005. Correlations for hydrocarbon-gas viscosity and gas density-validation and correlation of behavior using a large-scale database. SPEREE 561−572.

Lucas, K., 1981. Chem. Ing. Tech 53, 959.

Magoulas, K., Tassios, D., 1990. Thermophysical properties of n-alkanes from C1 to C20 and their prediction for higher ones. J. Fluid Phase Equilibria 56, 119–140.

Marruffo, I. Maita, J. Him, J. Rojas, G., 2001. Statistical forecast models to determine retrograde dew pressure and C7 + percentage of gas condensates on basis of production test data of eastern Venezuelan reservoirs. In: SPE Latin American and Caribbean Petroleum Engineering Conference. Paper No. 69393, Buenos Aires, Argentina.

McCain, W.D., 1991. Reservoir-fluid property correlations-state of the art (includes associated papers 23583 and 23594). Soc. Pet. Eng. Available from: https://doi.org/10.2118/18571-PA.

Nassar, I.S., El-Banbi, A.H., Sayyouh, M.H.M., 2013. Modified black oil PVT properties correlations for volatile oil and gas condensate reservoirs. Soc. Pet. Eng. Available from: https://doi.org/10.2118/164712-MS.

Nemeth, L.K., Kennedy, H.T., 1967. A correlation of dewpoint pressure with fluid composition and temperature. SPEJ 7, 99–104. Trans., AIME, 240.

Nishiumi, H., Saito, S., 1975. An improved generalized BWR equation of state applicable to low reduced temperatures. J. Chem. Eng. Jpn. 8 (5), 356–360.

Olds, R.H., Sage, B.H., Lacy, W.N., 1945. Volumetric and phase behavior of oil and gas from Paloma field. Trans. AIME 160. Available from: https://doi.org/10.2118/945077-G.

Olds, R.H., Sage, B.H., Lacey, W.N., 1949. Volumetric and viscosity studies of oil and gas from several San Joaquin Valley field. Trans. AIME 179, 287–302.

Ovalle, A.P., Lenn, C.P., McCain Jr., W.D., 2007. Tools to manage gas/condensate reservoirs; novel fluid-property correlations on the basis of commonly available field data. SPE Res. Eval. Eng. 10 (06). Available from: https://Doi.org/10.2118/112977-PA.

Papay, J., (1985). A Termelestechnologiai Parameterek Valtozasa a Gazlelepk Muvelese Soran. OGIL MUSZ, Tud, Kuzl. [Budapest]: 267–273.

Papp, I., (1979). Uj m6dszer Ioldgazok elteresi tenvezojenek szamirasara. koola] es Foldgaz, November, pp. 345–47.

Piper, L.D., McCain Jr., W.D., Corredor, J.H., 1993. Compressibility factors for naturally occurring petroleum gases (1993 version). In: Paper SPE 26668 Presented at the SPE Annual Technical Conference and Exhibition, Houston, TX., United States, Oct. 3-6. Doi: 10.2118/26668-MS.

Rayes, D.G., Piper, L.D., McCain, W.D., Poston, S.W., 1992. Two-phase compressibility factors for retrograde gases. Soc. Pet. Eng. Available from: https://doi.org/10.2118/20055-PA.

Said, S. and El-Banbi, A.H., 2005. A new and simple z-factor correlation. In: Paper Presented at the 9th International Mining, Petroleum, and Metallurgical Engineering Conference, Cairo University, Cairo, Egypt, Feb. 21–24.

Sancet, G.F., 2007. Heavy fraction C7 + characterization for PR-EOS. In: SPE-113026-STU Paper Presented at the SPE International Student Paper Contest, SPE Annual Technical Conference and Exhibition, November 11–14, Anaheim, California, USA.

Sanjari, E., Lay, E.N., 2012. An accurate empirical correlation for predicting natural gas compressibility factors. J. Nat. Gas Chem. 21, 184–188. Available from: https://doi.org/10.1016/S1003-9953(11)60352-6.

Sarem, A.M., 1961. Z-factor equation developed for use in digital computers. In: OGL Sept. 18, 1961, 118.

Shokir, E.M., 2008. Dewpoint pressure model for gas condensate reservoirs based on genetic programming. Energy Fuels 22 (5), 3194–3200. Available from: https://doi.org/10.1021/ef800225b.

Standing, M.B., 1947. A pressure-volume-temperature correlation for mixtures of California oils and gases, drilling and production practice, American Petroleum Institute, 1st January, New York, United States.

Standing, M.B., 1962. Oil-system correlation, Petroleum production handbook. McGraw-Hill Book Co, New York City, United States.

Standing, M.B., 1977. Volumetric and phase behavior of oil field hydrocarbon systems. Society of Petroleum Engineers, ISBN-13: 978-0895203007.

Stewart, W.F., Burkhardt, S.F., and Voo, D., 1959. Prediction of pseudo-critical parameters for mixtures. In: Presented at the AIChE Meeting, Kansas City, Missouri, USA.

Sutton, R.P., 2005. Fundamental PVT calculations for associated and gas/condensate natural gas systems. In: Paper SPE 97099 Presented at the 2005 SPE Annual Technical Conference and Exhibition, Dallas, TX (Oct. 9−12).

Sutton, R.P., 2007. Fundamental PVT calculations for associated and gas/condensate natural-gas systems. SPE Reservoir Eval. Eng 10 (3), 270−284.

Sutton, R.R., 1985. Compressibility factors for high-molecular weight reservoir gases. In: Paper No. SPE14265, SPE Annual Technical Meeting and Exhibition, Las Vegas.

Vasquez, M., Beggs, H.D., 1980. Correlation for fluid physical property predictions. JPT 32, 968−970.

Watansiri, S., Owens, V.H., Starling, K.E., 1985. Correlations for estimating critical constants, accentric factor, and dipole moment for undefined fractions. Ind. Eng. Chem. Process Des. Dev. 24 (2), 294−296.

Wichert, E., Aziz, K., 1972. Calculate Z's for sour gases. Hydrocarbon Process. 51, 119−122 (May).

Appendix C

Oil Correlations Range of Applicability

This appendix contains the working ranges of different oil correlations. The correlations are presented in alphabetical order. The references to all these correlations are given in the list of references for Appendix A, Oil Correlations Formulae.

BUBBLE POINT PRESSURE CORRELATIONS

ID	Correlation	Origin	No. of Samples	No. of Data Points	Data Range									
					P_b, psia		T, °F		R_s, scf/STB		API		Gas Sp.Gr.	
					Min	Max	Min	Max	Min	Max	Min	Max	Min	Max
1	Al-Marhoun (1985)	Saudi Arabia		200	107	4315	75	240	24	1901	14.3	44.6	0.752	1.367
2	Al-Marhoun (1988)	Saudi Arabia		160	20	3573	74	240	26	1602	19.4	44.6	0.75	1.3
3	Al-Mehaideb (1997)	United Arab Emirates		62	501	4822	190	306	128	3871	30.9	48.6	0.75	1.12
4	Al-Najjar et al. (1988)	Iraq		223	580	4181	160	242	139	2408	17	44	0.572	1.319
5	Al-Shammasi (1999)			1243	32	7127	74	342	6	3299	6	63.7	0.51	3.44
6	Asgarpour et al. (1988)	Canada		1500	291	3997	94	224	135	2107	23.2	51	0.655	1.586
7	Bolondarzadeh et al. (2006)	NA		NA	NA	NA	NA	NA	NA	NA	NA	NA	NA	NA
8	De Ghetto et al. (1995)	Africa North Sea		195	107	6614	81	342	9	3299	6	56.8	0.624	1.789
9	Doklah and Osman (1992)	United Arab Emirates		51	590	4640	190	275	181	2266	28.2	40.3	0.8	1.29
10	Elam (1957)	Texas		231	168	4595	85	245	11	1597	19	53	0.56	1.95
11	El-Banbi et al. (2006)	NA		NA	NA	NA	NA	NA	NA	NA	NA	NA	NA	NA
12	El-Mabrouk et al. (2010)	NA		NA	NA	NA	NA	NA	NA	NA	NA	NA	NA	NA

(Continued)

ID	Correlation	Origin	No. of Samples	No. of Data Points	Data Range									
					P_b, psia		T, °F		R_s, scf/STB		API		Gas Sp.Gr.	
					Min	Max	Min	Max	Min	Max	Min	Max	Min	Max
13	El-Sharkawy and Alikhan (1997)	Kuwait		175	302	4375	39	1586	130	250	19.9	42.76	0.663	1.268
14	Farshad et al. (1996)	Columbia		43	32	4138	95	260	6	1645	18	44.9	0.66	1.73
15	Glasso (1980)	North Sea		41	165	7142	80	280	90	2637	22.3	48.1	0.65	1.28
16	Hanafy et al. (1997)	Egypt		NA	36	5003	107	327	86	6539	17.8	47.7	0.633	1.627
17	Hassan (2011)	Indonesia		95	55	4765	118	320	1	1520	22.3	57.6	0.66	1.798
18	Hemmati and Kharrat (2007)	NA		NA	NA	NA	NA	NA	NA	NA	NA	NA	NA	NA
19	Ikiensikimama (2008)	NA		NA	NA	NA	NA	NA	NA	NA	NA	NA	NA	NA
20	Ikiensikimama and Ogboja (2009)	NA		NA	NA	NA	NA	NA	NA	NA	NA	NA	NA	NA
21	Karimnezhad et al. (2014)	NA		NA	NA	NA	NA	NA	NA	NA	NA	NA	NA	NA
22	Kartoatmdjo and Schmidt (1991)	Indonesia, North America,		5392	15	6055	75	320	0	2890	14.4	58.9	0.38	1.71
23	Kartoatmdjo and Schmidt (1994)	NA		NA	NA	NA	NA	NA	NA	NA	NA	NA	NA	NA

(Continued)

ID	Correlation	Origin	No. of Samples	No. of Data Points	P_b, psia Min	Max	T, °F Min	Max	R_s, scf/STB Min	Max	API Min	Max	Gas Sp.Gr. Min	Max
24	Khairy et al. (1998)	Egypt		39	236	4930	120	282	16	4569	30.7	54.3	0.675	1.417
25	Khamehchi and Ebrahimian (2009)	NA	NA	NA	NA	NA	100	306	83	1708	33.4	124	0.554	0.858
26	Khazam et al. (2016)	NA		NA	NA	NA	NA	NA	NA	NA	NA	NA	NA	NA
27	Kuparuk	NA		NA	NA	NA	NA	NA	NA	NA	NA	NA	NA	NA
28	Labedi (1982)	Libya, Nigeria & Angola		247	121	6557	100	306	12	3366	22.9	52	0.579	1.251
29	Labedi (1990)	NA		NA	NA	NA	NA	NA	NA	NA	NA	NA	NA	NA
30	Lasater (1958)	Midcontinent U.S.		158	48	5780	82	272	3	2905	17.9	51.1	0.57	1.2
31	Levitan and Murtha (1999)	NA		NA	NA	NA	NA	NA	NA	NA	NA	NA	NA	NA
32	Macary and El-Batanony (1992)	Gulf of Suez Egypt		90	1200	4600	130	290	200	1200	25	40	0.7	1
33	Mazandarani and Asghari (2007)	NA		NA	NA	NA	NA	NA	NA	NA	NA	NA	NA	NA
34	McCain (1991)	NA		NA	NA	NA	NA	NA	NA	NA	NA	NA	NA	NA
35	Mehran et al. (2006)	NA		NA	NA	NA	NA	NA	NA	NA	NA	NA	NA	NA
36	Moradi et al. (2010)	NA		NA	NA	NA	NA	NA	NA	NA	NA	NA	NA	NA

(Continued)

(Continued)

ID	Correlation	Origin	No. of Samples	No. of Data Points	Data Range									
					P_b, psia		T, °F		R_s, scf/STB		API		Gas Sp.Gr.	
					Min	Max	Min	Max	Min	Max	Min	Max	Min	Max
37	Movagharnejad and Fasih (1999)	NA		NA	NA	NA	NA	NA	NA	NA	NA	NA	NA	NA
38	Obomanu and Okpobiri (1987)	Nigeria		503	NA	NA	NA	NA	NA	NA	15	43	NA	NA
39	Okoduwa and Ikiensikimama	Nigeria		NA	NA	NA	NA	NA	NA	NA	NA	NA	NA	NA
40	Omar and Todd (1993)	Malaysia		93	790	3851	125	280	142	1440	26.6	53.2	0.61	1.32
41	Ostermann and Owolabi (1983)	NA		NA	NA	NA	NA	NA	NA	NA	NA	NA	NA	NA
42	Owolabi (1984)	Alaska Cook Inlet		7	515	1802	152	180	140	309	32.7	35.3	0.853	1.094
43	Petrosky (1990)	Gulf of Mexico		90	1574	6523	114	288	217	1406	16.3	45	0.58	0.86
44	Petrosky and Farshad (1993)	NA		NA	1574	6523	114	288	217	1406	16.3	45	0.578	0.852
45	Petrosky and Farshad (1998)	NA		NA	NA	NA	NA	NA	NA	NA	NA	NA	NA	NA
46	Standing (1947)	California		105	130	7000	100	258	20	1425	16.5	63.8	0.59	0.95
47	Standing (1981)	NA		NA	NA	NA	NA	NA	NA	NA	NA	NA	NA	NA
48	Steve (1993)	NA		NA	NA	NA	NA	NA	NA	NA	NA	NA	NA	NA
49	Vasquez and Beggs (1980)	Worldwide	600	6004	15	6055	75	294	0	2199	15.3	59.3	0.51	1.35
50	Velarde et al. (1997)	NA		728	70	6700	74	327	10	1870	12	55	0.556	1.367

SOLUTION GAS–OIL RATIO CORRELATIONS

ID	Correlation	Origin	No. of Samples	No. of Data Points	Data Range									
					R_s, scf/STB		T, °F		P_b, psia		API		Gas Sp.Gr.	
					Min	Max	Min	Max	Min	Max	Min	Max	Min	Max
1	Al-Marhoun (1988)	Saudi Arabia		200	24	1901	75	240	107	4315	14.3	44.6	0.752	1.367
2	Al-Shammasi (1999)	Worldwide		1243	6	3299	74	342	32	7127	6	63.7	0.51	3.44
3	Asgarpour et al. (1988)	Canada		1500	135	2107	94	224	291	3997	23.2	51	0.655	1.586
4	Casey and Cronquist (1992)	NA	NA	NA	NA	NA	NA	NA	NA	NA	NA	NA	NA	NA
5	De Ghetto et al. (1995)	Africa North Sea		195	9	3299	81	342	107	6614	6	56.8	0.624	1.789
6	Dindrouk and Christman (2001)	NA	NA	NA	133	3050	117	276	926	12,230	14.7	40	0.602	1.027
7	Doklah and Osman (1992)	United Arab Emirates		51	181	2266	190	275	590	4640	28.2	40.3	0.8	1.29
8	El-Sharkawy and Alikhan (1997)	Kuwait		175	130	250	39	1586	302	4375	19.9	42.76	0.663	1.268
9	Farshad et al. (1996)	Columbia		43	6	1645	95	260	32	4138	18	44.9	0.66	1.73
10	Glasso (1980)	North Sea		41	90	2637	80	280	165	7142	22.3	48.1	0.65	1.28

(Continued)

(Continued)

ID	Correlation	Origin	No. of Samples	No. of Data Points	Data Range									
					R_s, scf/STB		T, °F		P_b, psia		API		Gas Sp.Gr.	
					Min	Max	Min	Max	Min	Max	Min	Max	Min	Max
11	Hanafy et al. (1997)	Egypt		NA	86	6539	107	327	36	5003	17.8	47.7	0.633	1.627
12	Hassan (2011)	Indonesia		95	1	1520	118	320	55	4765	22.3	57.6	0.66	1.798
13	Hemmati and Kharrat (2007)	NA		NA	NA	NA	165	290	348	5156	18	49	0.523	1.415
14	Ikiensikimama (2008)	NA		NA	NA	NA	NA	NA	NA	NA	NA	NA	NA	NA
15	Kartoatmdjo and Schmidt (1991)	Indonesia, North America		5392	0	2890	75	320	15	6055	14.4	58.9	0.38	1.71
16	Kartoatmdjo and Schmidt (1994)	NA		NA	NA	NA	NA	NA	NA	NA	NA	NA	NA	NA
17	Khamehchi and Ebrahimian (2009)	NA		NA	83	1708	100	306	NA	NA	33.4	124	0.554	0.858
18	Khazam et al. (2016)	NA		NA	NA	NA	NA	NA	NA	NA	NA	NA	NA	NA
19	Knopp and Ramsey (1960)	NA		NA	NA	NA	NA	NA	NA	NA	NA	NA	NA	NA
20	Kuparuk (1990)	NA		NA	NA	NA	NA	NA	NA	NA	NA	NA	NA	NA
21	Labedi (1982)	Libya, Nigeria, and Angola		247	12	3366	100	306	121	6557	22.9	52	0.579	1.251
22	Labedi (1990)	NA		NA	NA	NA	NA	NA	NA	NA	NA	NA	NA	NA
23	Lasater (1958)	NA		158	3	2905	82	272	48	5780	17.9	51.1	0.57	1.2

(Continued)

(Continued)

ID	Correlation	Origin	No. of Samples	No. of Data Points	R_s, scf/STB Min	R_s, scf/STB Max	T, °F Min	T, °F Max	P_b, psia Min	P_b, psia Max	API Min	API Max	Gas Sp.Gr. Min	Gas Sp.Gr. Max
		Midcontinent U.S.												
24	Macary and El-Batanony (1992)	Gulf of Suez Egypt		90	200	1200	130	290	1200	4600	25	40	0.7	1
25	Mazandarani and Asghari (2007)	NA		NA	NA	NA	NA	NA	NA	NA	NA	NA	NA	NA
26	McCain (1991)	NA		NA	NA	NA	NA	NA	NA	NA	NA	NA	NA	NA
27	Petrosky (1990)	Gulf of Mexico		90	217	1406	114	288	1574	6523	16.3	45	0.58	0.86
28	Petrosky and Farshad (1993)	NA		NA	217	1406	114	288	1574	6523	16.3	45	0.578	0.852
30	Standing (1947)	California		105	20	1425	100	258	130	7000	16.5	63.8	0.59	0.95
31	Standing (1981)	NA		NA	NA	NA	NA	NA	NA	NA	NA	NA	NA	NA
32	Vasquez and Beggs (1980)	Worldwide	600	6004	0	2199	75	294	15	6055	15.3	59.3	0.51	1.35
33	Velarde et al. (1997)	NA		728	10	1870	74	327	70	6700	12	55	0.3556	1.367

SATURATED OIL FORMATION VOLUME FACTOR CORRELATIONS

ID	Correlation	Origin	No. of Samples	No. of Data Points	Data Range												
					B_o, rb/STB		T, °F		R_s, scf/STB		API		Gas Sp.Gr.				
					Min	Max	Min	Max	Min	Max	Min	Max	Min	Max			
1	Abdul-Majeed and Salman (1988)	Iraq		420	1.028	2.042	75	290	0	1664	9.5	59.5	0.51	1.35			
2	Ahmed (1989)	NA		NA	NA	NA	NA	NA	NA	NA	NA	NA	NA	NA			
3	Al-Marhoun (1985)	Saudi Arabia		200	1.02	2.42	75	240	24	1901	14.3	44.6	0.752	1.367			
4	Al-Marhoun (1988)	Saudi Arabia		160	1.032	1.997	74	240	26	1602	19.4	44.6	0.75	1.37			
5	Al-Marhoun (1992)	Middle East and North America		4012	1.01	2.96	75	300	0	3265	9.5	55.9	0.575	2.52			
6	Al-Marhoun (2003)	Saudi Arabia		NA	NA	NA	NA	NA	NA	NA	NA	NA	NA	NA			
7	Al-Mehaideb (1997)	United Arab Emirates		62	1.142	3.562	190	306	128	3871	30.9	480.6	0.75	1.12			
8	Al-Najjar et al. (1988)	Iraq		223	NA	NA	160	242	139	2408	17	44	0.572	1.319			
9	Al-Shammasi (1999)	worldwide		1345	1.02	2.916	74	342	6	3299	6	63.7	0.51	3.44			
10	Arps (1962)	NA		NA	NA	NA	NA	NA	NA	NA	NA	NA	NA	NA			
11	Asgarpour et al. (1989)	Canada		1500	1.05	2.3	94	224	135	2107	23.2	51	0.655	1.577			
12	Bolondarzadeh et al. (2006)	NA		NA	NA	NA	NA	NA	NA	NA	NA	NA	NA	NA			

(Continued)

ID	Correlation	Origin	No. of Samples	No. of Data Points	Data Range									
					B_o, rb/STB		T, °F		R_s, scf/STB		API		Gas Sp.Gr.	
					Min	Max	Min	Max	Min	Max	Min	Max	Min	Max
13	Casey and Cronquist (1992)	NA		NA	1.057	3.396	136	327	95	2395	19.6	43.5	NA	NA
14	Dindrouk and Christman (2001)	Gulf of Mexico		99	1.085	2.738	117	276	133	3050	14.7	40	0.602	1.027
15	Doklah and Osman (1992)	United Arab Emirates		51	1.216	2.493	190	275	181	2266	28.2	40.3	0.8	1.29
16	Elam (1957)	Texas		205	1.019	2.115	85	245	11	1597	19	53	0.562	1.95
17	El-Banbi et al. (2006)	NA		NA	NA	NA	NA	NA	NA	NA	NA	NA	NA	NA
18	El-Mabrouk et al. (2010)	NA		NA	NA	NA	NA	NA	NA	NA	NA	NA	NA	NA
19	El-Sharkawy and Alikhan (1997)	Kuwait		171	1.076	1.969	130	250	39	1586	19.9	42.76	0.663	1.268
20	Farshad et al. (1996)	Columbia		107	1.06	2.064	95	260	6	1645	18	44.9	0.66	1.7
21	Glasso (1980)	North Sea		41	1.032	2.588	80	280	90	2637	22.3	48.1	0.65	1.28
22	Hanafy et al. (1997)	Egypt		NA	NA	NA	NA	NA	NA	NA	NA	NA	NA	NA
23	Hemmati and Kharrat (2007)	NA	NA	NA	NA	NA	NA	NA	NA	NA	NA	NA	NA	NA
24	Ikiensikimama and Ogboja (2009)	NA	NA	NA	NA	NA	NA	NA	NA	NA	NA	NA	NA	NA

(Continued)

ID	Correlation	Origin	No. of Samples	No. of Data Points	Data Range									
					B_o, rb/STB		T, °F		R_s, scf/STB		API		Gas Sp.Gr.	
					Min	Max	Min	Max	Min	Max	Min	Max	Min	Max
25	Karimnezhad et al. (2014)	NA	NA	NA	NA	NA	NA	NA	NA	NA	NA	NA	NA	NA
26	Kartoatmdjo and Schmidt (1991)	Indonesia, North America,		5392	1.022	2.747	75	320	0	2890	14.4	58.9	0.38	1.71
27	Kartoatmdjo and Schmidt (1994)	NA		NA	NA	NA	NA	NA	NA	NA	NA	NA	NA	NA
28	Khairy et al. (1998)	Egypt		39	1.064	4.42	120	282	16	4569	30.7	54.3	0.675	1.417
29	Khazam et al. (2016)	NA		NA	NA	NA	NA	NA	NA	NA	NA	NA	NA	NA
30	Knopp and Ramsey (1960)	NA		NA	NA	NA	NA	NA	NA	NA	NA	NA	NA	NA
31	Labedi (1982)	Libya, Nigeria, and Angola		331	1.047	2.925	100	306	10	3366	22.91	52	NA	NA
32	Levitan and Murtha (1999)	NA		NA	NA	NA	NA	NA	NA	NA	NA	NA	NA	NA
33	Macary and El-Batanony (1992)	Gulf of Suez Egypt		90	1.2	2	130	290	200	1200	25	40	0.7	1
34	Mazandarani and Asghari (2007)	NA		NA	NA	NA	NA	NA	NA	NA	NA	NA	NA	NA
35	McCain (1991)	NA		NA	NA	NA	NA	NA	NA	NA	NA	NA	NA	NA
36	Mehran et al. (2006)	NA		NA	NA	NA	NA	NA	NA	NA	NA	NA	NA	NA

(Continued)

(Continued)

ID	Correlation	Origin	No. of Samples	No. of Data Points	Data Range									
					B_o, rb/STB		T, °F		R_s, scf/STB		API		Gas Sp.Gr.	
					Min	Max	Min	Max	Min	Max	Min	Max	Min	Max
37	Moradi et al. (2013)	NA		NA	NA	NA	NA	NA	NA	NA	NA	NA	NA	NA
38	Obomanu and Okpobiri (1987)	Nigeria		503	NA	NA	NA	NA	NA	NA	15	43	NA	NA
39	Omar and Todd (1993)	Malaysia		93	1.085	1.954	125	280	142	1440	26.6	53.2	0.612	1.32
40	Ostermann and Owolabi (1983)	NA		NA	NA	NA	NA	NA	NA	NA	NA	NA	NA	NA
41	Owolabi (1984)	Alaska Cook Inlet		7	1.129	1.205	152	180	140	309	32.7	35.3	0.853	1.094
42	Petrosky (1990)	Gulf of Mexico		90	1.118	1.623	114	288	217	1406	16.3	45	0.58	0.85
43	Petrosky and Farshad (1993)	NA		NA	NA	NA	NA	NA	NA	NA	NA	NA	NA	NA
44	Petrosky and Farshad (1998)	NA		NA	NA	NA	NA	NA	NA	NA	NA	NA	NA	NA
45	Standing (1947)	California		105	1.024	2.15	100	258	20	1425	16.5	63.8	0.59	0.95
46	Standing (1981)	NA		NA	NA	NA	NA	NA	NA	NA	NA	NA	NA	NA
47	Sulaimon et al. 2014	NA		NA	1.092	1.954	125	280	142	1440	26.9	53.2	0.038	1.315
48	Vasquez and Beggs (1980)	Worldwide		6004	1.028	2.226	75	294	0	2199	15.3	59.3	0.511	1.35
49	Velarde et al. (1997)	NA		2097	1.04	2.082	70	307	102	1808	11.6	53.4	0.561	1.101

SATURATED OIL COMPRESSIBILITY CORRELATIONS

ID	Correlation	Origin	No. of Samples	No. of Data Points	Data Range									
					C_o, psi^{-1}		T, °F		R_s, scf/STB		API		Gas Sp.Gr.	
					Min	Max	Min	Max	Min	Max	Min	Max	Min	Max
1	Ahmed (1989)			245	NA	NA	NA	NA	NA	NA	NA	NA	NA	NA
2	Al-Marhoun (1992)			3711	3.45	31.11	75	240	1	3113	10.4	49.2	0.657	1.588
3	Al-Mehaideb (1997)			244	NA	NA	190	306	128	3871	30.9	48.6	0.746	1.116
4	Calhoun (1947)			NA	NA	NA	NA	NA	NA	NA	NA	NA	NA	NA
5	Chew and Connally (1959)			NA	NA	NA	NA	NA	NA	NA	NA	NA	NA	NA
6	De Ghetto et al. (1995)			195	3.02	43	81	342	8	2986	6	56.8	NA	NA
7	Dindoruk and Christman (2004)			99	5.02	31.91	117	276	133	3050	14.7	40	0.602	1.027
8	El-Sharkawy and Alikhan (1997)			423	3.62	29.18	120	243	367	1568	24.5	39.8	0.807	1.234
9	Farshad et al. (1996)			726	2	34	95	260	6	1758	13.7	46.5	0.59	1.731
10	Ikiensikimama (2008)			NA	NA	NA	NA	NA	NA	NA	NA	NA	NA	NA
11	Kartoatmdjo and Schmidt (1991)			3588	2.701	127.4	75	320	0	2890	14.4	59	NA	NA
12	Kartoatmdjo and Schmidt (1994)			NA	NA	NA	NA	NA	NA	NA	NA	NA	NA	NA
13	Khazam et al. (2016)			NA	NA	NA	NA	NA	NA	NA	NA	NA	NA	NA
14	Labedi (1982)			267	5.81	57.04	128	306	128	306	NA	NA	NA	NA
15	McCain (1991)			NA	NA	NA	NA	NA	NA	NA	NA	NA	NA	NA

(Continued)

(Continued)

ID	Correlation	Origin	No. of Samples	No. of Data Points	Data Range									
					C_o, psi^{-1}		T, °F		R_s, scf/STB		API		Gas Sp.Gr.	
					Min	Max	Min	Max	Min	Max	Min	Max	Min	Max
16	McCain et al. (1988)			NA	NA	NA	NA	NA	NA	NA	NA	NA	NA	NA
17	Petrosky (1990)			304	3.507	24.65	114	288	217	1406	16.3	45	0.578	0.852
18	Petrosky and Farshad (1993)			NA	3.5	24.64	114	288	217	1406	16.3	45	0.578	0.852
19	Petrosky and Farshad (1998)			NA	NA	NA	NA	NA	NA	NA	NA	NA	NA	NA
20	Standing (1947)			NA	NA	NA	NA	NA	NA	NA	NA	NA	NA	NA
21	Standing (1981)			NA	NA	NA	NA	NA	NA	NA	NA	NA	NA	NA
22	Steve (1993)			NA	NA	NA	NA	NA	NA	NA	NA	NA	NA	NA
23	Vasquez and Beggs (1980)			4486	NA	NA	170	170	9	2199	15.3	59.5	0.511	1.351

SATURATED OIL DENSITY CORRELATIONS

ID	Correlation	Origin	No. of Samples	No. of Data Points	Data Range									
					Rho, lb/cf		T, °F		R_s, scf/STB		API		Gas Sp.Gr.	
					Min	Max	Min	Max	Min	Max	Min	Max	Min	Max
1	Ahmed (1989)			253	0.096	28.5	0.093	20.5	NA	NA	NA	NA	NA	NA
2	Hanafy et al. (1997)			210	0.093	7.139	NA	NA	NA	NA	NA	NA	NA	NA
3	McCain (1991)			328	NA	NA	NA	NA	NA	NA	NA	NA	NA	NA
4	Standing (1947)			26	0.16	315	0.142	127	NA	NA	NA	NA	NA	NA
5	Standing (1981)			NA	NA	NA	NA	NA	NA	NA	NA	NA	NA	NA

STOCK-TANK GOR CORRELATIONS

ID	Correlation	Origin	No. of Samples	No. of Data Points	Data Range							
					GOR scf/STB		T, °F		API		Gas Sp.Gr.	
					Min	Max	Min	Max	Min	Max	Min	Max
1	El-Mabrouk and Shirif (2003)			253	0.096	28.5	0.093	20.5	NA	NA	NA	NA
2	Okeke and Sylvester (2016)			210	0.093	7.139	NA	NA	NA	NA	NA	NA
3	Rollins et al. (1990)			328	NA	NA	NA	NA	NA	NA	NA	NA
4	Valko et al. (2003)			26	0.16	315	0.142	127	NA	NA	NA	NA

UNDERSATURATED OIL FORMATION VOLUME FACTOR CORRELATIONS

ID	Correlation	Origin	No. of Sample	No. of Data Points	Data Range									
					B_o, rb/STB		T, °F		R_s, scf/STB		API		Gas Sp. Gr.	
					Min	Max	Min	Max	Min	Max	Min	Max	Min	Max
1	Ahmed (1992)			253	0.096	28.5	0.093	20.5	NA	NA	NA	NA	NA	NA
2	Al-Marhoun (1992)			210	0.093	7.139	NA	NA	NA	NA	NA	NA	NA	NA
3	General equation			328	NA	NA	NA	NA	NA	NA	NA	NA	NA	NA

SATURATED OIL VISCOSITY CORRELATIONS

ID	Correlation	Origin	No. of Samples	No. of Data Points	Data Range									
					M_o, cp		R_s, scf/STB		T, °F		API		Gas Sp.Gr.	
					Min	Max	Min	Max	Min	Max	Min	Max	Min	Max
1	Abdul-Majeed et al. (1990)			NA	NA	NA	NA	NA	NA	NA	NA	NA	NA	NA
2	Abu-Khamsim and Al-Marhoun (1991)			NA	NA	NA	NA	NA	NA	NA	NA	NA	NA	NA
3	Al-Khafaji (1987)			72	NA	NA	0	2100	NA	NA	NA	NA	NA	NA
4	Al-Mehaideb (1997)			57	NA	NA	128	3871	NA	NA	NA	NA	NA	NA
5	Beggs and Robinson (1975)			2073	NA	NA	20	2070	NA	NA	NA	NA	NA	NA
6	Bergman (2004)			NA	NA	NA	NA	NA	NA	NA	NA	NA	NA	NA
7	Bergman (2007)			NA	NA	NA	NA	NA	NA	NA	NA	NA	NA	NA
8	Chew and Connally (1959)			2257	NA	NA	51	3544	NA	NA	NA	NA	NA	NA
9	De Ghetto et al. (1995)			195	0.07	295.9	9	3299	NA	NA	NA	NA	NA	NA
10	Dindrouk and Christman (2001)			95	0.161	8.7	133	3050	NA	NA	NA	NA	NA	NA
11	El-Sharkawy and Alikhan (1997)			254	0.05	20.89	10	3600	NA	NA	NA	NA	NA	NA
12	Glasso (1980)			NA	NA	NA	NA	NA	NA	NA	NA	NA	NA	NA
13	Hanafy et al. (1997)			NA	NA	NA	NA	NA	NA	NA	NA	NA	NA	NA
14	Hossain et al. (2005)			NA	NA	NA	NA	NA	NA	NA	NA	NA	NA	NA
15	Ikiensikimama (2008)			NA	NA	NA	NA	NA	NA	NA	NA	NA	NA	NA
16	Kartoatmdjo and Schmidt (1991)			NA	NA	NA	NA	NA	NA	NA	NA	NA	NA	NA

(Continued)

(Continued)

ID	Correlation	Origin	No. of Samples	No. of Data Points	Data Range									
					M_o, cp		R_s, scf/STB		T, °F		API		Gas Sp.Gr.	
					Min	Max	Min	Max	Min	Max	Min	Max	Min	Max
17	Kartoatmdjo and Schmidt (1994)			5321	0.096	586	0	2890	NA	NA	NA	NA	NA	NA
18	Khamehchi and Ebrahimian (2009)			NA	0.04	3	83	1708	NA	NA	NA	NA	NA	NA
19	Khan et al. (1987)			1841	0.13	77.4	24	1901	NA	NA	NA	NA	NA	NA
20	Khazam et al. (2016)			NA	NA	NA	NA	NA	NA	NA	NA	NA	NA	NA
21	Labedi (1982)			91	0.115	3.72	NA	NA	NA	NA	NA	NA	NA	NA
22	Labedi (1992)			NA	NA	NA	NA	NA	NA	NA	NA	NA	NA	NA
23	McCain (1991)			NA	NA	NA	NA	NA	NA	NA	NA	NA	NA	NA
24	Naseri et al. (2005)			NA	NA	NA	NA	NA	NA	NA	NA	NA	NA	NA
25	Osorio (1990)			NA	NA	NA	NA	NA	NA	NA	NA	NA	NA	NA
26	Petrosky (1990)			864	0.21	7.4	21	1885	NA	NA	NA	NA	NA	NA
27	Petrosky and Farshad (1993)			864	0.21	7.4	21	1885	NA	NA	NA	NA	NA	NA
28	Petrosky and Farshad (1995)			NA	NA	NA	NA	NA	NA	NA	NA	NA	NA	NA
29	Petrosky and Farshad (1998)			NA	NA	NA	NA	NA	NA	NA	NA	NA	NA	NA
30	Standing (1947)			NA	NA	NA	NA	NA	NA	NA	NA	NA	NA	NA
31	Standing (1981)			NA	0.3	35	0	3544	NA	NA	NA	NA	NA	NA
32	Steve (1993)			NA	NA	NA	NA	NA	NA	NA	NA	NA	NA	NA

UNDERSATURATED OIL COMPRESSIBILITY CORRELATIONS

ID	Correlation	Origin	No. of Samples	No. of Data Points	Data Range									
					C_o, psi^{-1}		T, °F		R_s, scf/STB		API		Gas Sp.Gr.	
					Min	Max	Min	Max	Min	Max	Min	Max	Min	Max
1	Ahmed (1989)			253	0.096	28.5	0.093	20.5	NA	NA	NA	NA	NA	NA
2	Al-Marhoun (1992)			210	0.093	7.139	NA	NA	NA	NA	NA	NA	NA	NA
3	Al-Marhoun (2003)			328	NA	NA	NA	NA	NA	NA	NA	NA	NA	NA

UNDERSATURATED OIL VISCOSITY CORRELATIONS

ID	Correlation	Origin	No. of Samples	No. of Data Points	Data Range									
					M_o, cp		Mob, cp		P, psi		P_b, psi		Gas Sp.Gr.	
					Min	Max	Min	Max	Min	Max	Min	Max	Min	Max
1	Abdul-Majeed et al. (1990)			253	0.096	28.5	0.093	20.5	NA	NA	NA	NA	NA	NA
2	Al-Khafaji et al. (1987)			210	0.093	7.139	NA	NA	NA	NA	NA	NA	NA	NA
3	Al-Mehaideb (1997)			328	NA	NA	NA	NA	NA	NA	NA	NA	NA	NA
4	Beal (1946)			26	0.16	315	0.142	127	NA	NA	NA	NA	NA	NA
5	Beggs and Robinson (1975)			NA	NA	NA	NA	NA	NA	NA	NA	NA	NA	NA
6	Bergman (2004)			NA	NA	NA	NA	NA	NA	NA	NA	NA	NA	NA
7	Bergman and Sutton (2006)			NA	NA	NA	NA	NA	NA	NA	NA	NA	NA	NA

(Continued)

ID	Correlation	Origin	No. of Samples	No. of Data Points	Data Range									
					M_o, cp		Mob, cp		P, psi		P_b, psi		Gas Sp.Gr.	
					Min	Max	Min	Max	Min	Max	Min	Max	Min	Max
8	De Ghetto et al. (1995)			195	0.13	354.6	NA	NA	NA	NA	NA	NA	NA	NA
9	Dindoruk and Christman (2001)			95	0.211	10.6	0.161	8.7	NA	NA	NA	NA	NA	NA
10	El-Sharkawy and Alikhan (1999)			254	0.2	5.7	NA	NA	1287	10,000	NA	NA	NA	NA
11	Glasso (1980)			NA	NA	NA	NA	NA	NA	NA	NA	NA	NA	NA
12	Hossain et al. (2005)			NA	NA	NA	NA	NA	NA	NA	NA	NA	NA	NA
13	Ikiensikimama (2008)			NA	NA	NA	NA	NA	NA	NA	NA	NA	NA	NA
14	Kartoatmdjo and Schmidt (1994)			3588	0.168	517.03	0.168	184.9	25	6015	NA	NA	NA	NA
15	Kartoatmdjo and Schmidt (1994)			NA	NA	NA	NA	NA	NA	NA	NA	NA	NA	NA
16	Khan et al. (1987)			1503	0.13	71	0.13	77.4	NA	NA	NA	NA	NA	NA
17	Khazam et al. (2016)			NA	NA	NA	NA	NA	NA	NA	NA	NA	NA	NA
18	Kouzel et al. (1965)			95	1.78	202	1.22	134	423	6015	NA	NA	NA	NA
19	Kouzel et al. (1997)			1279	NA	NA	NA	NA	NA	NA	NA	NA	NA	NA
20	Labedi (1982)			91	NA	NA	0.115	3.72	NA	NA	NA	NA	NA	NA
21	Labedi (1992)			NA	NA	NA	NA	NA	NA	NA	NA	NA	NA	NA
22	McCain (1991)			NA	NA	NA	NA	NA	NA	NA	NA	NA	NA	NA
23	Naseri et al. (2005)			NA	NA	NA	NA	NA	NA	NA	NA	NA	NA	NA

(Continued)

(Continued)

ID	Correlation	Origin	No. of Samples	No. of Data Points	Data Range									
					M_o, cp		Mob, cp		P, psi		P_b, psi		Gas Sp.Gr.	
					Min	Max	Min	Max	Min	Max	Min	Max	Min	Max
24	Petrosky (1990)			404	0.22	4.09	0.211	3.546	1600	10,250	NA	NA	NA	NA
25	Petrosky and Farshad (1993)			404	0.22	4.09	0.211	3.546	1600	10,250	NA	NA	NA	NA
26	Petrosky and Farshad (1995)			NA	NA	NA	NA	NA	NA	NA	NA	NA	NA	NA
27	Petrosky and Farshad (1998)			NA	NA	NA	NA	NA	NA	NA	NA	NA	NA	NA
28	Standing (1947)			0.16	315	0.14	127	NA	NA	NA	NA	NA	NA	NA
29	Standing (1981)			NA	NA	NA	NA	NA	NA	NA	NA	NA	NA	NA
30	Twu 1985			NA	NA	NA	NA	NA	NA	NA	NA	NA	NA	NA
31	Vasquez and Beggs (1980)			3593	0.117	148	NA	NA	126	9500	NA	NA	NA	NA

UNDERSATURATED OIL DENSITY CORRELATIONS

ID	Correlation	Origin	No. of Samples	No. of Data Points	Data Range									
					Rho, lb/cf		T, °F		R_s, scf/STB		API		Gas Sp.Gr.	
					Min	Max	Min	Max	Min	Max	Min	Max	Min	Max
1	General equation			253	0.096	28.5	0.093	20.5	NA	NA	NA	NA	NA	NA

FREE GAS SPECIFIC GRAVITY CORRELATIONS

ID	Correlation	Origin	No. of Samples	No. of Data Points	Data Range									
					Gas Sp.Gr		T, °F		R_s, scf/STB		API		Gas Sp.Gr.	
					Min	Max	Min	Max	Min	Max	Min	Max	Min	Max
1	McCain and Hill (1995)			253	0.096	28.5	0.093	20.5	NA	NA	NA	NA	NA	NA

TOTAL FORMATION VOLUME FACTOR CORRELATIONS

ID	Correlation	Origin	No. of Samples	No. of Data Points	Data Range									
					Bt, rb/STB		T, °F		Rs, scf/STB		API		Gas Sp.Gr.	
					Min	Max	Min	Max	Min	Max	Min	Max	Min	Max
1	Al-Marhoun (1988)			253	0.096	28.5	0.093	20.5	NA	NA	NA	NA	NA	NA
2	Glasso (1980)			210	0.093	7.139	NA	NA	NA	NA	NA	NA	NA	NA
3	Standing (1947)			328	NA	NA	NA	NA	NA	NA	NA	NA	NA	NA

DEAD OIL VISCOSITY CORRELATIONS

| ID | Correlation | Origin | No. of Samples | No. of Data Points | Data Range | | | | | | | | | | |
| | | | | | Mod, cp | | T, °F | | API | | R_s, scf/STB | | Gas Sp.Gr. | |
					Min	Max	Min	Max	Min	Max	Min	Max	Min	Max
1	Al-Khafaji et al. (1987)			350	NA	NA	60	300	15	51	NA	NA	NA	NA
2	Andrade (1930)			NA	NA	NA	NA	NA	NA	NA	NA	NA	NA	NA
3	Beal (1946)			753	0.865	1550	98	250	10.1	52.5	NA	NA	NA	NA
4	Beggs and Robinson (1975)			460	NA	NA	70	295	16	58	NA	NA	NA	NA
5	Bennison (1998)			16	6.4	8396	39	300	11.1	19.7	NA	NA	NA	NA
6	Bergman (2004)			NA	NA	NA	NA	NA	NA	NA	NA	NA	NA	NA
7	Bergman and Sutton (2006)			NA	NA	NA	NA	NA	NA	NA	NA	NA	NA	NA
8	De Ghetto et al. (1995)			195	0.46	1386.9	81	342	6	56.8	NA	NA	NA	NA
9	Dindrouk and Christman (2001)			95	0.896	62.63	121	276	17.4	40	NA	NA	NA	NA
10	El-Sharkawy and Alikhan (1999)			254	0.6	33.7	100	300	19.9	48	NA	NA	NA	NA
11	El-Sharkawy and Gharbi (2001)			NA	NA	NA	NA	NA	NA	NA	NA	NA	NA	NA
12	Fitzgerlad (1997)			7267	0.3	30,000	NA	NA	NA	NA	NA	NA	NA	NA
13	Glasso (1980)			29	0.616	39.1	50	300	20.1	48.1	NA	NA	NA	NA
14	Hossain et al. (2005)			NA	NA	NA	NA	NA	NA	NA	NA	NA	NA	NA
15	Ikiensikimama (2008)			NA	NA	NA	NA	NA	NA	NA	NA	NA	NA	NA
16	Kartoatmdjo and Schmidt (1991)			661	0.506	682	80	320	14.4	59	NA	NA	NA	NA
17	Kartoatmdjo and Schmidt (1994)			NA	NA	NA	NA	NA	NA	NA	NA	NA	NA	NA

(Continued)

(Continued)

ID	Correlation	Origin	No. of Samples	No. of Data Points	Data Range									
					Mod, cp		T, °F		API		R_s, scf/STB		Gas Sp.Gr.	
					Min	Max	Min	Max	Min	Max	Min	Max	Min	Max
18	Kaye (1985)			NA	NA	NA	NA	NA	NA	NA	NA	NA	NA	NA
19	Khan et al. (1987)			NA	NA	NA	75	240	14.3	44.6	NA	NA	NA	NA
20	Khazam et al. (2016)			NA	NA	NA	NA	NA	NA	NA	NA	NA	NA	NA
21	Labedi (1982)			91	0.66	4.79	100	306	32.2	48	NA	NA	NA	NA
22	Labedi (1992)			NA	NA	NA	NA	NA	NA	NA	NA	NA	NA	NA
23	McClain 1991			NA	NA	NA	NA	NA	NA	NA	NA	NA	NA	NA
24	Naseri et al. (2005)			NA	0.75	54	100	300	20	48	NA	NA	NA	NA
25	Naseri et al. (2012)			NA	NA	NA	NA	NA	NA	NA	NA	NA	NA	NA
26	Ng and Egbogah 1983			394	NA	NA	59	176	5	58	NA	NA	NA	NA
27	Oyedeko and Ulaeto (2011)			NA	NA	NA	NA	NA	NA	NA	NA	NA	NA	NA
28	Petrosky (1990)			118	0.725	10.25	114	288	25.4	46.1	NA	NA	NA	NA
29	Petrosky and Farshad (1993)			118	0.725	10.25	114	288	25.4	46.1	NA	NA	NA	NA
30	Petrosky and Farshad (1995)			NA	NA	NA	NA	NA	NA	NA	NA	NA	NA	NA
31	Petrosky and Farshad (1998)			NA	NA	NA	NA	NA	NA	NA	NA	NA	NA	NA
32	Sattarina et al. (2007)			NA	NA	NA	NA	NA	NA	NA	NA	NA	NA	NA
33	Standing (1947)			NA	0.865	1550	100	220	10.1	52.5	NA	NA	NA	NA
34	Standing (1981)			NA	NA	NA	NA	NA	NA	NA	NA	NA	NA	NA
35	Ulaeto and Oyedeko (2014)			NA	NA	NA	NA	NA	NA	NA	NA	NA	NA	NA
36	Whitson and Brule (2000)			NA	NA	NA	NA	NA	NA	NA	NA	NA	NA	NA

Appendix D

Gas Correlations Range of Applicability

This appendix contains the working ranges of different gas correlations. The correlations are presented in alphabetical order. Their references are covered in the list of references given in Appendix B, Gas Correlations Formulae.

DEW POINT PRESSURE CORRELATIONS

ID	Correlation	P (psi)		T (°F)		GCR (scf/STB)		API		Gas Sp.Gr.		C7+ Sp.Gr.		C7+ Mo. Wt.	
		C1		C2		C3		C4		C5		C6		C7+	
		Min	Max	Min	Max	Min	Max	Min	Max	Min	Max	Min	Max	Min	Max
1	Ahmadi and El-Sharkawy (2016)	NA	NA	40	340	NA	NA	NA	NA	0.65	1.89	0.72	0.85	106	253
2	Alzahabi et al. (2015)	0.19	0.95	0.02	0.21	0.00	0.12	0.00	0.51	0.00	0.12	0.00	0.09	0.00	0.17
3	El-Sharkawy (2001)	NA	NA	40	340	NA	NA	NA	NA	0.65	1.89	0.72	0.85	106	253
4	El-Sharkawy (2011)	0.19	0.95	0.02	0.21	0.00	0.12	0.00	0.51	0.00	0.12	0.00	0.09	0.00	0.17
5	England (2002)	NA	NA	NA	NA	NA	NA	NA	NA	NA	NA	NA	NA	NA	NA
6	Godwin (2012)	NA	NA	NA	NA	NA	NA	NA	NA	NA	NA	NA	NA	NA	NA
7	Kamaria et al. (2016)	NA	NA	NA	NA	NA	NA	NA	NA	NA	NA	NA	NA	NA	NA
8	Nemeth and Kennedy (1967)	NA	NA	40	320	NA	NA	NA	NA	NA	NA	0.773	0.868	106	235

Data Range

(Continued)

(Continued)

ID	Correlation	Data Range													
		P (psi) C1		T (°F) C2		GCR (scf/STB) C3		API C4		Gas Sp.Gr. C5		C7 + Sp.Gr. C6		C7 + Mo. Wt. C7 +	
		Min	Max	Min	Max	Min	Max	Min	Max	Min	Max	Min	Max	Min	Max
9	Olds et al. (1945)	0.03	0.97	0.00	0.15	0.00	0.11	0.00	0.38	0.00	0.07	0.00	0.09	0.00	0.14
10	Organick and Golding (1952)	NA	NA	NA	NA	NA	NA	NA	NA	NA	NA	NA	NA	NA	NA
		NA	NA	NA	NA	NA	NA	NA	NA	NA	NA	NA	NA	NA	NA
11	Ovalle et al. (2005)	NA	NA	93	267	2289	113,346	35.6	76.5	0.61	1.62	NA	NA	NA	NA
12	Shokir (2008)	NA	NA	75	320	NA	NA	NA	NA	NA	NA	0	0.04	106	224
13	Humound and Al-Marhoun (2001)	60	1215	100	310	3400	$1.5E+05$	42	56	0.66	0.82	0	NA	0	NA
		0.33	0.97	0.01	0.15	0.001	0.11	0	0.08	0	0.06	0	0.04	0	0.14
		0.58	0.84	0.005	0.13	0.001	0.0393	0	0.093						

CRITICAL PROPERTIES CORRELATIONS

ID	Correlation	Data Range													
		P (psi) C1		T (°F) C2		Gas Sp.Gr. C3		H2S C4		CO_2 C5		N_2 C6		C7+	
		Min	Max	Min	Max	Min	Max	Min	Max	Min	Max	Min	Max	Min	Max
1	El-Sharkawy (2000)	90	12,000	40	327	0.566	1.895	0	0.739	0	0.672	0	0.252	0	0.17
		0.17	0.97	0.00	0.29	0.00	0.13	0.00	0.08	0.00	0.05	0.00	0.05		
2	Piper et al.	514	12,814	78	326	0.0613	1.821	0	0.514	0	0.672	0	0.157	0	0.149
		0.19	0.95	0.02	0.18	0.00	0.13	0.00	0.09	0.00	0.06	0.00	0.04		
3	Standing (1981)	15	8220	20	280	0.591	1.074	NA	NA	0	0.015	NA	NA	NA	NA
		NA	NA	NA	NA	NA	NA	NA	NA	NA	NA	NA	NA		
4	Sutton (1985)	200	12,500	100	360	0.571	1.679	NA	NA	0	0.017	0	0.03	NA	NA
		NA	NA	NA	NA	NA	NA	NA	NA	NA	NA	NA	NA		
5	Sutton (2005)	12	17,065	505.7	652	0.554	1.861	0	0.9	0	0.89	0.	0.33	NA	NA
		NA	NA	NA	NA	NA	NA	NA	NA	NA	NA	NA	NA		
6	Sutton (2007)	12	17,065	0	460	0.556	2.819	0	0.9	0	0.89	0	0.33	NA	NA
		NA	NA	NA	NA	NA	NA	NA	NA	NA	NA	NA	NA		
7	Wichert and Aziz	154	7026	40	300	NA	NA	0	0.74	0.	0.55	NA	NA	NA	NA
		NA	NA	NA	NA	NA	NA	NA	NA	NA	NA	NA	NA		

GAS VISCOSITY CORRELATIONS

ID	Correlation	Data Range											
		P (psi)		T (°F)		Gas Sp.Gr.		C7 + Sp.Gr		C7 + Mo. Wt.			
		Min	Max	Min	Max	Min	Max	Min	Max	Min	Max		
1	Al-Nasser and Al-Marhoun (2012)	NA	NA	NA	NA	NA	NA	NA	NA	NA	NA		
2	Bicker and Katz (1943)	NA	NA	NA	NA	NA	NA	NA	NA	NA	NA		
3	Carr et al. (1954)	NA	NA	NA	NA	NA	NA	NA	NA	NA	NA		
4	De Ghetto et al. (1995)	NA	NA	NA	NA	NA	NA	NA	NA	NA	NA		
5	Dean and Stiel (1965)	NA	NA	NA	NA	NA	NA	NA	NA	NA	NA		
6	Dempcy (1965)	1	20 (Pr)	1.2	3 (Tr)	NA	NA	NA	NA	NA	NA		
7	El-Sharkawy (2006)	NA	NA	NA	NA	NA	NA	NA	NA	NA	NA		
8	Lee et al. (1966)	100	8000	100	340	NA	NA	NA	NA	NA	NA		
9	Londono et al. (2002)	NA	NA	23	NA	NA	NA	NA	NA	NA	NA		
10	Londono et al. (2005)	NA	NA	23	NA	NA	NA	NA	NA	NA	NA		
11	Lucas (1981)	NA	NA	NA	NA	NA	NA	NA	NA	NA	NA		
12	McCain (1991)	NA	NA	NA	NA	NA	NA	NA	NA	NA	NA		
13	Standing (1947)	NA	NA	NA	NA	NA	NA	NA	NA	NA	NA		
14	Sutton (2007)	14.7	20,305	45.7	1112	0.55	1.86	NA	NA	NA	NA		

Appendix E

Artificial Neural Network (ANN) Models Range of Applicability

This appendix contains the working ranges of the artificial neural network (ANN) models for oil and gas PVT properties. The ANN models are presented in alphabetical order. The references used in this appendix are given in Chapter 10, Artificial Neural Network Models for PVT Properties.

SINGLE-PHASE Z-FACTOR ARTIFICIAL NEURAL NETWORK MODELS

z Factor	T		P		H₂S		CO₂		N₂		C1		C2		C3		C4	
	Min	Max	Min	Max	Min	Max	Min	Max	Min	Max	Min	Max	Min	Max	Min	Max	Min	Max
Shateri et al. (2015)	40	300	NA	NA	0.0	73.9	0.0	54.5	0.0	25.2	17.3	97.5	0.0	28.7	0.0	13.2	0.0	5.3

z Factor	C5		C6		C7+		MW		Sp.Gr.		Tr		Pr		z Factor	
	Min	Max	Min	Max	Min	Max	Min	Max	Min	Max	Min	Max	Min	Max	Min	Max
Shateri et al. (2015)	0.0	3.6	0.0	2.7	0.0	8.2	0.0	150.0	0.0	0.9	1.0	2.0	0.2	10.2	0.4	1.2

DEW POINT PRESSURE ARTIFICIAL NEURAL NETWORK MODELS

Dew Point Pressure	T		H₂S		CO₂		N₂		C1		C2		C3	
	Min	Max	Min	Max	Min	Max	Min	Max	Min	Max	Min	Max	Min	Max
Kamari et al. (2013)	40	418	0	0.299	0.0	0.9	0.0	0.5	0.0	1.0	0.0	0.2	0.0	0.1

Dew Point Pressure	C4		C5		C6		C7+		MW		Sp.Gr.	
	Min	Max	Min	Max	Min	Max	Min	Max	Min	Max	Min	Max
Kamari et al. (2013)	0.0	0.1	0.0	0.4	0.0	0.1	0.0	0.1	106.0	262.0	0.7	0.9

OIL FORMATION VOLUME FACTOR ARTIFICIAL NEURAL NETWORK MODELS

Oil Formation Volume Factor	Rsi		T		API		Gas SP.Gr.		Bo		Pb	
	Min	Max	Min	Max	Min	Max	Min	Max	Min	Max	Min	Max
Elsharkawy (1998)	0.0	1568	130.0	245.0	0.6	0.9	0.63	1.20	1.03	1.93	NA	NA
Arabloo et al. (2014)	7.1	3299	74.0	360.9	6.0	56.8	0.52	3.44	1.02	2.92	NA	NA
Alimadadi (2011)	255.2	1998	140.0	277.0	18.9	34.8	0.88	1.00	1.17	2.39	NA	NA
Ikiensikimama (2012)	NA	NA	NA	NA	NA	NA	NA	NA	NA	NA	58	7127
Gharbi and Elsharkawy (1997)	13.0	3366	74.0	306.0	19.4	50.1	0.59	1.37	1.03	2.93	400	3362
Osman and Abdel-Aal (2002)	24.0	1453	75.0	240.0	17.5	44.6	0.75	1.82	1.03	1.89	58	7127

SOLUTION GAS OIL RATIO ARTIFICIAL NEURAL NETWORK MODELS

Solution Gas Oil Ratio	Rsi		T		API		Gas SP.Gr.		Bo		Pb	
	Min	Max	Min	Max	Min	Max	Min	Max	Min	Max	Min	Max
Elsharkawy (1998)	0.0	1568	130.0	245.0	0.6	0.9	0.63	1.20	1.03	1.93	NA	NA
Dutta and Gupta (2010)	10.0	1999	144.0	298.0	15.0	45.0	0.63	1.54	1.00	1.99	NA	NA
Oloso (2009)	193.0	1400	130.0	252.0	24.2	48.0	NA	NA	NA	NA	400	3362

BUBBLE POINT PRESSURE ARTIFICIAL NEURAL NETWORK MODELS

Bubble Point Pressure	Rs		N2		T		C1		API		C2		Gas Sp.Gr.		C3		Bo		C4		Pb		C7+		H2S		MW		CO2		Sp.Gr.	
	Min	Max	Min	Max	Min	Max	Min	Max	Min	Max	Min	Max	Min	Max	Min	Max	Min	Max	Min	Max	Min	Max	Min	Max	Min	Max	Min	Max	Min	Max	Min	Max
Dutta and Gupta (2010)	10	1999	NA	NA	144	298	NA	NA	15.0	45.0	NA	NA	0.6	1.5	NA	NA	1.0	2.0	NA	NA	100	4338	NA	NA	NA	NA	NA	NA	NA	NA	NA	NA
Ahmadi et al. (2015)	NA	NA	0.0	1.7	128	324	5.6	74.2	NA	NA	0.8	12.5	NA	NA	0.4	11.9	NA	NA	1.0	8.4	313	6880	10.7	83.2	0.0	3.7	134.0	324.0	0.0	9.1	0.7	0.9
Adeeyo (2016)	30	1600	NA	NA	110	275	NA	NA	14.0	56.5	NA	NA	0.5	1.0	NA	NA	1.0	2.2	NA	NA	550	5984	NA	NA	NA	NA	NA	NA	NA	NA	NA	NA
Rafiee-Taghanaki et al. (2013)	7.08	2496	NA	NA	74	360.9	NA	NA	15.7	56.5	NA	NA	0.5	3.4	NA	NA	1.0	2.9	NA	NA	58	5152	NA	NA	NA	NA	NA	NA	NA	NA	NA	NA
Al-Shammasi (2001)	6	3299	NA	NA	74	341.6	NA	NA	6.0	63.7	NA	NA	0.5	3.4	NA	NA	1.0	2.9	NA	NA	32	7127	NA	NA	NA	NA	NA	NA	NA	NA	NA	NA
Gharbi and Elsharkawy (1997)	13	3366	NA	NA	74	306	NA	NA	19.4	50.1	NA	NA	0.6	1.4	NA	NA	1.0	2.9	NA	NA	121	7127	NA	NA	NA	NA	NA	NA	NA	NA	NA	NA
Al-Marhoun and Osman (2002)	24	1453	NA	NA	75	240	NA	NA	17.5	44.6	NA	NA	0.8	1.8	NA	NA	1.0	3.6	NA	NA	90	3331	NA	NA	NA	NA	NA	NA	NA	NA	NA	NA
Osman and Abdel-Aal (2002)	24	1453	NA	NA	75	240	NA	NA	17.5	44.6	NA	NA	0.8	1.8	NA	NA	1.0	1.9	NA	NA	90	3331	NA	NA	NA	NA	NA	NA	NA	NA	NA	NA

OIL DENSITY AT THE BUBBLE POINT PRESSURE ARTIFICIAL NEURAL NETWORK MODELS

Density	Rsi		T		API		Gas SP.Gr.		Bo	
	Min	Max	Min	Max	Min	Max	Min	Max	Min	Max
Elsharkawy (1998)	0.0	1568	130.0	245.0	0.6	0.9	0.63	1.20	1.03	1.93
Alimadadi (2011)	255.2	1998	140.0	277.0	18.9	34.8	0.88	1.00	1.17	2.39

OIL VISCOSITY AT THE BUBBLE POINT PRESSURE ARTIFICIAL NEURAL NETWORK MODELS

Bubble Point Oil Viscosity	Rsi		T		API		Gas SP.Gr.		MuoD	
	Min	Max	Min	Max	Min	Max	Min	Max	Min	Max
Al-Gathe et al. (2015)	7.9	2988	75.0	342.0	15.0	53.0	0.6	1.8	0.2	164.0
Elsharkawy (1998)	0.0	1568	130.0	245.0	0.6	0.9	0.6	1.2	0.2	16.0
Oloso (2009)	193.0	1400	130.0	252.0	24.2	48.0	NA	NA	0.9	5.7

DEAD OIL VISCOSITY ARTIFICIAL NEURAL NETWORK MODELS

Dead Oil Viscosity	Rsi		T		API		Gas SP.Gr.		P	
	Min	Max	Min	Max	Min	Max	Min	Max	Min	Max
Al-Gathe et al. (2015)	NA	NA	68.0	342.0	5.9	135.0	NA	NA	NA	NA
Elsharkawy and Gharbi (2001)	NA	NA	130.0	243.0	24.5	39.8	0.81	1.23	0	9900
Elsharkawy (1998)	0.0	1568	130.0	245.0	0.6	0.9	0.63	1.20	0	9900
Oloso (2009)	193.0	1400	130.0	252.0	24.2	48.0	NA	NA	NA	NA

UNDER-SATURATED OIL VISCOSITY ARTIFICIAL NEURAL NETWORK MODELS

Under-Saturated Oil Viscosity	Rsi		T		API		Gas SP.Gr.		P		Pb		MuoD		Muob	
	Min	Max	Min	Max	Min	Max	Min	Max	Min	Max	Min	Max	Min	Max	Min	Max
Al-Gathe et al. (2015)	8	7804	NA	NA	6.0	57.0	NA	NA	2	15,305	64	6614	0	1387	0	296
Elsharkawy and Gharbi (2001)	NA	NA	130.0	243.0	24.5	39.8	0.81	1.23	0	9900	NA	NA	NA	NA	NA	NA
Elsharkawy (1998)	0	1568	130.0	245.0	0.6	0.9	0.63	1.20	0	9900	NA	NA	0	16	0	16
Dutta and Gupta (2010)	10	1999	144.0	298.0	15.0	45.0	0.63	1.54	1286	5940	100	4338	0	25	0	16

SATURATED OIL VISCOSITY ARTIFICIAL NEURAL NETWORK MODELS

Saturated Oil Viscosity	Rsi / MuoD		T / Muob		API / Muo		Gas SP.Gr. / P		Pb	
	Min	Max	Min	Max	Min	Max	Min	Max	Min	Max
Al-Gathe et al. (2015)	1.53	7804	75	342	13.5	56.8	0	7	64	6614
	0.2	163.6	0.1	75.0	NA	NA	NA	NA	NA	NA
Elsharkawy and Gharbi (2001)	NA	NA	130	243	24.5	39.8	1	NA	NA	NA
	NA	NA	NA	NA	NA	NA	NA	NA	NA	NA
Elsharkawy (1998)	0	1568	130	245	0.6	0.9	1	NA	NA	NA
	0.2	16.0	NA	NA	NA	NA	NA	NA	NA	NA

(Continued)

(Continued)

Saturated Oil Viscosity

	Rsi (MuoD)		T (Muob)		API (Muo)		Gas SP.Gr. (P)		Pb	
	Min	Max	Min	Max	Min	Max	Min	Max	Min	Max
Hajizadeh Y (2007)	157.1	4006	42	232	0.5	0.9	NA	NA	NA	NA
	NA	NA	0.1	9.0	NA	NA	97	8001	NA	NA
Dutta and Gupta (2010)	10	1999	144	298	15.0	45.0	1	2	100	4338
	0.3	25.0	0.0	15.5	0.1	27.8	1286	5940	NA	NA
Oloso (2009)	193	1400	130	252	24.2	48.0	NA	NA	400	3362
	0.9	5.7	0.3	1.9	NA	NA	NA	NA	NA	NA

UNDER-SATURATED OIL COMPRESSIBILITY ARTIFICIAL NEURAL NETWORK MODELS

Under-Saturated Oil Compressibility	Rsi		T		API		Gas SP.Gr.		P		Pb		MuoD		Muob	
	Min	Max	Min	Max	Min	Max	Min	Max	Min	Max	Min	Max	Min	Max	Min	Max
Elsharkawy (1998)	0	1568	130.0	245.0	0.6	0.9	0.24	16.00	0	9900	NA	NA	0	16	0	16

OIL FORMATION VOLUME FACTOR AT THE BUBBLE POINT PRESSURE ARTIFICIAL NEURAL NETWORK MODELS

Oil Formation Volume Factor at Bubble Point	Rsi		T		API		Gas SP.Gr.		Bob		Pb	
	Min	Max	Min	Max	Min	Max	Min	Max	Min	Max	Min	Max
Adeeyo (2016)	30	1600	110.0	275.0	14.0	56.5	0.53	0.99	1.0	2.2	550	5984
Al-Shammasi (2001)	6	3298	74.0	341.6	6.0	63.7	0.51	3.44	1.0	2.9	32	7127
El-Sebakhy (2007)	26	1602	74.0	240.0	19.4	44.6	0.74	1.37	1.0	2.0	130	3573
Al-Marhoun and Osman (2002)	24	1453	75.0	240.0	17.5	44.6	0.75	1.82	1.0	3.6	90	3331
Rafiee-Taghanaki et al. (2013)	7	2496	74.0	360.9	15.7	56.5	0.52	3.44	1.0	2.9	58	5152
Osman and Abdel-Wahhab	9	3617	58.0	341.6	11.0	63.7	0.51	1.79	1.0	3.6	107	7127

Appendix F

Worksheets for Oil PVT Correlations Selection

This appendix contains the work sheets for oil PVT (pressure, volume, and temperature) correlations selection. The first section of the appendix contains the input data applicability ranges for oil PVT correlations. These input data ranges are given in the tables of Appendix C, Oil Correlations Range of Applicability. The first four figures are graphical forms for applicable ranges of each of the four main inputs to oil correlations (e.g., reservoir temperature, API gravity of stock-tank oil, initial producing gas-oil ratio, and specific gravity of separator gas).

The following set of work sheets can be used to select PVT correlations for different oil properties based on basic information. The required inputs are initial producing gas—oil ratio, API gravity of stock-tank oil, specific gravity of separator gas, and temperature. The sheets should be used in the order they are presented in this appendix. The first correlation to be selected is the bubble point pressure correlation, followed by solution gas correlation, then oil formation volume factor, etc. The ranges for all PVT properties follow the work done in the appendix references. The sheets are limited to a subset of PVT correlations. An example calculation using these sheets is given in Chapter 9, Selection of PVT Correlations. The correlations references are listed in the Appendix A, Oil Correlations Formulae.

Reservoir Temperature (°F) Range of Applicability for Oil PVT Correlations

Temperature, F

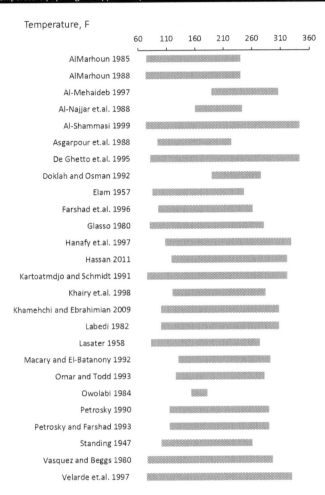

Stock-Tank Oil API Gravity Range of Applicability for Oil PVT Correlations

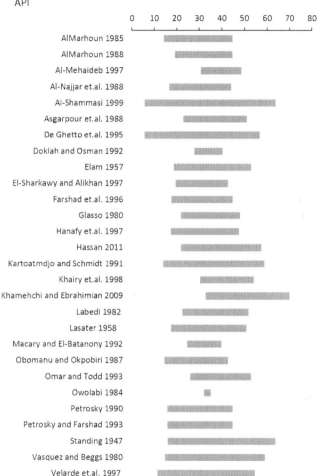

Initial Producing Gas-Oil Ratio Range of Applicability for Oil PVT Correlations

Initial Producing Gas Oil Ratio

Gas Specific Gravity Range of Applicability for Oil PVT Correlations

Gas Sp. Gr., Fraction

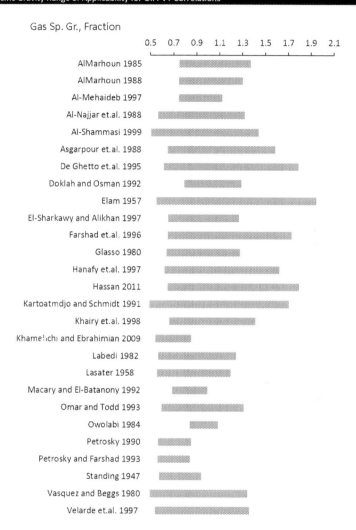

Worksheets for Al-Gathe (2015) correlations selection method.

Selection of Bubble Point Pressure Correlations

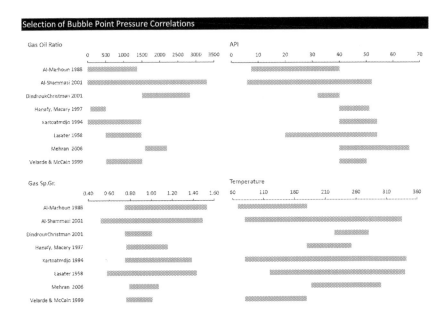

Selection of Solution Gas-Oil Ratio Correlations

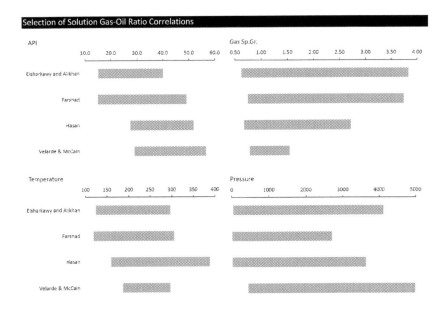

Selection of Oil Formation Volume Factor Correlations

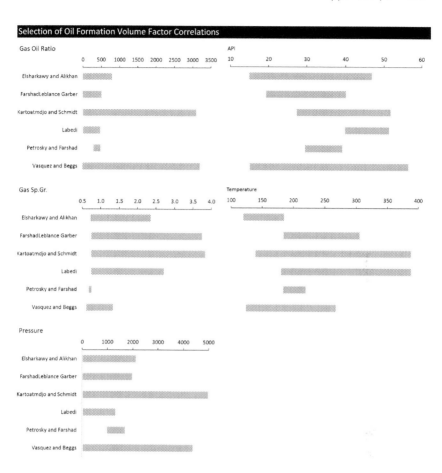

Selection of Saturated Oil Density Correlations

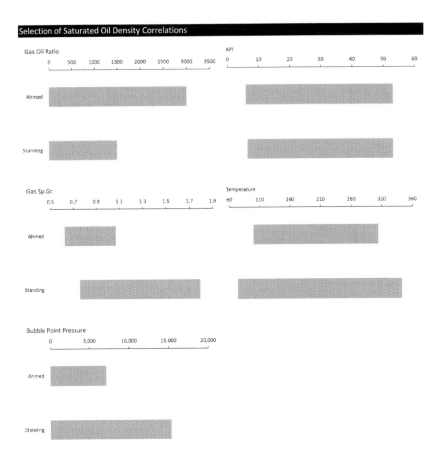

Selection of Under-Saturated Oil Compressibility Correlations

Selection of Dead Oil Viscosity Correlations

Selection of Under-Saturated Oil Viscosity Correlations

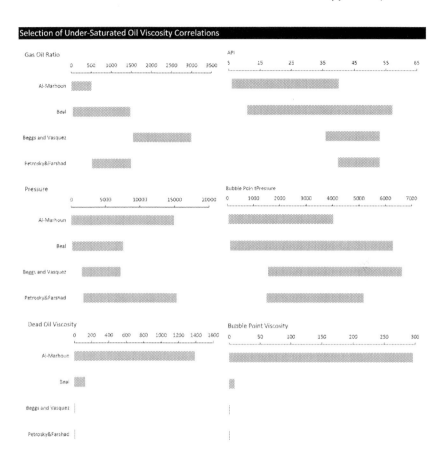

Selection of Viscosity at the Bubble Point Pressure Correlations

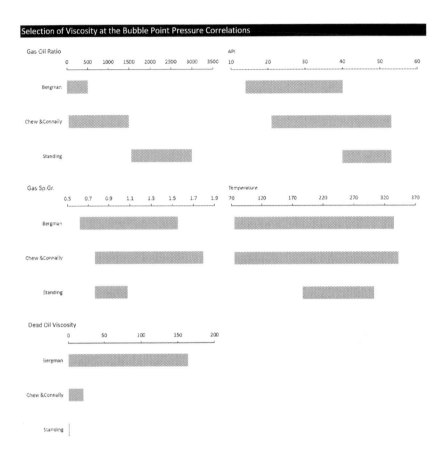

Selection of Saturated Oil Viscosity Correlations

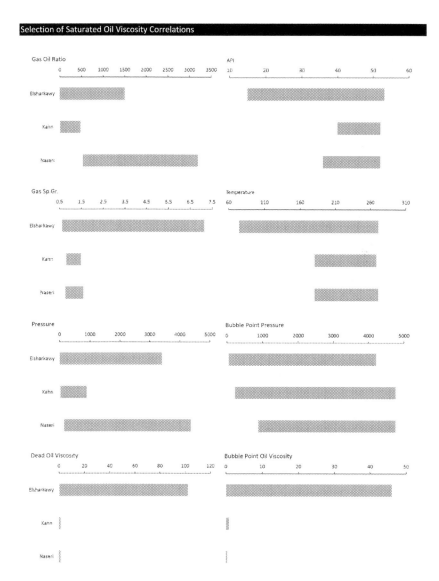

FURTHER READING

Al-Gathe, A.A., (2015). Development of Artificial Intelligence Systems for Black Oil PVT Properties Prediction. PhD Thesis. Cairo University, Cairo, Egypt.

Index

Printed in the United States
By Bookmasters